Water and Water Pollution Handbook

VOLUME 1
(in four volumes)

Water and Water Pollution Handbook

VOLUME 1

(in four volumes)

Edited by LEONARD L. CIACCIO
GTE LABORATORIES INCORPORATED
BAYSIDE, NEW YORK

MARCEL DEKKER, INC., New York 1971

MARCEL DEKKER, INC.
95 Madison Avenue, New York, New York 10016

LIBRARY OF CONGRESS CATALOG CARD NUMBER 78-134780

ISBN 0-8247-1104-1

PRINTED IN THE UNITED STATES OF AMERICA

Preface

The acceleration of interest in pollution among the general population in the past several years has at last affected that development of our instant communication age, the public opinion poll. This evident and enduring concern has no doubt been responsible for the actions by the legislative and administrative branches of the local and federal governments in pollution abatement.

Many disturbing environmental problems have arisen and they have evoked notable activities. The advent of mercury pollution in Lake Erie and the St. Clair River, first signaled by high content of that metal in fish found by Norvald Nimreite, a graduate student at the University of Western Ontario, has triggered a relatively fast reaction by the Canadian government and its Food and Drug directorate, the State of Michigan, the U.S. Food and Drug Administration, and the FWQA. Although some people see sluggishness on the part of the government, a limit on allowed mercury levels in fish has been set, fishing for sport has been banned, polluting industrial plants have been closed by court order, and an investigation of toxic pollutants has commenced. The solution to problems such as the removal of mercury from bottom sediments and the entrance of mercury and its transmission up the ecological food chain, as well as a comprehensive understanding of the effects of mercury on the health of man, are in need of elucidation. The tragic effects of teratogenicity and neurological damage due to the acute and chronic toxic effects of mercury poisoning in Minamata, Japan and upon the Huckelby family of Alamogordo, New Mexico portray an unpleasant result of our global carelessness.

The Suffolk County, Long Island legislature has banned the use of most existing detergents as of March 1, 1971. Their concern relates to the contamination by undegraded detergents of their drinking water, which is obtained from wells. Since the county's disposal system consists almost exclusively of septic tanks for private homes, the problem is quite serious. Although this is a bold measure, it is a stop-gap action. The installation of a sewer system must eventually be considered in light of the increasing population in that section of Long Island. All over the world interest has risen because of the pollution of Lake Baikal in Russia, the effect of the Aswan Dam on the ecology of the eastern Mediterranean, the concern about the eutrophication of the Baltic Sea, the rise in pollution of the rivers and lakes in Italy, the international concern about oil pollution in the world's oceans, etc.

The ability of biological organisms to concentrate low levels of dissolved, undesirable substances (e.g., lead, insecticides, mercury, PCB) and their transmission through the biological food chain is a situation for real concern. Although the complete mechanisms of concentration and transmission and their subsequent biological effects are unknown, some mitigating action is necessary. Professor Hutchinson points out that additive changes to the environment require a serious study of their effects on the biosphere. It does not appear to him that their presence would be the "basis for a revolutionary step forward in the evolution of the biosphere." We will have to change our way of living and ultimately, using our science and technology, control the side effects of the production of our goods and services.

The process of eutrophication is threatening our water resources and posing broad ecological changes. The cause and mechanism of this process has led to an important controversy: the respective roles of phosphates and BOD in the cause of algal blooms is being debated. In an attempt to ameliorate the situation, demands to ban the use of phosphates in detergents and cleaning formulations are increasing. Detergents containing lower or no phosphates are being formulated and marketed. The consumer is demanding such products and evidently is recognizing the nature of the problem of pollution. However, the recent response to replacing phosphates with other detergent builders may ultimately be more troublesome. The use of substances such as NTA, whose fate in and effect on the environment is unknown, does not guarantee an improvement over phosphates and may indicate a panic reaction.

It has been estimated that about 26 billion dollars will have to be spent to abate water pollution in the U.S. Since a complete description of ecological effects on the world and regional ecological systems may be many decades in coming, selective guidance from incomplete information will have to be used. One good general principle is to disturb the ecology as little as possible. This principle, however, does not help if limits are not set to the magnitude of "little." Whatever the actions, the task will be extensive and expensive. In his recent editorial in *Science*, Professor Etzioni claims that the elimination of pollution is "the wrong top priority."

His criticism is that the concern about pollution, "... has many features of a fad ..." and that there is much exaggeration about the demise of the biosphere being circulated in the media to focus on the water and air pollution problems. There is much scientifically based evidence that our environment is changing and not for the better. Some actions are urgent. However, Professor Etzioni casts environmental problems in a more general and logical light. The ugliness and crowding of cities, the deterioration of schools, hunger and malnutrition, racial and sexual inequality, the death of 57,000 Americans in traffic during 1970, are some of the unfashionable human problems that should be given top priority rather than the elimination of water and air pollution. He is referring to the lack of planning for human beings in our society by placing production and consumption of consumer goods as the objective of our society. It is easier, but it is a repudiation of the responsibility of leadership, to solve physical problems, to clean our effluents, or to put a man on the moon, but not to attack the "unfashionable" human problems.

World society will have to face this problem, because the developing countries vis-à-vis the industrial nations will of necessity become polluters as they increase their food production and industrial capabilities. Of course, they are utilizing the systems provided by the industrial nations with their attendant side effect—pollution.

Also important in this whole concept is the size of the world population. More people means more pollution; however, the limitation to population may be the effects of crowding on the sociological and psychological behavior of man. The need for materials and energy by a world population [as opposed to the present where 10 per cent (the U.S.) utilizes a large part of the world's resources] must be solved. This has its implications on pollution levels. Clearly the global manifestations of pollution and their widespread causes cannot be solved solely by bigger and better engineering and more and more money. The objective of society must be human needs, not the production of consumer goods. Until this happens, man in his biosphere will not be preserved biologically, socially, or psychologically.

A number of scholars in the last several years have speculated on the sociological causes for our environmental degradation. There is no universal agreement. Christianity is cited as the cause by White. However, Yi-Fu Tuan points out that the supposed idyllic relationship of ancient man and nature is not unspotted by ecological catastrophies. Moncrief and Goldman point out the effects of industrialization, technology, and the sociological system in exacerbating the pollution problem. With the large population increase since ancient times, the nature of ecological decay can be defined as "the commons problem." Some people are realizing in a practical way that they are the polluters and that individual actions will decide our direction in the battle to save the environment. Thus, the appeal of low phosphate detergents is gaining among a limited, albeit increasing number of consumers. The youth with their hope and idealism are demanding and working to save the environment not only through Earth Day programs but through positive action: from committees and taxpayer suits to demands in medical schools for courses related to environmental health. They are also concerned with people and eradication of human problems as goals of mankind and no doubt youth will help in changing our priorities as advocated by Professor Etzioni.

The nature of environmental science is somewhat analogous to the medical, nuclear, and space sciences. A number of social and scientific disciplines are necessary to define the complex system or phenomena that are the subjects of these fields. This multidisciplinary effort not only requires the cooperation of groups of scientists representing various disciplines, but demands a multidisciplinary scientist. Trained investigators with broad backgrounds must be supplied by the educational institutions. However, the mode of providing these new scientists can, in an unimaginative way, supply a sweep of courses in a number of the relevant subjects or more creatively provide scientifically integrated subject matter. Thus the educational effort can serve as a real means of providing new courses and programs to bring about an integration of the sciences, and more importantly, to train a different type of scientist. In this context, the link with the social and political sciences and the law can be strengthened. The educational effort of providing knowledge of current advances and new technologies to those in the scientific work force has not been adequately undertaken and could be benefitted by this integration. And more critically, those people, technical and nontechnical, white collar and blue collar, suddenly denied employment because of a recession or economic downturn, have the right to retraining programs resulting in meaningful work for their personal dignity and human necessities as well as the stability, viability, and peaceful future of our nation. Perhaps their reorientation via education toward the sociological and ecological tasks would provide a happy ending to a shocking economic and human event.

This integration, which definitely has a social aspect, may be the factor required to shore up the rapidly fading interest that students and the public have in the contributions of science. Of course, one great hurdle is unfreezing the classical structure of science teaching in the university. The parochialism of various science departments must be overcome so that interdisciplinary areas can be served. However, necessary emphasis in selective areas of the particular science must be preserved. The cause of interdisciplinary sciences can also be advanced, however, through other means such as publications. This is one of our objectives in the creation of this work.

The organization and preparation of a compendium of this type, focusing on a multidisciplinary area is an enlightening and humbling experience. The breadth of the subject requires the integration of many scientific areas. It is evident from the events in recent years that the chemical, physical, and biological definition of the environment and environmental problems is a most important activity at this time. There is need for specific analysis and identification of these materials in our water environment and their fate and effect on the ecological system, if one considers the nature of the system. Thus the analytical scientist, chemical, biological, and physical, is required for this task. Unfortunately, in many chemistry departments of our universities, analytical chemistry has been de-emphasized, although there are many who do maintain an excellent program. In all fairness, the fault cannot simply be attributed to internecine sniping among chemists, but rather to the failure of many analytical teaching staffs to adapt to the new teaching methods and the restructuring of courses consonant with the new and changing demands, particularly by the multidisciplinary sciences. However, analytical chemists in the universities appear to be responding positively. This is a propitious sign, because the future need for analytical scientists is great and will grow as a result of the increasing emphasis on multidisciplinary sciences.

The water environment can generally be characterized as a dilute, aqueous solution, containing a large variety of organic and inorganic chemical species, dissolved and in suspension, and including a variety of plant and animal life. A knowledge of the qualitative and quantitative composition of these systems is the first step toward revealing the nature of the particular environmental problem. Chemical and biological analyses are first called upon to supply information to the investigators. One may, at the outset, without regard to preventing perturbations in the system, obtain some very general analytical information. However, this type of information is at best only indicative of the identity of the constituents and gives no detailed information of the species present. To obtain knowledge of the specific species present and of their concentrations, the effect of the analytical treatment should in no way change the system with its unique equilibria and components. The analytical results should characterize the original system and not one that is a modification created by the analytical processing. This analytical treatment may cause decomposition or interactions of the components in the sample of the original system. However, analytical treatment can be chosen so that the energy of perturbation is small compared to that energy needed to cause chemical and physical changes in the system. Still, the disturbing question always remains about the possibilities of induced changes. Results obtained after modifications of the analytical approach along with consideration of the nature of the system should indicate whether induced changes have occurred. Throughout one's consideration of the determination of the composition of the system, it must be noted that most environmental systems are not constant in time; therefore, the effect of time on composition and analytical treatment on kinetics of occurring reactions must be evaluated. These difficult but fascinating considerations must be given thought in elucidating the composition of a system at a given time.

The main theme of these collected volumes was particularly this point, for at this stage in our investigation of the water environment we need good, specific, accurate analytical techniques to define from the static and dynamic states: what happened, what is happening, what may happen, and where it may take us. When we use the word, "analytical," we refer to the numerous disciplines necessary to completely evaluate the environmental situation. Thus we have endeavored to treat chemical, microbiological, viral, bioassay, and mathematical techniques as necessary ingredients in this most important task of defining our multifaceted problems. However, in order to intelligently consider analytical techniques to be utilized in defining the environmental problems, the physical, chemical, and biological characteristics of various parts of the water environment and associated systems are presented. Thus the formal structure of this four volume set consists of two parts: Part I, Environmental Systems and Part II, Chemical, Physical, Bacterial, Viral, Instrumental, and Bioassay Techniques.

The description of environmental systems in Part I includes chapters on the chemical, physical, and biological characteristics of water resources, estuaries, irrigation and soil waters, and wastes and waste effluents and in the areas of water purification, waste treatment, effects of pollution, self-purification, and mathematical modeling. After a definition of the systems, the analyst can utilize Part II effectively. In this main section a wide variety of analytical subjects are covered: Specific tests for BOD, residual chlorine, pesticides, herbicides, minor elements, and radionuclides; instrumental techniques such as infrared and mass spectroscopy, electrochemical, luminescence, gas chromatographic, and automated analyses; biological tests such as bacterial, viral, and bioassay techniques; other analytical subjects such as sampling, organic and inorganic analysis, concentration and separation techniques, insoluble matter, measurement system design, monitoring systems, and other chemical and physical phenomena such as the structure of water and chemical kinetics and dynamics in the water system.

The level of presentation is always a problem in a work such as this.

We must take on a pedagogic role for those who want to enter the field or may be trained only in certain aspects of the field, as well as the multidisciplinary scientist in the environmental field. The audience should represent quite a spectrum, in the both technical and nontechnical areas. Although it may sound as if we are quite ambitious about the group we hope to provide with information and knowledge, we think our book will fulfill this expectation.

Investigations into water pollution are moving rather rapidly, and since this book was assembled over a three-year period, we are bound to have a certain lag with respect to most recent developments. This we acknowledge quite freely, but at the same time we feel the basic information we are bringing forth will be useful for years to come. Our hope is to contribute to and influence the "human biological storage," alluded to by Sir Egerton, from which will spring the progress in this rapidly evolving science of the water environment.

During the generation of this work I have been fortunate to receive help and critical advice from a number of associates and friends. First I want to thank the management of the Research Laboratories of Wallace and Tiernan, Inc. (now Penn Walt, Inc.) and GTE Laboratories Incorporated, of Bayside, New York, for their understanding and generosity in the use of secretarial help in this undertaking. To Miss Virginia Aquino (Wallace & Tiernan Inc.) and Mrs. Lee Epstein (GTE Laboratories Incorporated), who generously contributed their secretarial skills, I am indebted. Dr. Richard Benoit and Mr. Harold Quis responded to my need for "brain storming" with wonderful results, which I appreciate very much. And to those who have been of invaluable aid in critical advice, Dr. Peter Cukor, Dr. Grace Vernon, Klaus Hameyer, Howard Madlin, Robert Rubino, James Cosgrove, and Danny Oblas, I am very grateful.

I would be remiss if I neglected to acknowledge two people who contributed to my professional development. Mr. Tom Grenfell, formerly Quality Control Director of Chas. Pfizer & Co., Inc., exemplified the creative, industrial, analytical, research chemist. Scientific realism and hope was impressed upon me by Professor Rudy Marcus who, while recognizing the complicated nature of basic physical chemical phenomena, labors assiduously in uncovering their characteristics.

To my most intimate human environment, that rare collection of individuals, my family, I must convey my most affectionate appreciation. To our children, Gloria, Luke, Dominic, Imelda and Rita, I acknowledge the lively atmosphere which stimulates, and to my wife, Eva Agostini Ciaccio, the civilizing spirit which keeps all pleasantly in bounds.

LEONARD L. CIACCIO

Glen Rock, New Jersey
December 10, 1970

Selected Bibliography

Mercury stirs more pollution concern, *Chem. Eng. News*, June 22, 1970, pp. 36, 37.

P. H. Abelson, Methyl mercury, *Science*, **169**, 237 (1970).

Mercury in the environment, *Environ. Sci. Tech.*, **4**, 890 (1970).

G. E. Hutchinson, The biosphere, *Sci. Am.*, **223**, 44 (1970).

A. Etzioni, The wrong top priority, *Science*, **168**, 921 (1970).

L. White, Jr., The historical roots of our ecological crisis, *Science*, **155**, 1203 (1967).

Yi-Fu Tuan, Our treatment of the environment in idea and actuality, *Am. Scientist*, **58**, 244 (1970).

G. Hardin, The tragedy of the commons, *Science*, **162**, 1243 (1968).

L. W. Moncrief, The cultural basis for our environmental crisis, *Science*, **170**, 508 (1970).

M. I. Goldman, The convergence of environmental disruption, *Science*, **170**, 37 (1970).

Contributors to Volume 1

Walter Abbott,* *Gulf Coast Research Laboratory, Ocean Springs, Mississippi*

Richard J. Benoit, *EcoScience Laboratory, Division of Environmental Research & Applications, Inc., Norwich, Connecticut*

Leon Bernstein, *U.S. Salinity Laboratory, U.S. Department of Agriculture, Riverside, California*

C. E. Dawson, *Gulf Coast Research Laboratory, Ocean Springs, Mississippi*

Walter H. Durum, *U.S. Geological Survey, Washington, D.C.*

Emil J. Genetelli, *Department of Environmental Sciences, Rutgers—The State University, New Brunswick, New Jersey*

Max Katz, *College of Fisheries, University of Washington, Seattle, Washington*

A. R. LeFeuvre, *Canada Centre for Inland Waters, Department of Energy, Mines and Resources, Burlington, Ontario*

C. H. Oppenheimer, *Florida State University, Tallahassee, Florida*

H. A. Painter, *Water Pollution Research Laboratory, Stevenage, Herts, England*

J. D. Rhoades, *U.S. Salinity Laboratory, U.S. Department of Agriculture, Riverside, California*

J. E. Singley, *College of Engineering, University of Florida, Gainesville, Florida*

*Present address: Environomics, Houston, Texas.

Contents

Contents of Volume 2

Contents of Volume 3

Contents to Volume 4

PROLOGUE

I will open rivers in the high hills, and
fountains in the midst of the plains.

I will turn the desert into pools of waters,
and the impassable land into streams of waters.

I will plant in the wilderness the cedar, and
the thorn, and the myrtle, and the olive tree.

I will set in the desert the fir tree, the elm,
and the box tree together,

That they may see, and know, and consider, and
understand together that the hand of the Lord
hath done this.

ISAIAH, 41; 18 & 19

Part I

Environmental Systems

Chapter 1 Chemical, Physical, and Biological Characteristics of Water Resources

Walton H. Durum
U.S. GEOLOGICAL SURVEY
WASHINGTON, D.C.

I. Introduction

We are witnessing what must surely be the greatest of all man's struggles—the struggle to control his environment. One of the most important environmental areas is the quality of life-giving water. What is it that we really are seeking to attain in water quality? Do we really understand the factors that constitute or control quality? Are we looking for the right parameters?

A familiar axiom is "What man cannot control he must learn to live with." These words best describe one inescapable fact—nature working through the forces of water sculptures the continent and leaves her indelible fingerprints on the landscape and in the solution products of that weathering action. Perhaps this is the real basis for concern among economists and water managers—that with the continuing change in our water resources problems, we cannot apply the old solutions to new situations. Love (*1*) calls for the scientific and engineering professions that deal with water to develop meaningful and understandable quality criteria for each kind of water use.

It is always vexing to those who look at water resources for the first time to discover that there is more to quantifying a river system than making a series of stream flow measurements, and that the water provided in nature's laboratory is seldom the pristine, sparkling, salt-free vintage about which the poets wrote. Indeed, nature has endowed us with water quality problems so vastly complicated and, in some areas of the United States, so uncontrollable that providing water of suitable quality will be the most formidable task of all to be faced by future water managers. Why is this so, and what knowledge have we gained about the situation?

II. Chemistry of Rainwater

Let's begin with the atmosphere, source of all precipitation recharge.

We are learning that rainwater has distinctive quality patterns and thus is a major factor in regional, national, and continental water quality. Investigations begun a decade or so ago by Egner and Ericksson (*2*) revealed that sea salt is the major source of chloride in rainfall. More recently Gambell (*3*), during carefully controlled field sampling and analyses for Na^+, Ca^{2+}, Mg^{2+}, NO_3^-, SO_4^{2-}, and Cl^-, has observed the composition of rainfall over a 34,000-sq-mile area in southeast Virginia and North Carolina.

Several features of Gambell's study of chemical composition of rainfall should be noted. First, it is apparent that rainfall contributes significant amounts of chemical substances to southeastern streams, which must, therefore, be considered in any serious "ionic" balance or geochemical studies involving dilute waters. Excluding bicarbonate, the quantity of dissolved solids supplied by rainfall amounted to almost one-half of the

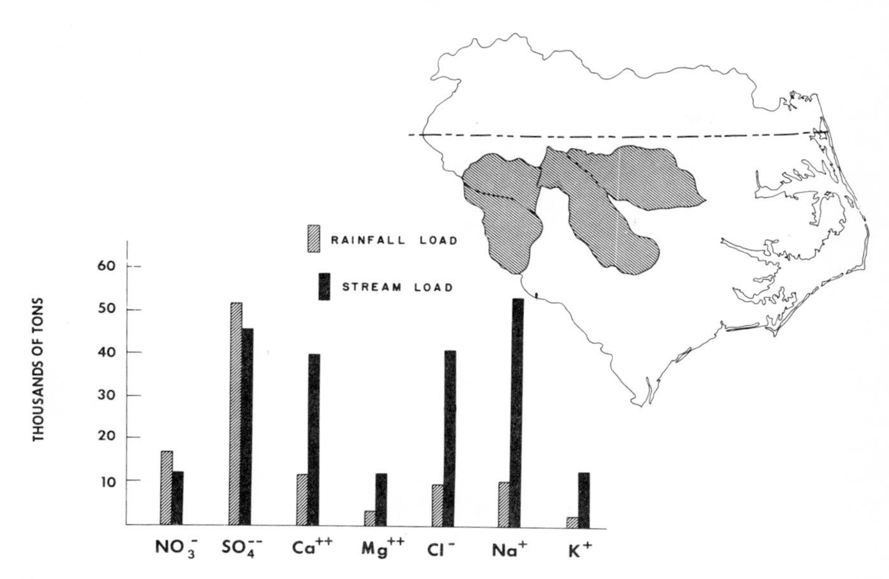

Fig. 1. Comparison of loads contributed by rainfall and streams, southeast Virginia and North Carolina, August 1962 to July 1963. [Reprinted from Ref. *3*.]

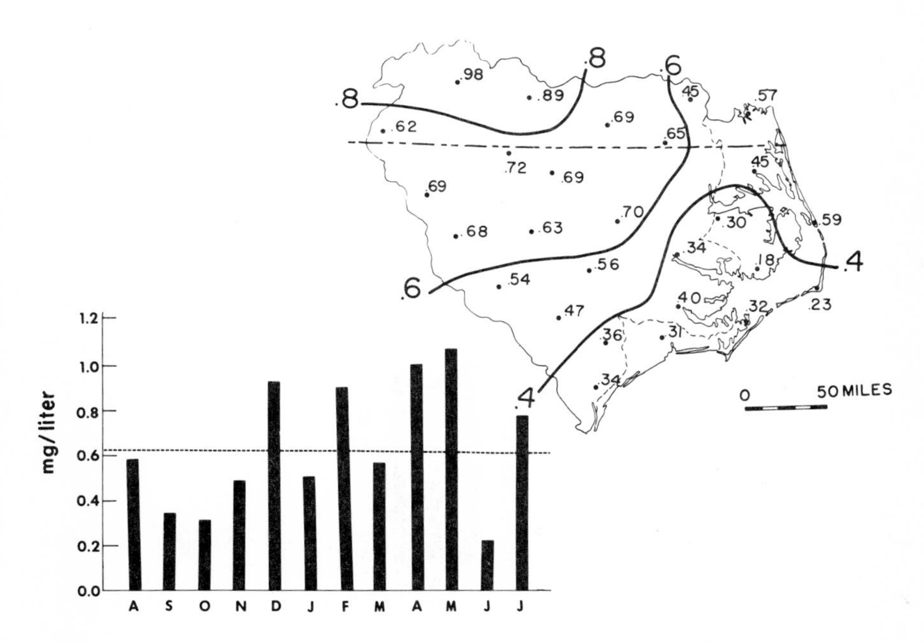

Fig. 2. Monthly average NO_3^- content of rainfall, southeast Virginia and North Carolina, August 1962 to July 1963. [Reprinted from Ref. *3*.]

dissolved solids load in the streams (Fig. 1); rainfall supplied more nitrates and sulfates than was carried by the streams, suggesting that these ions are removed by plants for nutritional purposes and are returned to the atmosphere or enter into other earth–water reactions (Figs. 2 and 3). It is significant, however, that precipitation can no longer be ignored as a major source of nutrients, at least judging by the results of a few studies of the Gambell(*3*) type; precipitation accounts for about one-quarter of the total chloride stream load.

The concentration of chloride derived from sea salt is highest along the coast and decreases sharply inland for about 200 miles (Fig. 4). There is a marked increase in chloride in winter, corresponding to the increased occurrence of coastal storms accompanied by onshore winds. Calcium is at a maximum in spring and summer and is apparently related to the soil dust associated with agricultural activity(*3*). Sodium and magnesium concentrations are slightly higher in winter, resulting from sea salt, but soil dust appears to be a significant source(*3*).

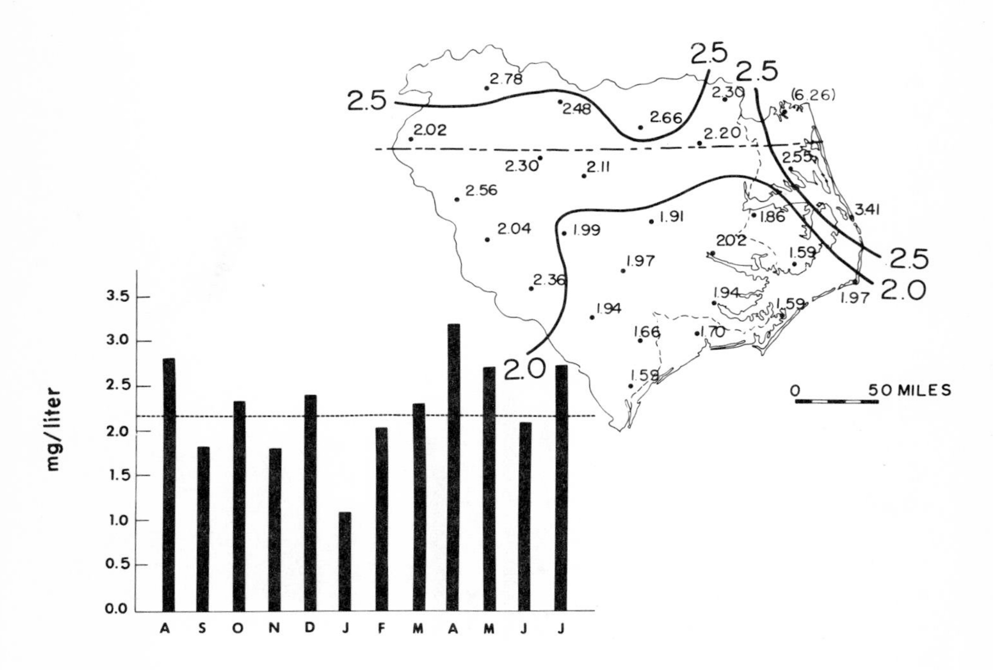

Fig. 3. Monthly average SO_4^{2-} concentration in rainfall, southeast Virginia and North Carolina, August 1962 to July 1963. [Reprinted from Ref. *3*.]

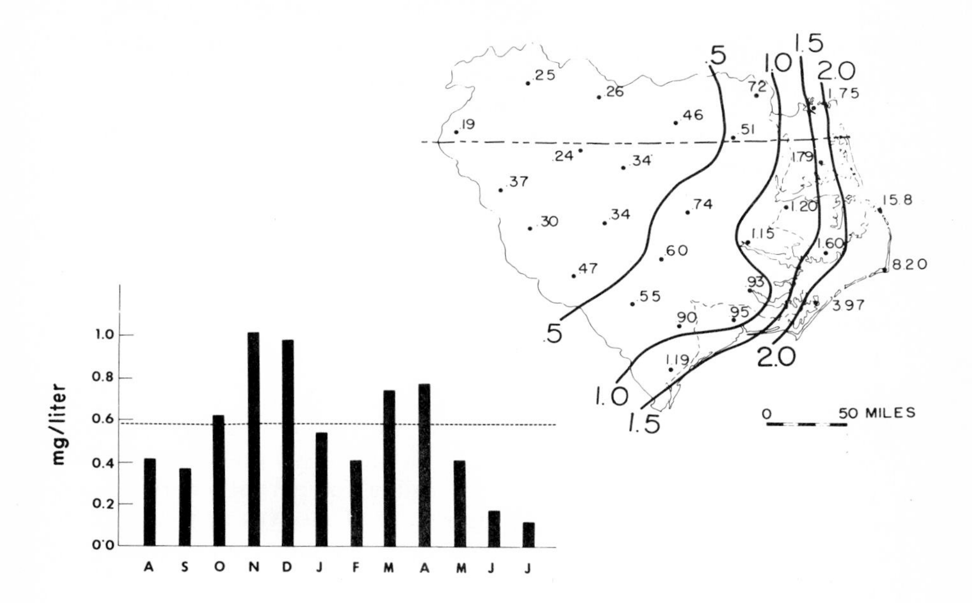

Fig. 4. Monthly average Cl^- content of rainfall in southeast Virginia and North Carolina, August 1962 to July 1963. [Reprinted from Ref. *3*.]

III. Nitrogen Compounds in Groundwater and Rainwater

In a survey of ammonium and nitrate in United States rainwater, Junge(*4*) reported a markedly uneven areal distribution of both. In general, values were low near coastlines and over the ocean at the few points. The major sources of fixed nitrogen, he concluded, are bound to certain geographical areas overland. Although some of the nitrogen found in rainwater is attributed to industrial activity and to fertilizers, Junge concluded that formation appears to be primarily bound to the soil.

Junge and Gustafson(*5*), reporting on data for a northern European network of precipitation stations, show a 10-fold range in the yield of total nitrogen in precipitation in various parts of northern Europe. Whitehead and Feth(*6*) reported on precipitation sampled at Menlo Park, California in 1957–1959.

Rankama and Sahama(*7*) give the nitrogen content of igneous rocks as 46 g/ton (which equals 46 mg/liter as nitrogen, or 186 mg/liter as nitrate) and that of sedimentary rocks as about 510 g/ton. During formation of sediments, those containing organic matter are enriched in nitrogen, but the fate of nitrogen during diagenesis and metamorphism is unknown.

We can conclude that nitrogen compounds reach surface and groundwaters from a great many diverse sources. Precipitation and soil seem to be of outstanding importance. Lithologic sources—especially those in unconsolidated rocks—remain virtually unexplored, their importance cannot now be evaluated.

IV. Groundwater and Equilibrium Chemistry

Observations of certain salt lakes in Egypt about the year 1800 indicated significant deposits of sodium carbonate. This led investigators to the conclusion that the high concentration of sodium chloride and dissolved calcium carbonate was responsible for a reaction that was opposite in direction to that in the laboratory. Thus was conceived the understanding that concentration of reactants can influence the extent of the reaction and composition of the products, refined to the law of mass action by Guldberg and Waage, about 1867. Today, we express the basic quantitative reactions of substances in water and earth materials as a reversible reaction: $A+B \leftrightharpoons C+D$. At equilibrium the rates of reactions to the right and left are equal and the equilibrium constant (K) can be calculated. For example, for the solution of calcite ($CaCO_3$) in water containing hydrogen ions:

$$K = \frac{[Ca^{2+}][HCO_3^-]}{[CaCO_3][H^+]} \quad (1)$$

When it was observed that K changes with temperature and is constant only in very dilute solutions, Lewis and Randall(*8*) introduced the concept of activity, subsequently placed on a strong theoretical basis by Debye and Hückel(*9*), to explain the decreasing effectiveness of the ionic material in solution as the concentration of the electrolytes increases. (See Chap. 16 for a discussion of kinetics.)

Various specialized types of equilibrium constants are the working tools of the geochemist concerned with basic reactions of water and its environment, especially groundwater. Feth(*10*) studied the chemical quality of water from granitic terrane in the Sierra Nevada and the composition of rocks and soils. He concluded from a study of perennial springs and seeps, mostly in granitic rocks, that the principal control is availability of carbon dioxide acquired during contact with the air and soil. Ephemeral spring water, for example, is continuously replenished with carbon dioxide and air and aggressively attacks the rock minerals, whereas perennial spring water, in a closed system, would have a limited amount of gaseous carbon dioxide available for reaction. Feth's hypothesis was based on the reaction:

$$H_2CO_3 \rightleftharpoons H^+ + HCO_3^- \tag{2}$$

Thus the hydrogen ion obtained from the reaction of carbon dioxide in water would control the ability of the water to dissolve constituents from the minerals.

The behavior of iron in underground water is another area of application of equilibrium chemistry. Each of us has experienced the perplexing problem of irrational behavior of iron—low iron content in water from one well, excessive iron in an adjacent well; unexplained variation in iron content with time, with continuous pumping of the same well; corrosion or encrustations of the well screen or casing.

The actual solubility of iron under different conditions at equilibrium can be readily calculated and shown graphically by means of the Eh-pH or stability field diagram. Garrels(*11*) and Hem and Cropper(*12*) have used the variables Eh and pH to define areas of stability for solid compounds, such as iron oxide or hydroxide, whose free energy is known. The only other variable is the activity of dissolved iron. Hem and Cropper's diagram of iron species stability fields for dissolved iron activity of 2×10^{-7} moles (Fig. 5) has also been modified to indicate the solubility of iron in the presence of other dissolved species such as sulfate and bicarbonate.

An example of these principles can be found in the Atlantic Coastal Plain sediments, where iron in water is a ubiquitous problem. Barnes and Back(*13*), reporting limonite to be a principal iron mineral, attributed the high iron content of groundwater to the contact of ferric hydroxide with groundwater having reducing characteristics. The reverse condition, i.e., highly oxidizing groundwater in contact with reduced minerals, such as pyrite, also occurs naturally in parts of the United States and can produce water high in dissolved iron.

Turning now to the Southeast, we see how departures from equilibrium can be demonstrated under field conditions. In discussing the chemistry of water samples of Ocala limestone and other formations of central Florida, Back(*14*) illustrated (Fig. 6) equilibrium conditions with respect to calcite as a 100% line: undersaturation by percentages under 100, and supersaturation by percentages over 100. The water is dissolving limestone in the areas designated as undersaturated, and may be precipitating calcite in supersaturated areas. From a practical viewpoint, where water is undersaturated, caverns and fissures can be enlarged hundreds of feet

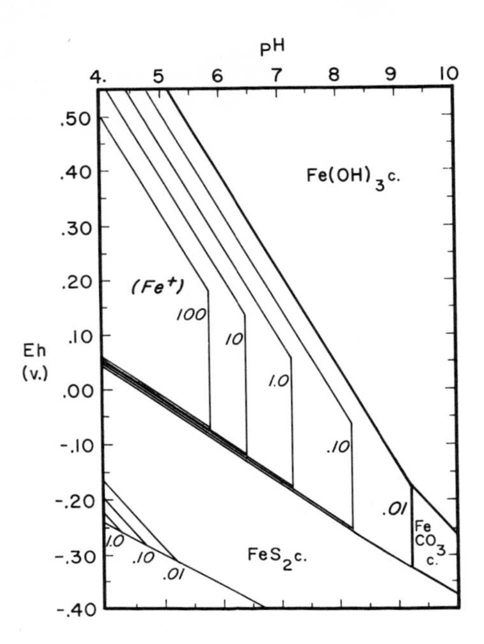

Fig. 5. Iron species stability fields. [Reprinted from Ref. *12*.]

below the water table, limited mainly by the circulation pattern of the water.

A third application of the principle of solution equilibrium (Fig. 7) is in the applied problems of introducing recharge through wells. If the water to be injected is substantially supersaturated with respect to calcite, and the native water has a pH the same as or higher than that of the water to be injected, calcium carbonate will be precipitated in the injection well and in the aquifer. Movement of water will be impaired. If the recharge water contains iron, any contact with air is likely to cause ferric hydroxide precipitation. According to Hem(*15*) the Eh of aerated water will generally range from about 0.35 to 0.45 V. It can be shown that 0.01 mg/liter of iron can remain in solution at an Eh of 0.45 V only if pH is less than 5.7. Hence, iron-bearing recharge water is almost certain to deposit iron in the recharge well. If the aquifer to be recharged contains iron-bearing water and iron minerals, the addition to or replacement of this water with aerated recharge water will precipitate ferric hydroxide where the two waters contact each other or are mixed together. Iron oxide precipitation commonly occurs during recharge operations at Grand Prairie, Arkansas.

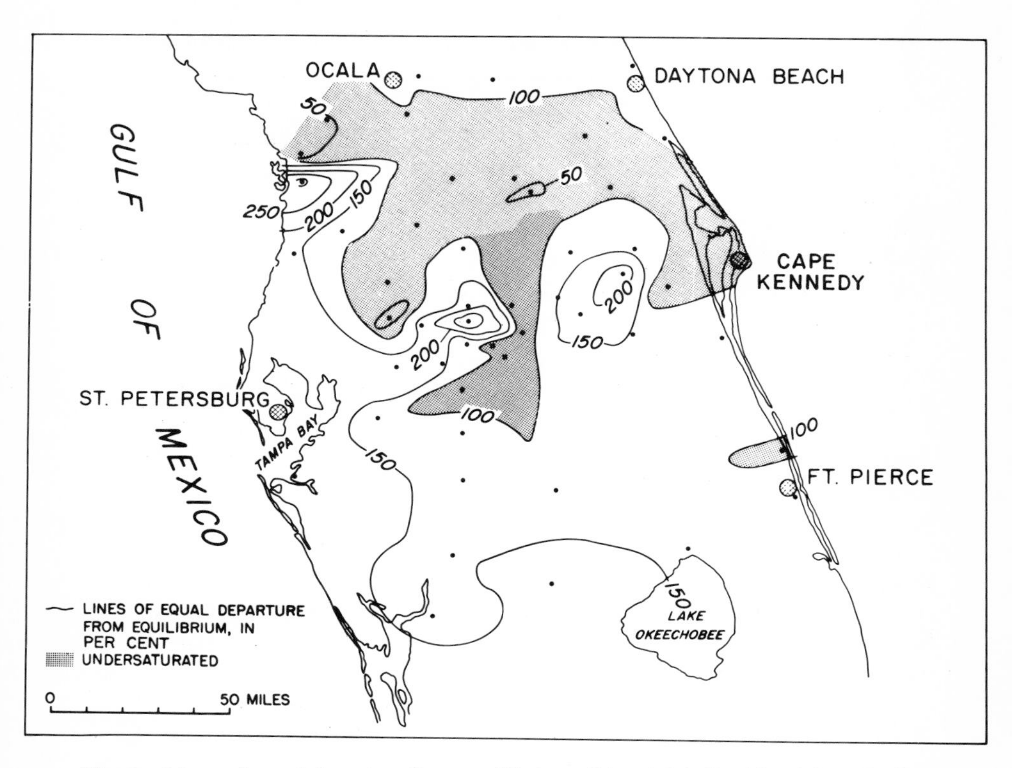

Fig. 6. Lines of equal departure from equilibrium with calcite. [Reprinted from Ref. *14*, p. 43, by courtesy of International Association of Scientific Hydrology.]

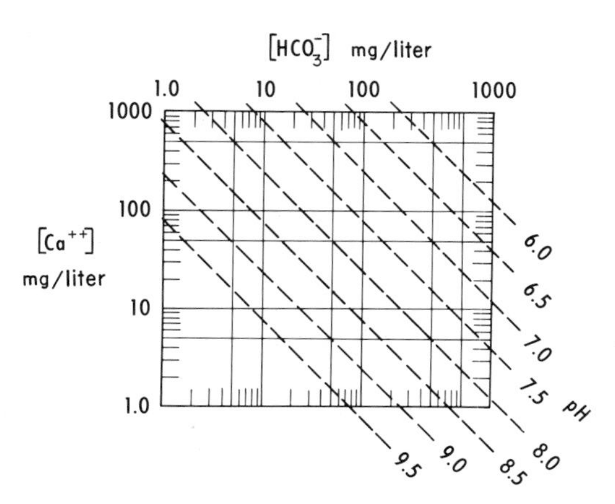

Fig. 7. Solution equilibrium: pH, calcium, bicarbonate, and calcite. [Reprinted from Ref. *15*, p. 19, by courtesy of International Association of Scientific Hydrology.]

The Geological Survey is now releasing a map that shows, by different patterns, the chemical concentration and composition of groundwater in the United States. Emphasis is given to waters with dissolved solids in the range 1000–3000 mg/liter at depths less than 500 ft. However, waters of mineral content greater than 3000 mg/liter and deeper than 500 ft are identified.

Pollutants or contaminants passing through the groundwater system seldom move with the velocity of the fluid. For this reason adjustments must be made to predict the flow of various ionic or non-ionic species. Two of the more important physical processes influencing these adjustments are dispersion and ion exchange. Leaching out of soluble material from a solid matrix may also have a drastic effect on water quality.

A. Dispersion

The dispersion of fluid streamlines or water particles during flow through porous media is characterized by a difference in fluid velocity as the water flows through cracks and around granules. The concentration front of tracer material or pollutant traveling down a column or through an aquifer usually develops a one-sided "normal" distribution due to this dispersion process. The degree of dispersion in porous media is characterized by D, the dispersion coefficient. Attempts to predict this coefficient for field situations have been totally unsuccessful. For well-sorted sands under carefully controlled laboratory conditions excellent results have been obtained, but when applying such results to large-scale field experiments errors of two to three orders of magnitude have resulted. Raffi et al.(*16*) have published an excellent discussion of dispersion phenomena.

B. Leaching

The passing of water through gypsum beds or any soluble material will increase its dissolved solids content. In order to express this increase in a quantitative manner an equilibrium-type relationship is usually assumed, and the amount dissolved is dependent on the degree of unsaturation of that particular species. In normal irrigation practice (see Chap. 3), leaching is used to remove accumulated salts and to maintain an adequate salt balance. This process produces a continual change in water quality of the reused water.

C. Ion Exchange

When ionic substances go into solution they usually dissociate into positive and negative ions called cations and anions, respectively. These ions have the ability to change places (exchange) with other cations or anions on the solid matrix (see Chap. 11).

The degree of this ion exchange is one of the most important facets concerning the movement of pollutants through an aquifer system. The molar (m) rate of ion exchange, $\partial m/\partial t$, can be expressed in a finite manner such as:

$$\frac{\partial m}{\partial t} = k[c - c^*] \tag{3}$$

where k is a rate constant and $[c - c^*]$ is the departure of the solute concentration from ion exchange equilibrium. However, the exchange reaction can be assumed to reach equilibrium at all times, i.e.,

$$m = K_d c \tag{4}$$

where K_d is a distribution coefficient (*17*) describing the ratio of the amount of ions absorbed on the solid to the amount dissolved in the liquid. It is usually necessary to determine the equilibrium constant of the ion exchange reaction and the exchange capacity of the solid in order to calculate the distribution coefficient. When the ion exchange adsorption isotherm (which depicts the ratio of the amount of solute in solution to the amount adsorbed on the solid) is linear, the distribution coefficient is constant. Also, the average velocity of the species under study is a fraction of the groundwater velocity, i.e., $1/1 + K_d$.

D. Transportation of Pollutants

The general equation that describes the concentration of an ionic species in space and time for one-dimensional flow is:

$$\frac{\partial c}{\partial t} = -v\frac{\partial c}{\partial x} + \frac{\partial^2 c}{\partial x^2} - \frac{\partial m}{\partial t} \tag{5}$$

For flow in the x direction, this equation has been solved for the concentration c of ion in solution as a function of x, space; t, time; and v, velocity. An increase in solute owing to leaching or any other unaccounted-for variable can be described in proper terms and added to the equation. As mentioned previously, Eq. (5) has been solved for simple cases where the following conditions hold: no dispersion, finite ion exchange reaction rates, and instantaneous equilibrium during the exchange reaction.

Solutions involving finite reaction rates have been obtained by Ogata(*18*) but are not in closed form, and a digital computer is still necessary to provide usable answers. Perhaps the most difficult task is the assignment of constants necessary to the solution of the equation. Dispersion coefficients, distribution coefficients, and ion exchange reaction rates vary widely in nature and are sometimes difficult to determine under controlled laboratory conditions, much less field conditions.

E. Biological Considerations

Groundwater stands in strong contrast to surface streams as a habitat for life. The fluctuations of terrestrial climate are muted and attenuated with increasing distance from the surface, and at still greater depth the effects of geophysical processes become dominant. Nevertheless, there is now no doubt that the chemical composition of groundwater may change not only by inorganic processes but also through the activity of microorganisms(*19*). In fact, deep-seated groundwaters contain a specific microflora which actively influences their chemical composition(*20*).

Bacteria are the significant forms of life in groundwaters. In general, these microorganisms are not uniformly distributed horizontally or vertically but depend, in part, on the distribution of organic matter, mineral species, sediment pore and particle size, and the intensity of water exchange. Also, there is an optimum temperature which favors bacterial activity, and the depth at which the optimum occurs depends upon the local geothermal gradient. The increase in geothermal heat with depth eventually precludes bacterial life at great depths.

Many of the most important chemical effects are caused by microorganisms whose life activities do not require the presence of preexisting organic matter. These so-called chemosynthetic, autotrophic microorganisms grow and reproduce by using inorganic food and the energy from the oxidation of inorganic compounds such as molecular hydrogen, methane, and reduced compounds of nitrogen, sulfur, and iron. Of particular interest are the anaerobic organisms, which are able to grow using both organic and mineral substances in the absence of oxygen.

Most of the readily available literature discusses biogenic transformations of sulfur, nitrogen, carbon, iron, and manganese, but the details of the inorganic and organic processes involved are very imperfectly known. Much of the work is in the early descriptive stage, as exemplified by Mekhtiyeva's study(*20*) of the distribution of microorganisms in the groundwaters of the Paleozoic strata of the Middle Volga region. He reported that groundwaters that were not in direct contact with petroleum contained active bacteria of several different physiological groups. The groundwaters in immediate contact with petroleum contained much more abundant and varied microfloras. The percentage of water samples containing bacteria diminished from the Permian to the Devonian aquifers. Sulfate-reducing bacteria were present in hydrogen sulfide waters but were absent from waters which were free of hydrogen sulfide. The following conditions were apparently unfavorable for bacteria: waters of the calcium chloride type, high salinity (800 meq/100 g), sulfate- and phosphate-free waters, waters with a high bromine content, and waters with a $(Cl + Na)/Mg$ ratio > 3. Bacteria were usually found in waters of low salinity, with $(Cl + Na)/Mg$ ratio < 3, high sulfate, hydrogen sulfide, and carbon dioxide contents.

Different events may occur with microorganisms as groundwaters move in sediments. Active or passive migration of the cells with or without multiplication may occur within the streaming water and the sediment; cells may grow through the sedimentary material, from surface films, attach to the surface of particles of the sediment, or be retained, adsorbed, or eluted within the sediment as reported by Wagner and Schwartz(*21*). Experiments by these authors with five species of bacteria and two types of sandy sediments showed that in transportation of bacteria through the sediment, movement and distribution in different layers of the sedimentary column were in general influenced by the size of the cells and the size of the grains of the sediment. More cells were retained with suspensions of cells in seawater than with suspensions in fresh water. Besides the physical laws which regulate retention and elution, biological differences were found among the different bacterial species. After eight days of flushing with cell-free water, the sediments still contained varying numbers of living bacteria.

This is a very brief review of groundwater biology which includes only the most general aspects of the ecosystem. However, the ecology of groundwaters is so poorly understood that even conceptual models of biological effects on water quality probably cannot be developed from existing data.

F. Future Sources of Groundwater

Preliminary studies by Feth(*10*) indicate that two-thirds of the country is underlain by rocks containing water which has more than 1000 mg/liter of dissolved solids. These water-bearing rocks, found at depths ranging from a few feet to more than 1000 ft, are potential sources of usable water in areas where freshwater conversion plants are determined to be economically feasible. As demand for fresh groundwater approaches the full capacity of available supplied, particularly in parts of the arid West, the need to develop mineralized water resources will grow rapidly.

V. Surface Water, Giant Vehicle for Transport

The variety of substances that determine surface water quality include inorganic, organic, and suspended matter, and products of the biologic system. These substances respond in different ways to natural and man-made environments. Water quality of a stream is determined by its environment—climate, geologic, hydrologic, physiographic, biological, and cultural. Thus, waters of similar quality can be expected from terranes of similar environment, and useful correlations between water quality and environment are probable. Some areas display a consistent pattern of overall similarity, but with an internal pattern of variability that gives a wider range of concentration than one would expect. This makes the setting of firm standards of natural quality for river segments and river systems difficult and oft times deceiving.

Let's look at the national picture. From maps adapted from Rainwater (*22*) we observe that in 50% of the country the prevalent concentration of dissolved solids in surface waters is less than 230 mg/liter. In 10% of the country, the prevalent concentration is more than 900 mg/liter, but these waters are generally associated with areas of incompletely leached rocks and the infrequent and often intense rainfall characteristic of the semiarid regions. This type of environment yields extreme ranges in the composition and mineral content of the water, sometimes a factor of 10 or more.

A general observation is that waters with dissolved solids concentration less than 120 mg/liter are soft; from 120 to 350 mg/liter they are moderately hard to hard; and with greater than 350 mg/liter of dissolved solids they are very hard. This is because calcium bicarbonate, which has limited solubility, tends to dominate lower concentrations, whereas the more soluble sulfate ion in association with calcium and magnesium is generally predominant at the higher concentrations (Fig. 8 and Tables 1 and 2).

As for the hydraulic mechanics of solute transport, the main emphasis of research has been on adapting and verifying Taylor's concepts (*23*) in an open channel. The basic convective diffusion equation for a two-dimensional uniform channel is

$$\frac{\partial c}{\partial t} + u\frac{\partial c}{\partial x} = \frac{\partial}{\partial x}\left(\epsilon_x \frac{\partial c}{\partial x}\right) + \frac{\partial}{\partial y}\left(\epsilon_y \frac{\partial c}{\partial y}\right) \tag{6}$$

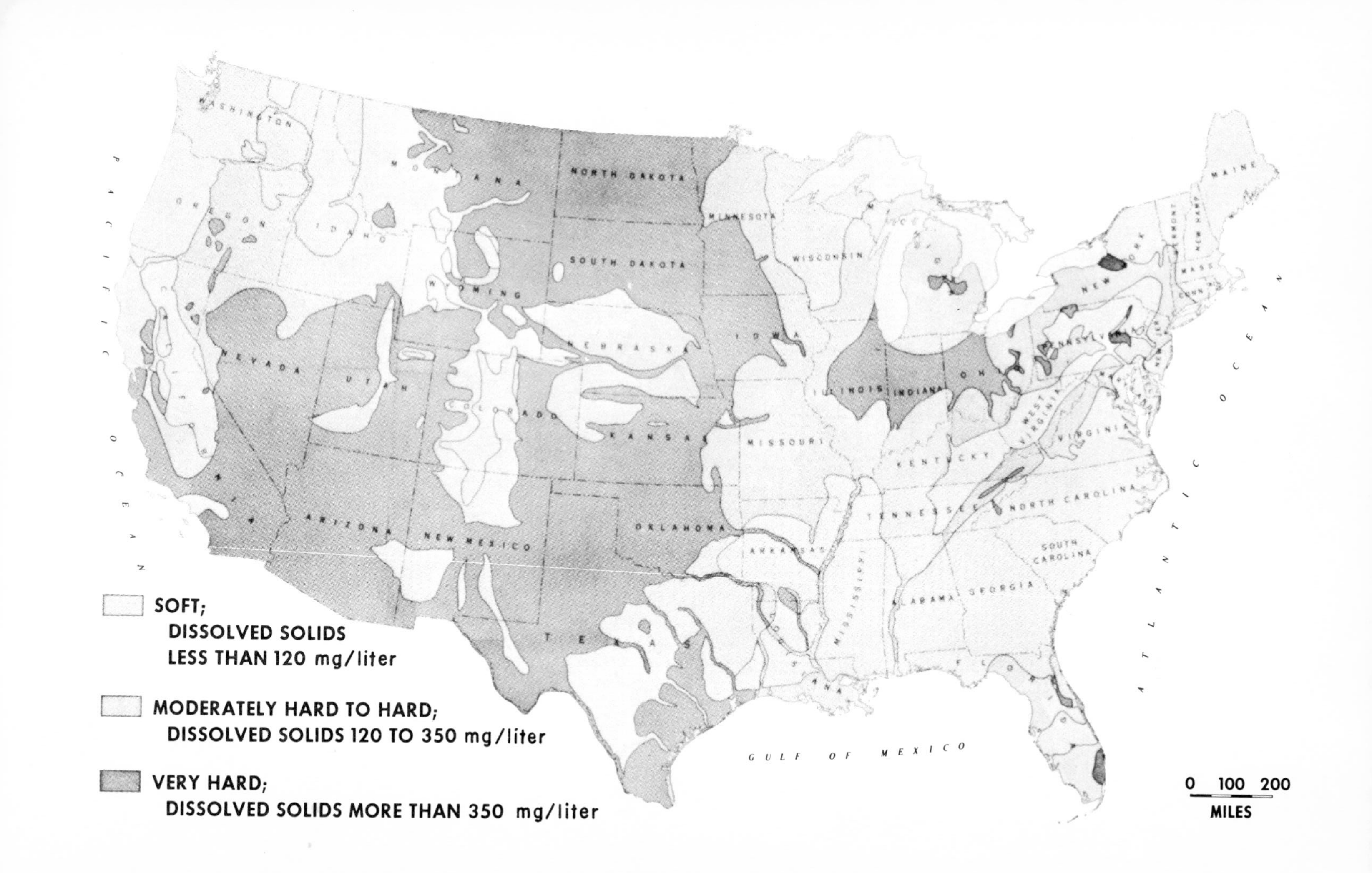
SOFT;
DISSOLVED SOLIDS
LESS THAN 120 mg/liter
MODERATELY HARD TO HARD;
DISSOLVED SOLIDS 120 TO 350 mg/liter
VERY HARD;
DISSOLVED SOLIDS MORE THAN 350 mg/liter
0 100 200
MILES
PACIFIC OCEAN
ATLANTIC OCEAN
GULF OF MEXICO
WASHINGTON
OREGON
IDAHO
MONTANA
NORTH DAKOTA
SOUTH DAKOTA
WYOMING
NEVADA
UTAH
COLORADO
NEBRASKA
KANSAS
ARIZONA
NEW MEXICO
OKLAHOMA
TEXAS
MINNESOTA
WISCONSIN
IOWA
MISSOURI
ARKANSAS
ILLINOIS
INDIANA
KENTUCKY
TENNESSEE
MISSISSIPPI
ALABAMA
GEORGIA
SOUTH CAROLINA
NORTH CAROLINA
VIRGINIA
WEST VIRGINIA
MAINE
MASS
CONN

where u and ϵ refer to the velocity and lateral diffusion coefficient, respectively, at a point (x, y), and c and t refer to concentration and time, respectively. Sayre and Chang(*24*), among others, concluded that the Fickian diffusion equation is a valid approximation of the dispersion. From an applicational viewpoint, however, the most significant outcome of this approach is probably due to Fischer(*25*), who proposed that the magnitude of the apparent dispersion coefficient K in natural streams is controlled predominantly by the lateral variation of longitudinal velocity

TABLE 1
Analyses of Typical Surface Waters and Groundwaters in the United States[a,b]

Analysis number[c]	1	2	3	4	5	6	7	8
Date of collection	8/22/61	9/1/61	10/19/61	1961	8/9/61	9/15/61	8/23/61	7/25/61
Silica (SiO_2)	5.9	2.1	5.5	8.7	25	8.3	9.1	21
Iron (Fe)	0.00	0.17	0.40	—	0.03	0.72	0.10	0.01
Manganese (Mn)	0.00	0.14	0.00	—	—	0.01	0.06	0.00
Calcium (Ca)	8.5	28	37	84	65	12	57	26
Magnesium (Mg)	2.6	7.0	8.9	28	23	6.1	32	6.2
Sodium (Na)	3.6	4.1	17	92	14	7.5	3.4	86
Potassium (K)	1.5	0.9	5.1	4.0	1.8	0.7	1.7	1.9
Bicarbonate (HCO_3)	25	92	128	140	179	78	332	254
Carbonate (CO_3)	0	0	0	1	0	0	0	0
Sulfate (SO_4)	9.0	18	48	285	112	3.8	12	11
Chloride (Cl)	5.5	8.0	10	83	18	3.0	3.5	40
Fluoride (F)	0.1	0.0	0.4	0.4	0.8	0.4	0.1	0.7
Nitrate (NO_3)	4.7	0.5	3.2	1.4	0.2	1.2	0.2	0.0
Dissolved solids	59	129	222	657	410	87	284	318
Hardness as $CaCO_3$	32	99	129	323	256	55	274	90
Noncarbonate hardness as $CaCO_3$	11	24	24	206	110	0	2	0
Specific conductance (μmhos at 25°C)	91	213	324	1040	535	137	498	533
pH	6.4	7.6	7.7	8.4	7.9	6.8	7.6	7.2
Color	5	3	5	—	5	5	1	0

[a]Reprinted from Ref. *22a*.
[b]All concentrations in milligrams per liter.
[c]Analyses numbers are identified as follows:

1. Baltimore, Md., North Branch Patapsco River (raw).
2. Detroit, Mich., Detroit River (raw).
3. St. Louis, Mo., Mississippi River (raw).
4. Los Angeles, Calif., Colorado River (raw).
5. Jacksonville, Fla., Composite of several wells (raw).
6. Memphis, Tenn., Allen Well Field. Composite of several wells 400–600 ft deep (raw).
7. Rockford, Ill., Well No. 15, 1355 ft deep (raw).
8. Houston, Texas, Southwest Well Field. Composite of 12 wells 490–2000 ft deep (raw).

rather than by the vertical variation. Fischer's formula for estimating the coefficient from the stream-gauging data has been verified in a number of streams (*26*).

As for the magnitudes of the lateral diffusion coefficient, ϵ, the above studies as well as many other measurements indicate that it can be adequately expressed by

$$\epsilon = \alpha R U^* \tag{7}$$

where α is the constant, R is the hydraulic radius, and U^* is the shear velocity. Thus, for example, ϵ in natural streams is known to vary from 0.2 to 0.6 RU^*, while K can range from 10 to 4000 RU^* in natural streams (*27,28*).

VI. River Chemical Composition and Loads

Rivers of the conterminous United States discharge annually to the oceans an average of about 225 million tons of dissolved solids (*29*). This is equivalent to about 131 mg/liter or 82 tons from each square mile of drainage. We observe in Fig. 9 that the rate of runoff from the lands varies greatly from place to place. For example, in the North Atlantic slope basins, runoff is about 19.3 in. annually, as compared to about 1.9 in. annually from the western Gulf of Mexico basins. Comparing further, we find that the annual chemical load from North Atlantic slopes is about 126 tons/sq mile of drainage, contrasting with about 31 tons/sq mile for western Gulf basins.

A. *Minor Elements*

Whereas major ions in natural waters may vary in concentration from a fraction of a milligram to several thousand milligrams per liter, the range of minor elements is for the most part a few tenths of a microgram per liter. In Table 3, only the observed range and median values for the three hydrosolic metals – aluminum, iron, and manganese – and for the alkaline earths – barium and strontium – are much over 100 μg/liter. Other elements detected occur at or below 100 μg/liter, all of which have median (or middle) values of 10 μg/liter or less (*30*).

This is an important consideration in our study, for we are dealing with quantities of an element which are often much less than its solubility product and which are constantly adjusting toward equilibrium with a new

TABLE
Analyses by U.S. Geological Survey of Typical

Surface waters[b]	1			2			3			4		
	Max	Min	Avg	Max	Min	Avg	Max	Min	Avg	Max	Min	Avg
Date of Collection	Jan. 1–10, 1962	May 1–10, 1962	Oct. 1961–Sept. 1962	Oct. 1–31, 1961	April 1–30, 1962	Oct. 1961–Sept. 1962	Dec. 1–9 1961	March 1–31, 1962	Oct. 1961–Sept. 1962	Dec. 13–28, 1961	March 29, April 5, 1962	Oct. 1961–Sept. 1962
Silica (SiO_2)	8.7	2.1	4.4	9.0	7.6	9.5	6.4	5.6	6.2	16	15	15
Iron (Fe)	0.20	0.00	0.01	0.02	0.10	0.07	—	—	—	—	—	—
Manganese (Mn)	0.14	0.00	0.03	—	—	—	—	—	—	—	—	—
Calcium (Ca)	20	8.2	16	8.7	7.5	7.5	46	29	33	83	35	61
Magnesium (Mg)	8.0	2.6	6.0	2.5	1.8	2.4	14	7.1	8.3	24	14	17
Sodium (Na)	10	2.2	7.3	9.0	5.9	7.4	26	7.6	10	64	13	40
Potassium (K)	2.5	1.2	1.9	2.9	1.6	2.1	3.0	2.0	2.0	9.2	4.0	7.6
Bicarbonate (HCO_3)	66	18	51	41	31	36	86	73	80	301	146	195
Carbonate (CO_3)	0	0	—	0	0	—	0	0	—	0	0	—
Sulfate (SO_4)	30	17	26	7.8	6.4	6.9	103	42	49	178	45	124
Chloride (Cl)	9.0	3.2	7.5	7.6	4.1	5.4	36	10	14	21	8.2	14
Fluoride (F)	0.6	0.0	0.2	0.1	0.1	0.1	0.4	0.1	0.2	0.5	0.3	0.4
Nitrate (NO_3)	5.7	2.6	4.1	4.0	1.7	2.3	5.8	4.2	4.4	4.6	3.0	4.0
Dissolved solids	125	46	104	81	59	69	296	154	175	554	220	394
Hardness as $CaCO_3$	83	32	65	32	26	28	173	102	116	307	145	223
Noncarbonate hardness as $CaCO_3$	31	17	24	0	1	0	102	42	50	61	26	66
Specific conductance (μmhos at 25°C)	198	86	168	113	84	96	475	249	288	867	346	607
pH	7.8	6.5	—	6.9	7.7	—	8.2	6.9	—	7.6	7.3	—
Color	8	2	4	15	32	25	20	5	11	—	—	—

[a]All concentrations in milligrams per liter.
[b]Surface waters are identified as follows:
1. Delaware River at Trenton, N.J. [U.S. Geol. Surv. Water-Supply Paper 1941 (WSP 1941)].
2. Roanoke River at Jamesville, N.C. (WSP 1941).
3. Ohio River at Lock and Dam 53, near Grand Chain, Ill. (WSP 1942).

environment. Further, it appears that some minor elements essential to the growth of living organisms in the stream may be removed at a rate faster than they are supplied to a stream.

B. *Time-Weighted Characteristics*

Let us consider some of the larger streams in Ohio and examine some selected water quality characteristics. First, we evaluate the continuous record, that is, data taken from frequent sampling, and charts and graphs from instruments mounted in the river. Table 4 gives percentage of time,

2
Surface Waters of the United States[a]

5			6			7			8		
Max	Min	Avg	Max	Min	Avg	Max	Min	Avg	Max	Min	Avg
Sept. 1–30, 1962	April 1–30, 1962	Oct. 1961– Sept. 1962	March 1–31, 1962	Nov 1–30, 1961	Oct. 1961– Sept. 1962	Sept. 27–30, 1962	May 8–23, 1962	Oct. 1961– Sept. 1962	March 9–28, 1962	June 1–30, 1962	Oct. 1961– Sept. 1962
8.7	7.4	9.5	11	9.5	11	19	18	18	18	6.5	13
0.16	0.00	0.03	—	—	—	0.01	0.00	0.00	—	—	—
46	32	40	48	39	43	230	55	77	22	18	20
14	8.0	10	18	18	19	51	12	20	6.4	3.9	4.1
26	11	18	37	34	34	212	28	68	10	3.6	6.2
3.1	2.1	2.7	3.9	3.8	3.7	4.3	3.2	4.1	1.8	0.8	1.7
160	101	128	193	172	177	239	170	195	101	61	76
0	0	—	0	0	—	0	0	—	0	0	—
47	34	41	37	29	33	730	77	156	20	11	15
36	11	19	54	51	57	200	20	58	5.0	1.0	3.0
0.2	0.0	0.2	0.3	0.2	0.3	0.5	0.3	0.4	0.3	0.1	0.2
1.8	1.7	2.2	1.8	0.5	1.6	8.9	2.3	3.6	1.5	0.4	1.0
279	157	218	330	268	293	1,620	314	531	140	69	99
171	113	140	194	172	186	785	185	271	88	54	67
40	30	35	36	30	40	585	46	111	5	4	4
434	275	352	540	464	516	2,180	477	803	221	125	161
7.6	7.2	—	7.5	7.2	7.3	7.9	7.5	7.7	7.8	7.7	—
10	5	18	—	—	—	—	—	—	—	—	—

4. Missouri River at Nebraska City, Nebr. (WSP 1943).
5. Mississippi River at St. Francisville, La. (WSP 1944).
6. Colorado River at Austin, Texas (WSP 1944).
7. Colorado River near Grand Canyon, Ariz. (WSP 1945).
8. Columbia River near The Dalles, Ore. (WSP 1945).

or durations, of water quality characteristics in Ohio streams, 1960–1964. Next, we show the water quality parameter, either temperature, specific conductance, or pH. An analysis of the data by using computer techniques gives the water quality data expressed as a percentage of time (or the number of days each year) when the individual parameter was equal to or less than the value shown for 1, 10, 25, 50, 75, or 90% of the time. For example, the Mahoning River at Levittsburg, Ohio, had a temperature of 33°F, at or near freezing, 10% of the time, or about 35 days a year. At the other extreme, the temperature was 74°F, or higher, 10% of the time. At the 50% duration, or about one-half of the days, the temperature was 54°F or less. In examining Table 4 further, we can see that 80% of the

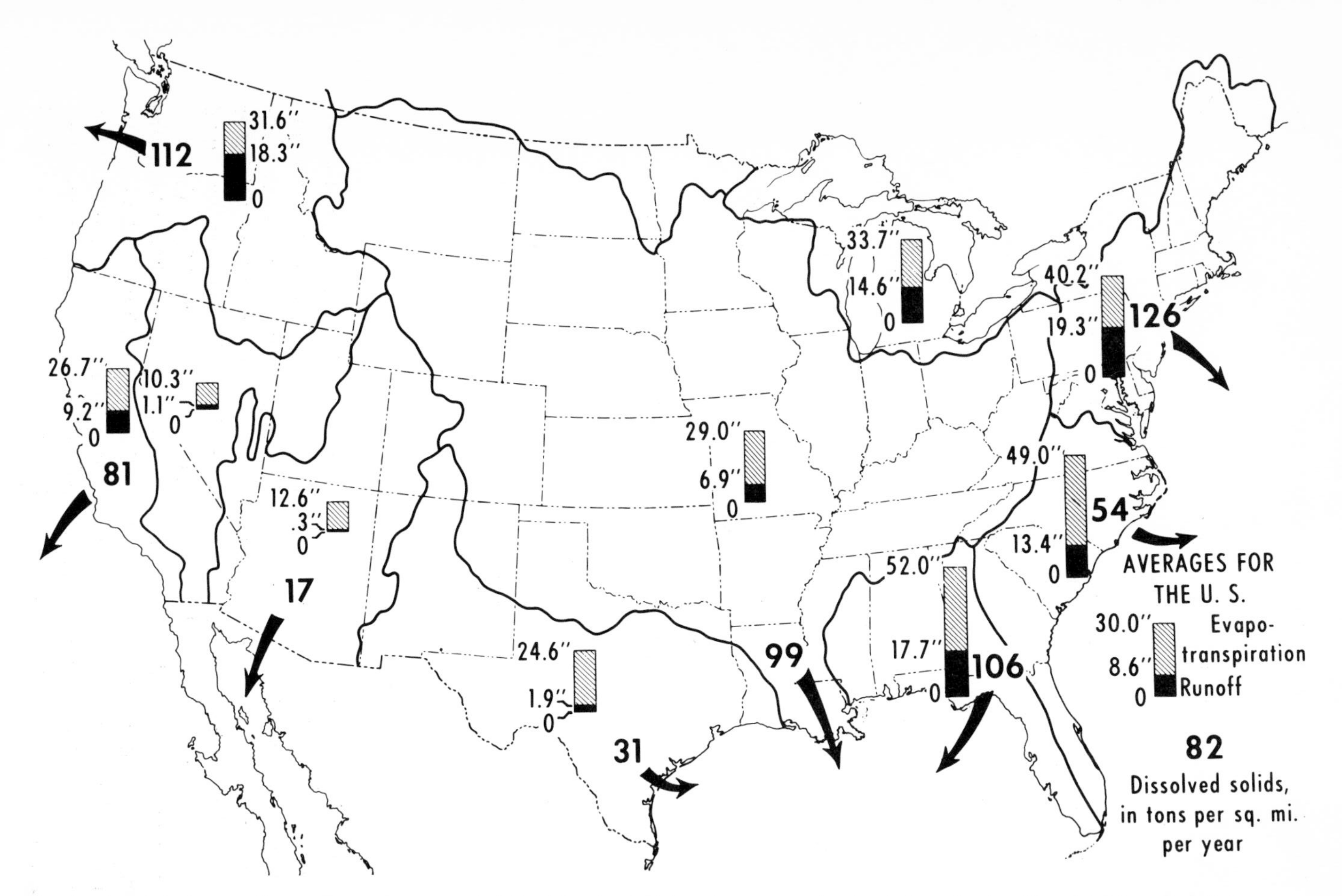

Fig. 9. Precipitation runoff, evapotranspiration, and yield of dissolved solids to the ocean and major drainage basins. [Reprinted from Ref. *29*.]

TABLE 3
Minor Elements in Large Rivers of North America[a]

Element	Median, μg/liter	Range, μg/liter	Atlantic Coast	Gulf Coast	Pacific Coast
Ag	0.09	0–0.94	+	0	0
Al	238	12–2550	0	+	0
B	10	1.4–58	0	0	0
Ba	45	9–152	—	+	—
Co	0	0–5.8			
Cr	5.8	0.72–84	+	—	—
Cu	5.3	0.83–105	0	+	0
Fe	300	31–1670	0	+	0
Li	1.1	0.075–37	—	+	0
Mn	20	0–185	+	0	0
Mo	0.35	0–6.9	+	0	+
Ni	10	0–71	+	0	0
Pb	4.0	0–55	0	—	+
Rb	1.5	0–8.0	0	+	—
Sr	60	6.3–802	+	—	0
Ti	8.6	0–107	+	+	—
V	0	0–6.7			
Zn	0	0–215			

[a]+ = Significant percentage of determinations above median.
0 = Approximately 50% of determinations above and below median.
— = Significant percentage of determinations below median.

TABLE 4
Duration Table of Water Quality Parameters of Ohio Streams, 1960–1964

Station number	Stream and location	Parameter[a]	Percentage of time that parameter was equal to or less than that shown: 1%	10%	25%	50%	75%	90%
1	Mahoning River at Leavittsburg	Temperature	32	33	36	54	69	74
2	Mahoning River at Lowellville	Specific conductance	305	490	530	705	955	1080
		pH	6.7	6.7	6.7	6.8	7.1	7.2
		Temperature	44	52	57	80	88	92
3	Ohio River at East Liverpool	Specific conductance	201	275	350	425	555	600
		pH	4.8	5.4	5.8	6.2	6.7	7.1
		Temperature	32	35	41	61	75	79
4	Ohio River at New Cumberland Dam, at Stratton	Specific conductance	205	220	280	320	440	550
		Temperature	32	37	41	55	74	80
5	Tuscarawas River at Newcomerstown	Specific conductance	260	735	1200	1900	3100	3800
		pH	6.4	6.7	6.9	7.1	7.5	7.9
		Temperature	32	33	41	57	72	78

TABLE 4 (continued)

Station number	Stream and location	Parameter[a]	1%	10%	25%	50%	75%	90%
			Percentage of time that parameter was equal to or less than that shown					
6	Killbuck Creek at Killbuck	Temperature	32	32	35	52	66	70
7	Salt Fork near Cambridge	Specific conductance	201	260	300	400	500	550
		pH	5.0	6.8	7.0	7.2	7.4	7.7
8	Licking River near Newark	Temperature	32	34	43	61	72	76
9	Licking River near Dillon	Temperature	35	38	42	55	71	76
10	Muskingum River at McConnelsville	Specific conductance	225	395	490	885	1450	1750
		pH	6.5	6.9	7.1	7.3	7.5	7.7
		Temperature	32	34	42	59	75	80
11	Muskingum River near Beverly	Specific conductance	260	530	690	1150	1180	1980
		pH	6.1	6.6	7.0	7.3	7.9	8.3
		Temperature	33	44	54	73	83	86
12	Hocking River at Athens	Specific conductance	50	410	705	970	1200	1360
		pH	5.0	6.2	6.6	6.9	7.1	7.4
		Temperature	32	32	38	54	69	74
13	Sandy Run near Lake Hope	Specific conductance	90	200	320	475	770	1240
		pH	3.2	3.4	3.5	3.7	3.9	4.2
14	Ohio River near Huntington, W. Va.	Specific conductance	160	210	340	420	640	720
		pH	6.1	6.2	6.5	6.9	7.3	7.5
		Temperature	33	37	45	59	77	82
15	Olentangy River near Delaware	Temperature	33	36	39	53	68	75
16	Olentangy River near Worthington	Temperature	32	33	38	54	69	76
17	Scioto River at Chillicothe	Temperature	32	34	41	60	74	79
18	Paint Creek near Bourneville	Temperature	32	35	41	55	69	73
19	Scioto River at Higby	Temperature	32	34	40	56	78	80
20	Scioto River at Lucasville	Specific conductance	215	445	605	730	840	912
		pH	7.0	7.0	7.3	7.5	7.8	8.0
		Temperature	32	34	40	54	69	75
21	Great Miami River at Taylorsville	Specific conductance	310	520	610	625	680	738
		pH	7.2	7.6	7.9	8.1	8.3	8.5
22	Stillwater River at Englewood	Specific conductance	240	405	540	610	670	692
		pH	7.1	7.8	7.9	8.1	8.4	8.5
23	Mad River at Dayton	Specific conductance	340	595	630	680	745	775
		pH	7.1	7.5	7.7	7.8	8.0	8.2
24	Great Miami River at Miamisburg	Temperature	34	40	45	62	77	84
25	Great Miami River near Miamisburg	Specific conductance	310	605	650	725	760	855
		pH	6.6	7.3	7.4	7.7	7.9	8.0

TABLE 4 (continued)

Station number	Stream and location	Parameter[a]	Percentage of time that parameter was equal to or less than that shown					
			1%	10%	25%	50%	75%	90%
26	Great Miami River at Middletown	Specific conductance	520	650	710	780	840	880
		pH	6.9	7.0	7.2	7.6	7.9	8.0
27	Great Miami River near Middletown	Specific conductance	520	680	750	830	910	960
		pH	6.7	6.9	7.0	7.2	7.5	7.6
28	Great Miami River at Hamilton	Temperature	34	39	44	60	77	83
29	Great Miami River near Hamilton	Specific conductance	520	610	720	820	916	970
		pH	6.7	6.9	7.0	7.4	7.8	8.2
30	Great Miami River at Elizabethtown	Specific conductance	475	607	640	695	760	800
		pH	6.6	7.1	7.4	7.8	7.9	8.2
		Temperature	32	39	45	64	77	83
31	Maumee River at Waterville	Temperature	32	33	37	54	73	79
32	Maumee River at Anthony Wayne Bridge, at Toledo	Specific conductance	501	515	540	580	625	640
		pH	6.7	7.1	7.2	7.3	7.4	7.9
33	Maumee River at Toledo	Specific conductance	300	450	510	560	595	830
		pH	5.0	6.7	6.9	7.0	7.5	7.9
34	Maumee River at Toledo	Specific conductance	255	408	445	507	565	688
		pH	6.5	6.7	7.0	7.3	7.5	7.6
35	Maumee River at Buoy 39, at Toledo	Specific conductance	201	305	320	360	385	450
		pH	6.9	6.9	7.1	7.2	7.3	7.4
36	Maumee River at Buoy 33, at Toledo	Specific conductance	205	220	240	290	345	370
		pH	6.8	6.8	7.0	7.3	7.5	7.8
37	Sandusky River near Fremont	Specific conductance	230	430	520	810	975	1020
		pH	6.7	6.8	7.2	7.5	8.0	8.2
		Temperature	32	33	40	57	73	80
38	Huron River at Milan	Temperature	32	32	35	54	70	76
39	Black River near Elyria	Specific conductance	405	505	760	875	1000	1160
		pH	5.6	6.8	7.0	7.2	7.4	7.6
		Temperature	32	32	34	50	68	70
40	Cuyahoga River at Independence	Temperature	32	36	39	57	70	75
41	Grand River at Painesville	Specific conductance	1000	3300	4100	8600	9950	12400
		pH	6.1	6.5	6.8	7.0	7.3	7.4
		Temperature	32	35	40	61	79	84

[a]Specific conductance in μmhos at 25°C, temperature in °F.

time (the difference between 10 and 90%) the temperature ranged between 33 and 74°F. Table 4 then allows us to examine the stream quality in its present condition and to select goals and set absolute criteria, or accept the present situation as it is. Many other things can be inferred from Table 4. For example, note that at station 2, Mahoning River at Lowellville, the 50% temperature value is 80°F, indicating that the annual average water temperature has increased about 26°F in a reach of about 40 miles from Levittsburg to Lowellville. Further, we observe a change at extreme values, i.e., temperature downstream has increased 19°F, from 33 to 52°F, about 10% of the time, and increased 18°F, from 74 to 92°F, about 90% of the time.

C. Biological Considerations

Concluding a 13-year study of the numbers and kinds of aquatic species found in rivers in the eastern United States, Patrick(*31*) summarizes aquatic life in natural and polluted areas of streams and rivers as follows:

> As a result of these studies of sections of rivers and streams, species numbering 350 to 400 in the larger groups of organisms such as protozoa insects, fish and algae usually vary less than 33 percent from the mean for areas in natural rivers in which all the diversified habitats have been collected.

This relative similarity in numbers of taxa is better understood if one considers the total number of taxa treated in this paper for the rivers draining into the Atlantic Ocean or Gulf of Mexico adjacent to the southeastern states. In algae the total number of systematic entities is approximately 311, whereas the number of entities in any one river varied between 58 and 105. In protozoa the total number of entities in all rivers was approximately 291, while the number in any one river varied from 38 to 86. In insects the total number of entities in all rivers was approximately 213, while in any one river it usually varied from 48 to 99. The exception was the Escambia River, which had 29 species probably because brackish water occasionally invades this area. In fish the total number of entities in all of these rivers was approximately 108, while in any one river it varied from 13 to 39. We refer to "approximate numbers" because in some cases, where entities were involved which could not be specifically determined, more than one entity may have been involved. This was particularly true for the insects.

This relatively similar number of species found at any one time (when one considers the total number of species which might be collected) is exemplified if we consider the studies that Patrick(*31*) made on the Savannah River. Between 1951 and 1956, within a distance of 35 miles, she studied two areas five times and two areas six times. As a result of these studies the following total numbers of species identified for the various groups were: algae, 394; protozoa, 289; insects, 241; and fish, 106. However, the number of species found at any one station at a single time varied as follows: algae, 49–131; protozoa, 30–59; insects, 34–73; fish, 10–34. It is interesting to note that the total numbers of species for each group found in this series of studies are very similar to the total numbers of species for the nine rivers considered in this paper.

If one compares river regions of similar structural types, the numbers of species of the various groups of organisms are not consistently more similar than when compared to regions representing different structural areas of a river. That is, the numbers of species in the North Anna, White Clay, North Fork of the Holston, and Ottawa are not more similar to each other than they are to those in the Escambia and Savannah. Nor are the numbers of species in the Potomac and Flint Rivers more similar to each other than to those found in other river regions. Thus the region of the river studied does not per se seem to influence the number of species present. However, there is a slightly higher number of species of protozoa and insects in hard water rivers(*31*) and of algae in soft water rivers, although the ranges of species numbers overlap in the rivers considered.

We see (Table 5) that 84 species of algae is the mean number of total species for any one river. The greatest deviation is 25%, which occurs in the Savannah River for 1954 and in the North Fork of the Holston.

In the protozoa the mean number of species for all rivers is 59 (Table 5) and the variation in species numbers is usually less than 33%. The small-

TABLE 5

Total Number of Systematic Entities for Each Study[a] in Each River

	Soft water rivers						Hard water rivers								
Entity	Escambia	Savannah, 1954	Savannah, 1955	North Anna	White Clay	Flint	N. Fork Holston	Rock Creek	Ottawa, 1955	Ottawa, 1956	Potomac, 1956	Potomac, 1957	Mean, all rivers	Mean, soft rivers	Mean, hard rivers
Algae	77	105	101	98	73	79	63	65	76	58	105	103	84	89	78
Protozoa	38	61	40	58	56	51	—	86	—	48	85	68	59	51	72
Insects	29	58	51	61	57	—	83	48	59	61	89	99	63	51	73
Fish	39	19	35	21	20	13	21	24	18	28	18	29	24	25	23

[a] Reprinted from Ref. *31*, by courtesy of Academy of Natural Sciences of Philadelphia.

est number of species was found in the Escambia River, which deviated 37% from the mean total number and 26% from the mean total number of species for soft water. The highest number of species, 86, was found in Rock Creek, which is a natural area a short distance below the area of recovery from organic pollution. It has often been observed that the number of species of protozoa may increase as a result of mild organic pollution. This number of species of protozoa is a deviation of 46% from the mean for all rivers of 19% from the mean for hard water rivers. Table 5 shows that the mean number of species is considerably greater in hard water rivers.

The mean number of species of insects for all rivers is 63 (Table 5) and the variation is usually less than 33%. The lowest number of species was found in the Escambia. This is probably due in part to the fact that this station is slightly brackish on rare occasions. The number of species in this area was 29, which deviates 54% from the mean for all rivers. If, however, we consider just soft water rivers this number is 43% of the mean because the number of species in soft water is less than in hard water. The highest number of species, 99, was found in the Potomac in 1957 (Table 5). This number is 57% higher than the mean for all rivers and 36% higher than the mean for hard water rivers.

The mean number of species of fish is very similar in hard and soft water, and the variation in species numbers is usually less than 33%. The lowest number of species was 13 in the Flint River, which deviated 46% from the mean. The highest number of species collected, 39, was in the Escambia River(*32*), which deviated 63% from the mean. The fish deviated more from the mean than any of the other groups studied. This is probably due in part to the fact that fish are the most difficult group to collect.

Of interest also is the fact that so many of the taxa have only been found in one area. When one examines the systematic lists, it is evident that the taxa are not endemic and have been found in other localities in the geographical area in which most of these rivers occur. It is probable that if the rivers had been studied in more areas and at many different times during the year more overlap would have occurred.

The main problem is the collectability of the species. Many taxa exist during periods which are not optimum for their development in very small populations. Other taxa, due to their life cycle, exist in such minute forms at certain times of the year that they are uncollectable. Due to variation in the environmental factors, the ecological conditions of a river may be very different on the same day of the same season in two different years and hence the collectable taxa may vary greatly. Likewise, the collectable populations may vary from month to month. In organisms with short life cycles even a few days can cause a considerable change in the relative population sizes of the taxa.

It is convenient to consider stream ecosystems under two subdivisions: (1) those with eroding (hence generally firm) substrates, and (2) those with depositing (hence generally soft) substrates. In many cases these stream bed types alternate in the pools and rapids of small streams. Aquatic communities are quite different in the two situations(*33*). The communities of pools resemble those of ponds or lakes, while the life of the hard-bottom rapids is composed of more specialized forms adapted in various ways to life in the current.

Streams are incomplete ecosystems in that much of the biological or metabolic energy flow is based on organic matter imported from the drainage basin. Organic matter may also leave the community by downstream export or by sedimentation below the active metabolic level of the community. Thus, according to Odum(*34*), the rate balance of organisms at any unit volume in the system is given as follows:

$$I_m + P = E_x + R \tag{8}$$

where I_m is the import rate, E_x the export rate, P the production rate, and R the community respiration rate. The quantity of consumers that can be supported is dependent on the primary production and the import rates.

Respiration is a function of the concentration of organic matter in the water and on the bottom, and primary production is a function of light and nutrients. Thus a stream in which $P/R > 1$ is designated autotrophic because production exceeds respiration, resulting in a net storage or export of organic matter. A stream in which $P/R < 1$ is designated heterotrophic and is characterized by a net loss or import of organic matter(*35*).

Algae, either floating or attached, are the chief primary producers in streams and, of course, are closely involved in waste assimilation and oxygen production. Therefore, the ecology of stream algae is of utmost importance to the subject of this report. In an extensive review of the ecology of river algae, Blum(*36*) discussed the following factors which influence algal abundance and distribution: size of stream, current rate, water level, depth, temperature, light, turbidity, dissolved gases, pH, calcium, phosphorus, nitrogen, salinity, and other solids. The conclusions were qualitative and disappointingly few in number: A phytoplankton bloom (population increase) seems to follow the period of highest concentration of nitrates in such a way that a decrease in nutrients precedes by a few days to a few weeks the maximum development of phytoplankton. For attached algae: (a) An extended period of low water with little change in hydrographic conditions is conducive to the full development of benthic organisms; (b) a period of several cloudy days can induce detachment and disappearance of certain algae; (c) an ice cover can bring about sharp changes in the vegetation; (d) periodicity of some species depends on that of some other species; (e) leafing out of bank vegetation has sharp repercussions in the stream. In general, "given sufficient warmth and a series of favorable days, the water chemistry in respective parts of various streams becomes the factor which determines the production of a dense plankton and of a dense benthic vegetation" (*36*).

The fact is that we know very few details about algal production in surface waters (see Chap. 4).

D. Pesticides

Pesticide analyses of samples of a water-suspended sediment mixture obtained during the first year of study by Brown and Nishioka(*37*) are partially summarized in Table 6 to indicate frequently occurring pesticides and place of occurrence. Based on data of the study(*37*), the following observations can be made:

(1) No herbicide was found at any time at any station. The absence of herbicides from any of the samples analyzed may be due at least in part to their susceptibility to degradation. Recent studies by Goerlitz and Lamar(*38*) on the stability of herbicides added to water samples indicate that 2,4-D may be degraded rapidly, especially in samples obtained from areas repeatedly treated with 2,4-D.

(2) All insecticides were found at one time or another, but not at all stations.

(3) The insecticide concentrations found were very small, amounting to 0.005 μg/liter in slightly more than 50% of all positive results.

(4) Although no definite pattern could be noted in pesticide occurrence, positive results were most frequently found in February, March, April, and May.

(5) The most frequently found insecticide was lindane, which occurred 46 times out of a total of 165 positive results. The most infrequently occurring insecticide was aldrin, which was observed only four times at all stations during the year.

(6) On a geographical basis, most frequent occurrence of pesticides was at the Rio Grande Station below Anzalduas Dam, Texas. Thirty-two positive results, about 20% of the total, were obtained at this station. The least numbers of positive results were noted at

TABLE 6
Number of Occurrences of Insecticides[a]

Pesticides	Missouri River at Nebraska City, Nebr.	Arkansas River below John Martin Reservoir, Colo.	Arkansas River at Van Buren, Ark.	Brazos River at Richmond, Texas	Colorado River at Wharton, Texas	Rio Grande below Anzalduas Dam, Texas	Colorado River at Yuma, Ariz.	Sacramento River at Verona, Calif.	Yakima River at Kiona, Wash.	Snake River at King Hill, Idaho	Columbia River near The Dalles, Ore.	Totals
Aldrin	1	—	1	—	—	—	—	—	—	1	1	4
DDD	—	1	2	2	—	5	—	1	2	—	—	13
DDE	—	3	2	2	2	4	2	—	3	—	—	18
DDT	2	1	2	3	2	1	1	—	1	1	—	14
Dieldrin	5	3	5	3	3	5	1	2	—	1	1	29
Endrin	1	—	—	—	1	3	1	—	—	1	—	7
Heptachlor	1	3	2	1	1	4	—	—	1	1	—	14
Heptachlor epoxide	2	3	1	2	2	2	2	2	2	1	1	20
Lindane	4	8	5	3	3	8	3	3	4	1	4	46
Totals	16	22	20	16	14	32	10	8	13	7	7	165

[a]Reprinted from Ref. *37*, by courtesy of Pesticides Monitoring Journal, Federal Committee on Pest Control.

the Snake and Columbia River stations, each of which recorded only seven pesticide occurrences.

(7) No relationship can be noted between pesticide occurrence and the various factors in pesticide application, since the latter parameters cannot be ascertained in most areas.

(8) No evaluation of other forms of pesticide movement, such as might be involved in sediment transport, can be made on the basis of the data obtained. Data presented will provide a basis for investigations into other areas of pesticide relationships, such as mode of transport, time of travel, and improved sampling techniques.

A discussion of various aspects of pesticides may be found in other volumes of this work (see in particular Chaps. 6 and 23).

VII. Water and Sediment Considerations

One finds that the streams of heavily watered areas of the United States are the carriers of the lowest concentrations of sediment. For example, in the great belt of heavily populated areas east of the Mississippi River, sediment concentrations in most surface waters average less than 270 mg/liter (Fig. 10). For the eastern United States we can say that sediment concentrations average less than 1900 mg/liter except for a belt along the east side of the Mississippi River extending south from Madison, Wisconsin, nearly to the Gulf Coast. In the tier of states from Virginia to Georgia one observes that the slopes of the east flanks of the Appalachian chain are being eroded.

In resolving sediment problems one needs information about concentration, size composition, and mode of transport. These parameters are controlled by the composite effect of various elements of the natural environment. But the more important elements controlling sheet erosion, or soil loss, appear to be (a) the steepness of the terrain, (b) the texture and erodibility of the soil mantle, (c) the amount of disturbance of the soil mantle by cultivation and the impact of rainfall, (d) the infiltration capacity of the soil mantle, and (e) the protection by vegetation.

Usually, computations of soil loss are based on the determined erodibility of a soil for certain standard conditions of land slope, precipitation, and vegetal cover such as those given by Musgrave (*39*).

Probably the steeper a land slope, the larger the fraction of precipitation that will run off over the ground if all other pertinent factors remain constant (*40*). Steep slopes also tend to cause high velocities of flow over the soil surface. The further the water flows over the soil surface, the

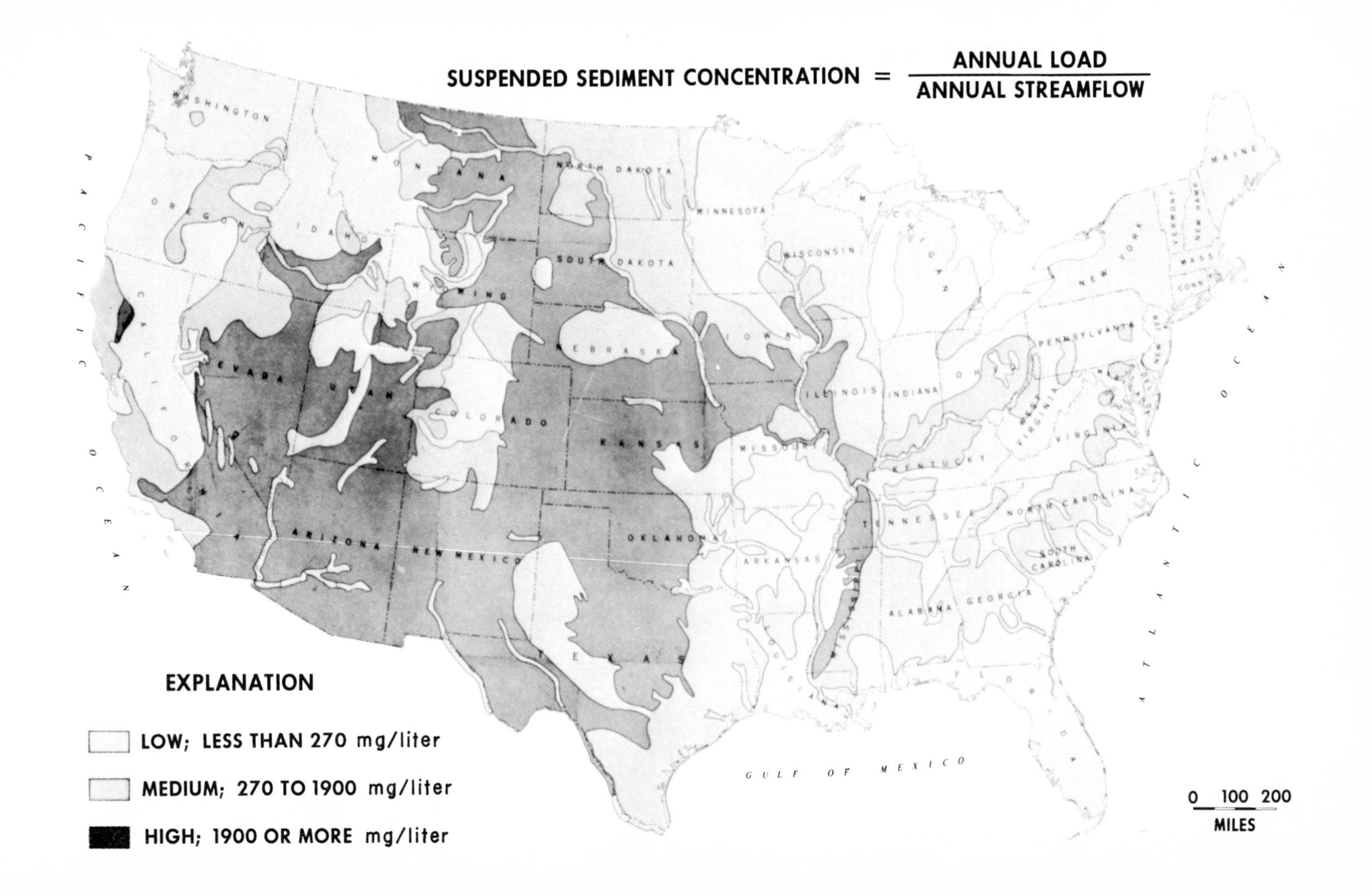

SUSPENDED SEDIMENT CONCENTRATION = ANNUAL LOAD / ANNUAL STREAMFLOW
EXPLANATION
LOW; LESS THAN 270 mg/liter
MEDIUM; 270 TO 1900 mg/liter
HIGH; 1900 OR MORE mg/liter
0 100 200
MILES
PACIFIC OCEAN
ATLANTIC OCEAN
GULF OF MEXICO
WASHINGTON
OREGON
CALIFORNIA
IDAHO
NEVADA
UTAH
ARIZONA
MONTANA
WYOMING
COLORADO
NEW MEXICO
NORTH DAKOTA
SOUTH DAKOTA
NEBRASKA
KANSAS
OKLAHOMA
TEXAS
MINNESOTA
IOWA
MISSOURI
ARKANSAS
LOUISIANA
WISCONSIN
ILLINOIS
MISSISSIPPI
MICHIGAN
INDIANA
KENTUCKY
TENNESSEE
ALABAMA
OHIO
GEORGIA
FLORIDA
WEST VIRGINIA
VIRGINIA
NORTH CAROLINA
SOUTH CAROLINA
PENNSYLVANIA
NEW YORK
MARYLAND
DEL.
NEW JER.
VERMONT
NEW HAMP.
MASS.
CONN.
MAINE

greater its velocity and depth become. Thus, the water has greater opportunity for picking up sediment. Precipitation on a soil surface tends to loosen soil particles and make them readily available for transportation.

In contrast to soil erosion, channel erosion (the entrainment of sediment from the bed or banks of a stream) occurs when the flow of a stream has the ability to transport much more of the available sediment than it is already carrying. Thus, reservoir planners are faced with the problem of increased channel erosion below reservoirs during periods when water is allowed to flow. Channel erosion is caused by the released water before it has been able to adjust its carrying ability to the available sediment.

Many streams erode locally even though their flows are in general equilibrium with available sediment loads. A winding stream tends to scour the outside of bends, but much of the eroded sediment, especially the larger particles, may be quickly deposited on the inside of the bends.

Langbein and Schumm(*41*) related sediment concentration to precipitation (Figs. 11 and 12). They also demonstrated that the presence of abundant vegetation in humid regions accounts for low to moderate sediment concentration and thus a smaller range of variability than in arid lands(*41*). Forested mountain areas in arid regions exhibit similar qualities.

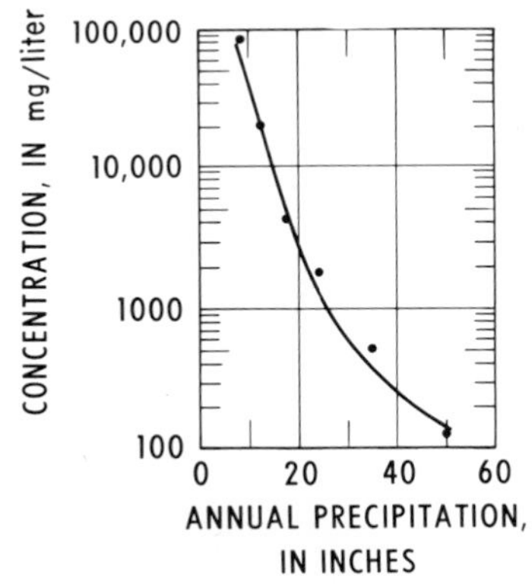

Fig. 11. Effect of annual precipitation on sediment concentration. [Reprinted from Ref. *41*, p. 1079, by courtesy of American Geophysical Union.]

VIII. Transport of Dissolved and Suspended Load

Leopold et al.(*32*) and Colby(*40*) have demonstrated that transport rate is related to several parameters, some independent and some dependent. The variables relate not only to available supply of sediment but also to

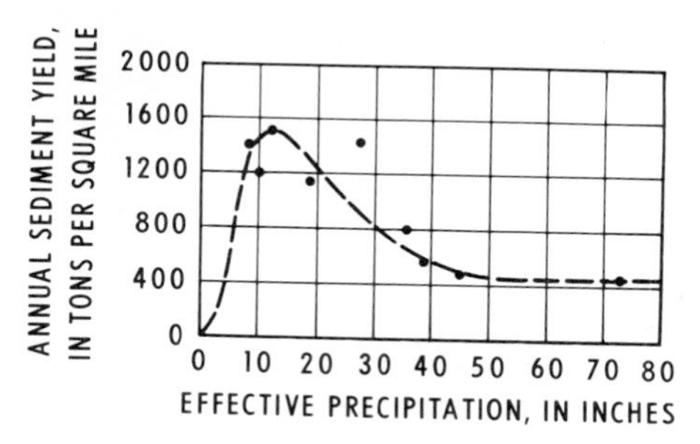

Fig. 12. Climatic variation of yield of sediment. [Reprinted from Ref. *41*, p. 1077, by courtesy of American Geophysical Union.]

sizes, shapes, and densities of the particles; velocities of flow; channel widths, depths, and slopes; bank roughness and bed configuration; and density, temperature, and at times the chemical composition of the water. The relation between transport rate of sands, mean velocity, and depth is an empirically useful one, but its basic mechanics are complex and poorly understood. Here, for low values of flow velocity [$<$ 2.5 fps (feet per second)], increasing depth is associated with decreasing rate of transport of sand, while at velocities of about 2.5 fps the transport rate of sand is independent of channel depth. However, at high values of mean flow velocity, higher transport rate of sand occurs with increases in water depth.

We observe that the dissolved load increases with increasing annual runoff up to about 10 in. For higher annual runoffs dissolved loads level off at about 125 tons/sq mile per year, as we observe for the North Atlantic slopes (see Fig. 9). In these wetter regions the dissolved load is a function of the availability of salts in the rocks; that is, it depends on rock composition and weathering rate and not on the available runoff. As pointed out by Leopold et al.(*32*), inasmuch as the dissolved load is less dependent than sediment load upon the quantity of flow, it follows that the larger the percentage of the load carrier in solution, the larger the percentage of material which is being removed from the basin by more frequent flows of smaller magnitude. The concentration of dissolved material in a river channel decreases with increasing magnitude of flow. Although the process of solution may be aided by floods and stream flow of high velocity, it is more dependent upon the presence of soluble, permeable rocks and abundant total precipitation to percolate through them.

If we now plot the dissolved load in tons per day against the discharge rate in cubic feet per second, we observe that the slope of each curve for most streams is less than 1. By combining this information with the frequency distribution of large and small discharges, we may compute the percentage of material transported as dissolved load by flows of varying magnitude occurring with varying frequency. As an example, for the Delaware River(*42*), the largest portion of the dissolved load is carried by relatively frequent flows.

The data indicate that the greater the percentage of the total load carried in solution, the more significant from an erosional viewpoint will be the flows of smaller magnitude. Leopold et al.(*32*) further point out that in a sample of about 70 rivers in different regions of the United States approximately 20% of the total measured load is carried in solution. Percentage of load in solution varies from 1% in the Little Colorado River in Arizona to 64% in the Juniata River in Pennsylvania.

The authors(*32*) conclude that this regional variation appears to be systematic and not random. In arid and semiarid regions where vegetation is sparse and sediment production high, dissolved load is a relatively small part of the total. Where precipitation is greater, runoff higher, and vegetation more abundant, the percentage of the total load which is dissolved increases. In areas of low runoff, the dissolved load constitutes only about 9% of the total load, whereas in areas of high runoff—where discharge is equal to or greater than 0.7 ft^3/sec per sq mile—the dissolved load constitutes about 37% of the total.

Langbein and Dawdy(*43*), in a study of published records for many streams, state:

> With no flow there is no load, and as runoff increases the dissolved load increases. When runoff exceeds 3 inches, additional runoff does not produce much additional dissolved load. At about 10 inches of runoff the load approaches 150 tons per sq mile per year, a maximum value. At about this value, and thus when the annual precipitation exceeds 25 inches annually, weathering proceeds at about the maximum rate and additional precipitation does not produce any greater quantity of soluble products.

Concentration of dissolved solids, on the other hand, decreases from an average of about 800 mg/liter in an arid climate, and where runoff exceeds about 10 in., concentration decreases with runoff as a straight dilution effect.

Leopold and co-workers(*32*) summarize the significance of these relations as follows:

> From the viewpoint of work done in erosion and transport, then, the evidence from measurements of the dissolved and suspended load carried by rivers suggests that in many drainage basins a large proportion of the work is performed by relatively frequent events of moderate magnitude.

The results of the example presented can be generalized as follows: To the extent that the movement of sediment by water or air is dependent upon shear stress, rate of movement can be described by the equation,

$$q = k(\nu - \nu_c)^n \tag{9}$$

where q is the rate of transport, k is a constant related to the characteristics of the material transported, ν is the shear stress per unit area, ν_c is the critical or threshold shear required to move the material, and n is an exponent. If the frequency distribution of such stresses, determined by climatic and meteorological events, follows a log-normal distribution as suggested by Leopold et al.(*32*), then the product of the weight of material moved times the frequency will always reach a maximum such that the greatest quantity of material is transported by frequent event rather than by the extremes.

Flow frequency distribution curves(*41*) are useful for the analysis of the flow values associated with principal transport of sediment and dissolved load. The Delaware River is a good example. It can be shown that the mean flow for the Delaware River at Trenton, 11,750 ft^3/sec, occurs at about the 35% position on the duration curve; that is, the river flows at a rate less than the mean flow during 65% of the time. This figure lies between 60 and 75% for many rivers. Similarly, many rivers flow at a rate less than half the mean flow about 25% of the time.

Flows of 35,000 ft^3/sec and above contribute only about 5% of the Delaware River runoff(*42*). All flows greater than 10,000 ft^3/sec contribute only 40% of the total runoff. Thus, we observe that floods provide large discharges but occur on so few days that they do not contribute as much water as lower flows, which occur much more often.

Next we observe the contribution of various ranges of discharge to the total dissolved load of the river(*42*). Low flows have larger concentrations of dissolved salts than do flood flows. The Delaware River at Trenton has an average concentration of 86 mg/liter. However, for about 5% of the time the dissolved solids concentration exceeds 127 mg/liter and about 5% of the time it is less than 53 mg/liter. In summary, 90% of the flows at Trenton range between 2000 and 35,000 ft^3/sec, and dissolved solids range narrowly between 53 and 127 mg/liter.

Leopold et al.(*32*) discuss the results of similar computations for Brandywine Creek in Pennsylvania and for the Rio Puerco in New Mexico. Despite climatic and physiographic differences, 50% of the total suspended load of both of these streams was transported by flows which occur on the average 1 day or more per year. Although many lower flows were required to transport the remaining 50% of the sediment, at least half of the suspended sediment is removed from these drainage basins by low and moderate flows.

A catastrophic flood may by itself transport an immense volume of sediment, but this type of flood recurs so infrequently that its cumulative effect is less than that of the more moderate, but much more frequent, annual and biannual floods.

IX. Temperature Characteristics

We have long ignored the significance of temperature, possibly the single most important water quality characteristic. Physical, chemical, and biological properties and reactions in water are functions of temperature. As we have seen, even sediment transport in streams is temperature dependent(*40*).

Computer studies(*44*) by the U.S. Geological Survey of 3-year continuous thermograph records at 200 key station sites, mostly in the eastern and northwestern United States, reveal that average maximum and minimum temperatures are reasonably consistent from year to year. For example, in records for about 200 stations, the average maximum and average minimum for 3 years at any site seldom varied more than 4°F and frequently varied no more than 1°F. Further, the spread between average maximum and average minimum is rarely more than 2°F. It is evident that annual extremes range widely following fluctuations in air temperature and in the number of days in each group of temperature ranges, e.g., 40–44°F. Statistical analysis is continuing in the preparation of a nationwide map of surface water temperature characteristics in which thermograph records will be supplemented by the much larger record of daily or more frequent single observations.

Ward(*45*), reporting on annual variation of stream water temperature in Arkansas, gives an empirical equation, namely:

$$T = a[\sin(bx + c)] + \bar{T} \tag{10}$$

where T is the temperature of the stream water in °F; a is the amplitude of the sine curve in °F; $b = 360°\text{F}/365\text{ days} = 0.987°\text{F/day}$; x is the number of days since October 1 ($x = 1$ for October 1); c is the phase coefficient, in degrees; and $\bar{T}$ is the arithmetic mean or average temperature in °F. Ward concludes (1) that the average temperature of a stream, at a specific point, does not vary appreciably from year to year; (2) that the characteristics of the temperature sine curve do not change materially from year to year; (3) that the analysis of 12 monthly averages yields results almost identical to those from the analysis of daily temperature readings; and (4) that daily temperatures seem to be reasonably well represented by an arithmetic frequency distribution.

X. Oxygen and Reaeration Capacity

We have seen that our rivers and groundwater channels are natural drainage ways, with an enormous capacity for transporting natural salts dissolved from rocks and the debris from weathering of soils. Even before the era of American settlement and expansion, our streams probably carried more than 200 million tons of dissolved solids annually to the oceans. This continuous process is not going to alter significantly, and it is this process— by precipitation, streams, and seepage and groundwater flow—that to a large extent is the basis of quality and composition of the water. Over and above this enormous chemical load, the waters also have a natural capacity for assimilating some of the waste products of municipalities, industries, mining, urban drainage, and agriculture. The question is how much and where? Few water managers are in agreement.

In recent years, the available assimilative capacity of the natural environment has been rapidly used up in pockets or sectors, and it is becoming more difficult to protect one environmental medium, such as water, without damaging another, such as air. In the past, when pollution control effects were undertaken, it was often assumed that if a liquid or gaseous waste stream was treated or if solid wastes were burned or hauled off the premises the pollution problem was solved. But in recent years we have gradually come to appreciate that air, water, and solid waste problems are closely interrelated and that their analysis and control are best viewed as a systems problem relating to the whole process of residuals generation and control.

A case in point is the well-established equation for expressing the rate of absorption of oxygen per unit of time in rivers:

$$dc/dt = k_2(c_s - c) \tag{11}$$

involves the concentration of oxygen, c; the concentration for saturation, c_s, at a given temperature, T; the time, t, in days; and the coefficient of reaeration, k_2, which is a function of hydraulic characteristics. American and British studies by Langbein and Durum(*46*) have fairly well delineated boundaries or oxygen uptake by streams.

Leopold(*47*) has listed the number, lengths, and drainage areas of streams in the United States and has classified them as to their order. The coefficients of reaeration and the assimilative capacity for each stream of given order have been computed by the formula given in footnote b of Table 7(*46*). The values for total assimilative capacity refer to the total load, in tons of oxygen per day, at mean flow that could be absorbed from the air by the river system for each unit (milligrams per liter) of oxygen less than the saturation value.

TABLE 7

Hydraulic Factions and Total Assimilative Capacity of Streams of Different Orders[a]

Stream order	Average discharge, ft^3/sec	Average depth, ft	Average velocity, ft/sec	Coefficient of reaeration, day^{-1}	Total length of streams, miles	Total assimilative capacity[b], tons per day per unit deficiency in dissolved oxygen	River representative of each order
1	0.6				1,570,000		
2	2.8				810,000		
3	14	0.55	1.2	9.3	420,000	16,300	
4	65	0.95	1.6	5.5	220,000	19,000	
5	310	1.8	1.8	2.6	116,000	20,000	Pecos
6	1,500	2.7	2.0	1.8	61,000	30,000	Shenandoah, Raritan
7	7,000	5	2.5	1.0	30,000	31,000	Allegheny, Kansas, Rio Grande
8	33,000	12	3.0	0.37	14,000	21,000	Tennessee, Wabash
9	160,000	25	4.0	0.19	6,200	18,000	Columbia, Ohio
10	700,000	45	5.0	0.10	1,800	9,400	Mississippi

[a]Reprinted from Ref. *46*.

[b]QLk_2/v, where k_2 is the proportion of natural logarithmic units. However, the values of k_2 given in the table are in terms of common logarithmic units, hence assimilative capacity equals 2.3 QLk_2/v. Since Q is given in cubic feet per second, L and v in miles, and k_2 in $days^{-1}$, the formula used is $(1/2700)(QL/v)k_2$. The quantity QL/v is the volume of water in the stream at mean flow.

One could compute similar values for, say, average low flow for the systems (lower 25% quartile) when oxygen levels are minimized, if the hydraulic characteristics of the channel are known. However, the main purpose of the computations is to give some order of magnitude of the capacity of the stream for "reconditioning" itself when oxygen-consuming substances are encountered.

The results indicate that, although the total assimilative capacity among the several orders is roughly of corresponding magnitude, most of the assimilative capacity occurs in streams of the sixth and seventh order, not in streams of the highest or lowest order (Table 7).

At average flows the approximate rate of recovery of dissolved oxygen by the U.S. river systems per unit (milligrams per liter) deficiency below oxygen saturation is about 165,000 tons/day. A sewered population of 140 million has a BOD (biochemical oxygen demand) daily loading of 8000 tons, based on a per capita loading of 0.12 lb/day. Although the assimilation capacity of U.S. river waters is enormous (20 times the BOD population demand), neither people nor stream systems are uniformly distributed to take full advantage of natural reaeration capacity.

XI. Design of Water Quality Data Programs

Water quality, unlike some other aspects of the hydrologic cycle, deals with a momentary condition which may never recur in a similar sequence of events. This state or condition is defined by the chemical, biological, and physical characteristics in relation to all other hydrologic characteristics. It is difficult and costly to obtain data on the many characteristics that comprise the quality of a water resource. The objective, however, of national water quality data programs is to obtain this wide spectrum of information for all surface water and groundwater resources.

A decade ago such a difficult and far-reaching objective would have appeared impractical. However, modern technology is providing new tools which place this objective well within the realm of practicality. Because of their extensive experience gained over a span of many years, and because of active programs in scientific research and statistical analysis, data information sources can face the burden of an ever-increasing number of "new things" to be measured and assessed and can readily hope to achieve more efficient operations with modern tools.

A basic and principal concept of national data information sources, such as the U.S. Geological Survey, is that water quality information to fulfill the requirements of all data users will eventually be needed on every stream and body of water. There is reason to believe that this ideal concept can be achieved under the present organizational structure.

A. Criteria

A comprehensive data collection program must be designed to satisfy a great number and variety of needs. Continuous sampling of every body of water would be the ultimate way to obtain data, but this approach is impractical and uneconomical. Thus, data needs must be met by a "limited" collection program, and to guarantee a maximum benefit from limited data, quantitative techniques should be used to extend data from measured parts of the hydrologic system to unmeasured areas, where feasible.

If we knew the value of data to a specific user or could apply some monetary worth factor, design of a basic data program would be greatly simplified. In the absence of these values, the following criteria should guide the development of a well-balanced water quality data program.

(1) Provide data judged to be of maximum value to national and regional planners; that is, provide a base line of selected inorganic, organic, and biological parameters in relation to flow characteristics.
(2) Provide data in the time frame required by regional and local water managers for solving both indigenous and water-user problems.
(3) Obtain such data as will adequately describe average national conditions and the frequency and extent of variation from the averages.
(4) Plan well-defined programs which will show the effects of man's developments on natural conditions.
(5) Develop and incorporate systems of interpolating and extrapolating data where possible, in order to approximate water quality characteristics from fixed points.
(6) Provide the technical guidance essential for sound social and economic decisions in national and regional water resources development.
(7) Provide for sensitivity testing of the data by modern methods, such as systems modeling, in order to improve network design.
(8) Strive for a balance between the details described in the record and the cost of obtaining the record.

Beyond the basic measurements of the various physical, chemical, and biological parameters, a water quality basic data program should provide the additional information required to understand water quality. This information includes statistical summaries, the relation of parameters to stream flow, and the relation of water quality to the functioning of the hydrologic cycle. The following elements are considered essential to a water quality basic data program: (1) Data collection activities, (2) statistical descriptions, and (3) simulation models. These are described in detail in the following pages (see also Chaps. 5 and 26).

B. Data Collection Activities

For optimum development and management of water for the major beneficial uses, current information is needed on the chemical and physical properties of the resource. Data requirements for the many users are sometimes compatible within limits, whereas others are at complete variance.

The environment is constantly changing, and these changes cause variations in the chemical, physical, and biological quality of the waters of the nation. As a first requirement, data should be obtained on rivers to cover all internal drainages in the United States and thereby form a highly flexible information system responsive to variations caused by physiography, climate, geology, and population.

C. Water Quality Parameters

Table 8 includes a cross section of parameters to be considered for inclusion in a comprehensive data collection program. Parameters to be observed in any basin should be selected with close consultation and cooperation with federal, state, and local agencies to form a manageable number of base line or index measurements. Although the listed parameters do not constitute an all-inclusive program, they include the important factors involved in water quality management. Furthermore, they comprise the mainstay of parameters included in river quality standards being adopted by states for pollution abatement. The proposed base line parameters are about equally divided between so-called conservative (stable) and nonconservative (unstable) parameters and emphasize basic data of utmost importance for overall assessment of inorganic, organic, and biological stream quality.

Suggested frequencies for measuring each parameter are included in Table 8 but should be considered only as a general guide. For each site, determination of the proper frequency of measurement will depend on the variability in concentration of each parameter and on how accurately one needs to describe this variability. If the variability in values is large throughout a year, many measurements may be needed to adequately describe a parameter. If variability is small, frequency could be altered seasonally to achieve desired accuracy. If values are negligible or do not constitute an immediate or foreseeable problem, measurements could be omitted except for periodic checks to see whether they remain negligible. Statistically we are concerned with how well each measurement in the sampling program represents the population sampled (flow segment). Each parameter has its separate population as well as a relation to all other parameters. Clearly, these are practical considerations in sampling (see Chap. 10). The subject of water quality parameters is covered in more detail in another part of this work (see Chap. 19).

TABLE 8
Water Quality Parameters

Parameter	Index factor	Suggested frequency[a]
Major and important cations and anions[b]	Salinity Inorganic variability Geochemical	C, M
Specific conductance or dissolved solids		C
Minor elements[c]		M
Hardness (Ca + Mg)		W
Nitrogen cycle (NO_2, NO_3, total N)	Nutrients	D
Phosphorus cycle (PO_4 and total P)	Nutrients	D
Dissolved oxygen	Oxidation Aquatic health	C, M
pH	Acidity and corrosiveness Geochemical Aquatic health	C, M
Radioactivity[d] (gross alpha, beta)	Radiochemical	Q
Carbon (total or dissolved C)	Organic	M
Pesticides[e]	Organic Ecology, public health	M

Parameter	Index factor	Suggested frequency[a]
Phytoplankton and benthic organisms (selected)	Biological integration	M
Turbidity or suspended sediment	Physical	C or D
Temperature	Physical	C
Color, odor	Physical	M
Biochemical oxygen demand (BOD)	Organic waste	W
Microorganisms (coliform)	Waste, human and animal	W

[a]C, continuous; D, daily; W, weekly; M, monthly; Q, quarterly.

[b]Includes ions from PHS drinking water standards – iron, manganese, chloride, sulfate, fluoride – and possibly others, such as alkalinity, acidity, or sodium.

[c]Includes those listed in PHS standards: copper, lead, arsenic, cyanide, chromium (hexavalent total), barium, cadmium, selenium, silver, zinc.

[d]Also includes specific nuclides radium 226 and strontium 90.

[e]Federal Committee on Pest Control Primary List includes aldrin, DDD, DDE, DDT, dieldrin, endrin, heptachlor, heptachlor epoxide, lindane, 2,4-D, silvex, 2,4,5-T.

D. *Data Characteristics and Levels of Precision*

Water quality data must meet a wide variety of users' requirements. In order to design a sampling program to yield all the necessary data, it is essential to define many data characteristics. Three of those most commonly needed are (1) instantaneous characteristics, (2) characteristics for a given time interval, and (3) patterns of fluctuation.

1. Instantaneous Characteristics

An instantaneous water quality measurement provides current information on concentration at a given place and time. The instantaneous measurement may be a field determination, a laboratory determination, or a value sensed and recorded by a monitor. It must be within verifiable limits, especially for meeting legal requirements associated with stream quality standards. It also provides information for many other management and operational needs. If a data user has a critical need for the information in day-to-day operations, relatively high precision in measurement is justified. Examples are data for public water supply treatment plants, industrial cooling, food processing, reservoir releases, and pollution control.

The precision attainable for an instantaneous measurement can be defined by limitations in the sampling technique and in the method of analysis—either field or laboratory. A suggested reasonable level of overall precision for *all streams* in physical and chemical quality measurements is 10%, based on developing quality controls and standard methods. Although experience in biological observations, such as the observation of coliform bacteria, is not yet sufficient to evaluate a 10% precision objective, such a level may be sought.

Electronic equipment, such as the multiple probe sentinel for measuring dissolved oxygen, temperature, specific conductance, turbidity, and pH, is subject to instrumental errors. Required accuracy in manufacturers' specifications for these instruments has been established as follows: dissolved oxygen, 2%; specific conductance, 2%; turbidity, 5%; temperature, 2%; and pH, 5%. Continuous sensing with multiple probe electronic units provides valuable information where interest is great, the problem extensive, variability large, and where real-time data are needed (see Chaps. 26 and 27).

2. Characteristics for a Given Time Interval

The finite time interval is the basic unit for quantitative evaluation and statistical inference for describing variations and trends, average conditions, and future conditions based on probability analysis. The best way to determine the data characteristics for a given time interval (1 hour, 1 day, 1 month, etc.) is to measure the water quality continuously and then combine the instantaneous characteristics measured in an appropriate statistical summary. However, intermittent sampling will be used much more extensively than continuous measurements because of economic constraints.

In intermittent sampling, the investigator attempts to space intermittent measurements to simulate a continuous record of the stream quality. Statistics on the behavior of the stream for the given interval are computed from the intermittent measurements; these computations should meet stated levels of precision. For intermittent measurements, a sampling schedule must be devised to provide statistics that are representative of the characteristics of the population (water segment) sampled. Value of the data, costs, efficiency, and other factors will dictate the pattern of sampling.

3. Patterns of Fluctuation

In addition to concentration at a specific time of flow, other statistical characteristics of water quality, such as means, extremes, and deviations, are becoming increasingly important in water resources planning and development. We do not yet have sufficient information and experience to define levels of precision for such statistics as magnitude, rates, and recurrent patterns of fluctuation. However, the need for such information may require the establishment of precision objectives.

Data characteristics described in Sec. XI.D, 1 and 2 should eventually lead to regionalization of the record; that is, the capacity to extrapolate data from one site to another site on the same stream. In addition, the synthesis of historical data by means of mathematical models is possible and should be used as a tool in data extrapolation. Mathematical models are discussed later in this paper (see also Chaps. 5 and 26).

Needs for data should be reviewed periodically and sampling schedules varied seasonally to ensure maximum usefulness, especially for the nonconservative elements. For example, water users often need to know if the dissolved oxygen concentration is maintained above minimal level continuously during the dry season, but they are less interested in the oxygen level during the wet season, at which time oxygen measurements could be reduced substantially. For all quality measurements, the total array of samples should include some collected during both low and high flow conditions.

It is recommended that the levels of accuracy listed in Table 9 be established for describing a water quality characteristic for a given time interval of stream flow for all streams. Accuracy is given in the table as standard error in per cent of mean of the item and is based on the equivalent of a 5-year record.

E. Short-Term Design and River Management Data

In addition to a network of relatively fixed locations, essential for assuring a uniform base of information for national accounting and surveillance, the objective of providing real-time short-term water quality data for river management problems is of national concern. Short-term measurements are essential to the support of the basic network because they provide information on (1) catastrophic occurrences or accidents resulting in downstream problems or hazards, (2) time of travel of pollutants and potential pollutants resulting from accidents, (3) droughts, with concurrent quality extremes in which refined measurements are needed for immediate decisions, and (4) changes in quality resulting from

TABLE 9

Suggested Statistical Summaries and Levels of Accuracy (accuracy sought for all stream quality data is based on a 5-year record)[a]

Chemical, biological
a. Concentration of properties:
Maximum observed concentration
Minimum observed concentration
Annual minimum 7-day
Mean and standard deviation of annual concentration
Mean and standard deviation of each calendar month concentration
Annual frequency distribution
Trend
b. Load of properties
Monthly mean
Monthly standard deviation
Annual frequency distribution
Annual mean load
Trend
Sediment
a. Concentration
Maximum
Minimum
Monthly mean
Monthly standard deviation
Annual frequency distribution
b. Load
Maximum
Minimum
Monthly mean
Monthly standard deviation
Annual frequency distribution

[a]Incomplete analyses show that the statistics of water quality vary greatly from stream to stream. However, using as a standard a systematic sample of 12 monthly observations, most inorganic constituents can be measured within 10–20% of the mean for the year. Accuracy can be improved by measurements over a longer period than a year.

Where the standard error, SE, of the mean, $\bar{x}$, is equal to the square root of its sample variance, $S^2(\bar{x})$, which is therefore

$$S^2(\bar{x}) = \frac{S^2}{n} \text{ and}$$

$$SE(\bar{x}) = S/\sqrt{n}$$

then the accuracy of the mean is expressed as $\bar{x} \pm SE(\bar{x})$ and is given as percentage of the mean:

$$\text{Accuracy of mean } (\%) = 100[SE(\bar{x})/\bar{x}]$$

water management decisions, which cause unexpected alterations in the network model and require relocation of stations.

Observations by the U.S. Geological Survey at hydrologic bench mark sites are providing a record of the variations in quality, essentially unaffected by man, as a base line for documenting the natural trends in water quality in a basin. The observations at the bench mark sites therefore provide a key to the net effects of management on the quality of the downstream river.

The water quality information subsystem has three functional components: (1) sample collection, (2) constituent analysis, and (3) data processing. Inputs to the subsystem include the hydrologic environment(s) being sampled, other hydrologic measurements made as part of this subsystem or as part of some other data program (e.g., stream flow), user information requirements, and various institutional and economic constraints. These components are linked together by lines of communication which in large part control the rate of information flow through the subsystem.

Water quality records are examined in order to determine (1) the mean and variability of each constituent measured for complete chemical analysis, (2) the statistical characteristics of the time series for each constituent, and (3) the effect of serial correlation on the variance of the estimated mean for each constituent.

F. Water Quality Modeling

The mathematical model is being used as a water quality planning and management tool to simulate time and spatial variations of flow and concentration of dissolved solids throughout a basin. Other parameters of water quality can be included in the model after more experience is gained. Several standard models are available for testing various basin quality and flow characteristics.

The mathematical model as a technique for analyzing the behavior of physical systems has a long history, but only in recent years has its use as a tool for problem solving become widespread. This has come about as part of the rapid development of operations research and systems and associated advances in electronic computers. In general, the model attempts to duplicate the essence of a system or activity without actually measuring the parameters in reality.

Known needs for water quality modeling are (1) a water quality planning and management tool to simulate time and spatial variations of flow and concentration of dissolved solids throughout a basin and (2) a network of sampling stations for determining basic qualities of surface waters (water quality information system).

REFERENCES

1. S. K. Love, *J. Am. Water Works Assoc.*, **53**, 1366 (1961).
2. H. Egner and E. Ericksson, *Tellus*, **7**, 134 (1955).
3. A. W. Gambell, *U.S. Geol. Surv. Water-Supply Paper 1535-K*, 1965.
4. C. E. Junge, *Trans. Am. Geophys. Union*, **39**, 241 (1958).
5. C. E. Junge and P. E. Gustafson, *Tellus*, **9**, 164 (1957).
6. H. C. Whitehead and J. H. Feth, *J. Geophy. Res.*, **69**, 319 (1964).
7. K. Rankama and T. G. Sahama, *Geochemistry*, Chicago Univ. Press, 1950, p. 911.
8. G. N. Lewis and M. Randall, *Thermodynamics and the Free Energy of Chemical Substances*, McGraw-Hill, New York, 1923.
9. P. Debye and E. Hückel, *Physik. Z.*, **24**, 185 (1923).
10. J. H. Feth, *U.S. Geol. Surv. Open-File Rep.*, 1964.
11. R. M. Garrels, *Mineral Equilibria at Low Temperature and Pressure*, Harper, New York, 1960, p. 254.
12. J. D. Hem and W. H. Cropper, *U.S. Geol. Surv. Water-Supply Paper 1459-A*, 1959.
13. I. Barnes and W. Back, *Geochim. Cosmochim. Acta*, **29**, 85 (1965).
14. W. Back, *Intern. Assoc. Sci. Hydrol.*, **VIII**, 43 (1963).
15. J. D. Hem, *Intern. Assoc. Sci. Hydrol.*, **19**, 45 (1960).
16. M. N. E. Raffi, W. J. Kaufman, and D. K. Todd, *Rept. No. 3, I.E.R. Ser. 90*, Sanitary Eng. Res. Lab., Univ. of California, Berkeley, Calif., 1956.
17. S. W. Mayer and E. R. Tompkins, *J. Am. Chem. Soc.*, **69**, 2866 (1947).
18. O. Ogata, *U.S. Geol. Surv. Prof. Paper 411-H*, 1964.
19. M. S. Gurevich, *Geologic Activity of Microorganisms* (Transl.), Consultants Bureau, New York, 1962, p. 65.

20. V. L. Mekhtiyeva, *Geochemistry*, **8**, 817 (1962).
21. M. Wagner and W. Schwartz, in *Symposium on Marine Microbiology*, C. C. Thomas, Springfield, Ill., 1963, p. 179.
22. F. H. Rainwater, *U.S. Geol. Surv. Atlas HA-61*, 1962.

22a. C. N. Durfer and E. Becker, *U.S. Geol. Surv. Water-Supply Paper 1812*, 1964.

23. G. I. Taylor, *Doc. Roy. Soc. (London)*, **223A**, 1954.
24. W. W. Sayre and F. M. Chang, *U.S. Geol. Survey Open-File Rept.*, 1967.
25. H. B. Fischer, *J. Hydraulics Div., Am. Soc. Civil Engrs.*, **93**, 187 (1967).
26. H. B. Fischer, *J. Sanit. Eng. Div., Am. Soc. Civil Engrs.*, **94**, 927 (1968).
27. H. B. Fischer, *U.S. Geol. Surv. Prof. Paper 575-D*, 1967.
28. N. Yotsokura, H. B. Fischer, and W. W. Sayre, *U.S. Geol. Surv. Circ.*, 1968.
29. W. H. Durum, S. G. Heidel, and L. J. Tison, *U.S. Geol. Surv. Res. Art. 266*, C326 (1961).
30. W. H. Durum and J. Haffty, *Geochim. Cosmochim. Acta*, **27**, 1 (1963).
31. R. Patrick, *Proc. Acad. Nat. Sci. Philadelphia*, **113**, 215 (1961).
32. L. B. Leopold, M. G. Wolman, and J. P. Miller, *Fluvial Processes in Geomorphology*, W. H. Freeman, San Francisco and London, 1964, p. 522.
33. H. B. N. Hynes, *The Biology of Polluted Waters*, Liverpool Univ. Press, Liverpool, England, 1960.
34. H. T. Odum, *Limnol. Oceanog.*, **1**, 102 (1956).
35. E. D. Odum, *Ecology*, Holt, Rinehart, and Winston, New York, 1963.

36. J. L. Blum, *Botan. Rev.*, **22**, 291 (1956).
37. E. Brown and Y. A. Nishioka, *J. Pesticides Monitoring*, **2**, 38 (1967).
38. D. F. Goerlitz and W. L. Lamar, *U.S. Geol. Surv. Water-Supply Paper 1817-C*, 1967.
39. G. W. Musgrave, *J. Soil Water Conserv. India*, **2**, 133 (1947).
40. B. R. Colby, *U.S. Geol. Surv. Bull. 1181-A*, 1963.
41. W. B. Langbein and S. A. Schumm, *Trans. Am. Geophys. Union*, **39**, 1076 (1958).
42. L. T. McCarthy, Jr., *U.S. Geol. Surv. Water-Supply Paper 1779-X*, 1964.
43. W. B. Langbein and D. R. Dawdy, "Some General Comments on the Occurrence of Dissolved Solids in Surface Waters of the United States," U.S. Geol. Surv., unpublished.
44. J. F. Blakey, *U.S. Geol. Surv. Atlas HA-235*, 1966.
45. J. C. Ward, *J. Sanit. Eng. Div., Am. Soc. Civil Engrs.*, **89**, 1 (1963).
46. W. B. Langbein and W. H. Durum, *U.S. Geol. Surv. Circ. 542*, 1967.
47. L. B. Leopold, *Am. Scientist*, **50**, 511 (1962).

Chapter **2**

Physical, Chemical, and Biological Characteristics of Estuaries

Walter Abbott and C. E. Dawson*
GULF COAST RESEARCH LABORATORY
OCEAN SPRINGS, MISSISSIPPI

SECTION ON BACTERIA

C. H. Oppenheimer
FLORIDA STATE UNIVERSITY
TALLAHASSEE, FLORIDA

*Present address: Environomics, Box 66805, Houston, Texas 77006

I. Introduction

Estuaries and the areas offshore to them are the sites of most of the great fisheries of the world(*1–4*). Concomitantly, the estuaries, being inshore, are more subject than any other marine environment to the influence of man. De Falco(*5*), for example, has recently designated them as the "septic tank of the megalopolis." The subjects of fisheries and waste disposal are interrelated, and a vast estuarine literature has arisen based on these two aspects.

This chapter deals primarily with what we regard as a "typical" estuary —an estuary associated with one of the alluvial deltaic pans that are scattered over most of the temperate and tropical coastlines of the earth at the mouths of rivers. Well-known examples are associated with such great rivers as the Amazon, the Mississippi, the Indus, and the Ganges, to mention only a few. For specialized treatments of all phases of estuarine research, reference should be made to the excellent treatises recently edited by Lauff(*6*) and Ippen(*7*).

Following the usage of Emery and Stevenson(*8*), but not that of Pritchard(*9*) or Russell(*10*), we exempt fjords from consideration. Fjords are special topographic features formed by glaciers and require special treatment.

Also exempted from discussion are many coastal indentations and seawater embayments not related to mouths of streams. Except for sporadic, unchanneled, local runoff and direct precipitation, these embayments are fed only by the sea. Atolls, coves, and bights are examples of such bodies of water. If rainfall is sufficiently great and oceanic exchange is somewhat restricted, measurable dilution can occur in these bodies, leading to an environment that is biologically comparable to an estuary. Conversely, if rainfall is lacking and exchange is restricted, a hypersaline environment may result. However, such basins usually tend to be isosaline with respect to the adjacent ocean. Chemical, physical, and biological differences from the open sea result from the stilling and protecting effect of enclosure plus the availability of specialized habitats.

Many attempts have been made to define the term estuary(*8,9,11–17*, and others), and there is considerable disagreement among the definers. Presumably, these disputes relate to the fact that proposed definitions tend to be oriented toward particular specialties. Pritchard(*9*) and Bowden(*15*), for example, define the term in a physical oceanographic context and insist upon a semienclosed basin with a free, permanent connection to the sea; Day(*11*) and Scott et al.(*17*) define the term in an ecological context without requiring the permanent sea connection. The restrictions imposed by the physical oceanographic definitions simplify computational and investigational problems in studies of circulation, mixing, and diffusion, but automatically eliminate such coastal features as sounds and lagoons from the estuarine category.

Perhaps the most useful definition is the generalized one of Ketchum (*12*): "An estuary is . . . a region where river water mixes with, and measurably dilutes, sea water." Under this definition, the vast, freshened sheets of pelagic water that cover the continental shelves opposite the mouths of great rivers, notably, the Amazon, the Orinoco, and the Mississippi, can be regarded as estuarine areas, and there are sound biological reasons for so regarding them.

From time to time, legislative definitions of estuaries have been proposed relative to specific bills under consideration. These definitions, although subject to final interpretation by the courts, show the diverse meanings attached to the word "estuary."

Public Law 89–753, 89th Congress, the Clean Water Restoration Act of 1966, defines as follows:

> . . . the term 'estuary' means all or part of the mouth of a navigable or interstate river or stream or other body of water having unimpaired natural connection with open sea and within which the sea water is measurably diluted with fresh water derived from land drainage.

From the technical standpoint, this is a rather tight definition, oriented toward the physical oceanographic viewpoint. Unfortunately, such rigor is not always employed.

The Estuarine Areas Bill (H.R. 25, 90th Congress, first session), considered during 1967 by the Committee on Merchant Marine and Fisheries, U.S. House of Representatives, offered the following definition:

> The terms 'estuary' or 'estuaries' mean all or part of the tidal portion of the navigable waters in the United States up to the mean high water line, including, but not limited to, any bay, sound, lagoon, or channel, and the lands underlying all such waters; . . .

Even a conservative judicial interpretation might assume that this definition embraced the entire coastline of the United States.

Inland to the main basin of an estuary, there is a portion of the feeder stream which responds physically to changes in the estuarine water level. As tidal inflow raises the water level in the estuary, an upstream point is reached where the incoming estuarine water physically forces back the seaward-flowing fresh water of the river. Beyond this point, whose location depends on river flow, there is a length of stream in which flood tide effects produce reversal of flow. These effects are rapidly damped by forces related to channel configuration, river discharge volume, and similar factors. As one progresses upstream, increasingly shorter flood tide periods and correspondingly longer ebb tide periods ensue.

Ultimately, a point is reached at which downstream flow exactly compensates for upstream tidal thrust at peak flood tide. At this point and time, a brief period of stagnation occurs, followed by resumption of downstream flow. Above the stagnation point, the river does not exhibit stoppage or reversal of flow related to normal tidal influences, but shows discharge velocity and water level variations correlated with tidal cycles. Finally, with rise of channel bed above sea level, no further influences of tide on stream flow are evident.

Those portions of the stream which receive no saline water from the estuary during flood tide, but which do exhibit tidally correlated effects, such as flow reversal, stagnation, or flow variation, are regarded herein as the tidal river.

The foregoing description presents an idealized situation. Not all estuary feeder streams show all of these tidal phenomena, and some may show none of them.

II. Physical Characteristics

A. Morphology and Classification

Estuaries have occasionally formed as the result of subsidence and maritime flooding of a coastal, river-fed area. San Francisco Bay, the Gulf of California, and the Gulf of Aqaba are examples of this type of formation(*9,10*). However, the typical coastal plain estuaries were created by inundation of river valleys as sea level rose during postglacial thaws.

At the height of the Wisconsin and Würm glaciations, sea level was 400–500 ft (120–150 m) lower than at present. Coastal rivers ran much farther seaward across dry land (now submerged continental shelf). With glacial retreat, the sea rose and flooded the valleys farther and farther upriver, ultimately reaching the coastline as we know it now. In many instances, multiple-river systems were inundated so that branched estuaries, with several freshwater influxes, were created. Such was apparently the case with Chesapeake Bay on the Atlantic Coast(*10*) and with the Baffin Bay-Alazan Bay system and the Aransas Bay-Copano Bay complex on the Texas coast [cf. (*18*)].

Russell(*10*), reviewing estuarine origins, concluded that the most recent rise in sea level may have started about 50,000 years ago. Thus, a maximum age limit is placed on existing estuaries, and most are certainly much younger than the 50,000-year maximum. Morgan(*19*) has documented the formation and disappearance of small estuarine systems due to scouring and sedimentation over periods of 100–150 years in the active lower delta area of the Mississippi River.

This suggests that individual estuaries are geological transients. During the Cenozoic, a number of periods of glaciation were separated by intervals of warming climate and rising sea level. Since rivers presumably existed throughout this era, estuaries must have formed as the sea rose during warming periods, and vanished as ocean level dropped during succeeding cold periods.

Even older epochs may have included estuarine periods(*20*). Siler(*21*) points out that the Del Rio formation of the Texas Cretaceous contains a dwarfed biota including ammonites that are the same species as those found at full size in distinctly marine deposits of equivalent age. The fossil oysters in the Del Rio, however, are not dwarfed. Reasoning from the estuarine preferences of modern oysters, he suggests that the dwarfing of characteristically marine forms may be the result of life in an adverse, presumably estuarine, habitat. Hedgpeth(*22*) discussed similar data but considered that no definite conclusion was possible.

The premise that estuaries have been and are sites of rapid geomorphological change is widely accepted. The exact shape of a newly formed estuary would depend on the conformation of the river valley walls at any particular sea level. Viewed from above, however, the overall impression would be that of an elongate triangle with its base at the coastline and its apex at the river. Immediately upon stabilization of water level, changes in basin conformation would begin.

As seawater passed across the continental shelf and into the estuary, it would be slowed by frictional effects on the bottom and by opposition from outflowing river water. As a result, kinetic energy, which maintains particulate matter in suspension, would be dissipated. Depending upon whether the area involved was a high energy or low energy coast (i.e., a coast with large or small tidal fluctuation, rapid or slow inshore currents, narrow or broad shelf, etc.), suspended material would settle out either to form a spit across part of the estuary mouth (high energy coast) or to form offshore barrier islands (low energy coast) [cf. (*23*)]. Figure 1 is a diagram of these two alternatives.

Meanwhile, at the inland end of the estuary, similar effects would result from slowing of the river water as it entered the broad estuarine basin. Bar and delta formation would begin, thereby further altering basin topography. Behrens(*18*), using echo-sounding techniques, has demonstrated the depth of burial and general conformation of the drowned river

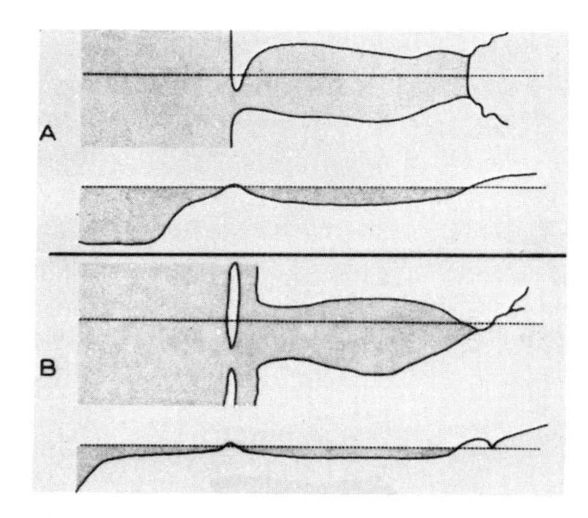

Fig. 1. Diagrams of bar formation at the mouth of an estuary viewed from above and in longitudinal section. A: high energy coast; B: low energy coast.

valleys which represented the primary basins of several Texas estuaries. His findings substantiate the general description just given.

Along a low energy coast, the formation of offshore barrier islands creates, parallel to the coastline, a protected coastal lagoon which is itself estuarine in nature. The Biloxi Bay-St. Louis Bay-Mississippi Sound complex of the central Gulf Coast of the United States is an example of such a situation, as are the bay systems of the Texas coast.

The complexity of estuary dynamics points up the difficulty inherent in any attempt at rigid classification. Pritchard(*9*) has proposed several broad geomorphological categories that may be conceptually useful. These are drowned river valleys, fjords (excluded here), bar-built estuaries, and tectonically produced estuaries.

Another type of estuarine classification has been based upon evaporation: runoff ratios(*9, 14, 24, 25*). This approach leads to three classes of estuaries:

(1) Positive estuary—one in which average freshwater influx exceeds evaporative water loss. A net export of water from the estuary to the sea occurs during each tidal cycle.

(2) Neutral estuary—one in which evaporation equals freshwater influx so that, on the average, neither net water import from the sea nor export to the sea occurs.

(3) Negative estuary—one in which evaporation exceeds freshwater influx. Seawater is imported and concentrated so that "salinities" severalfold greater than those of the ocean may be attained. This is the "hypersaline" estuary of Hedgpeth(*22,25–27*). The Laguna Madre of

Texas and the Laguna Madre of Mexico are well-known North American examples of this estuarine category.

The neutral estuary is a rare item, and the negative estuary is a highly specialized system, which is treated in the final section of this chapter. Our immediate concern relates to the positive estuary—the "typical" coastal plain estuary.

B. Circulation and Mixing

As glacial retreat slowed and sea level stabilized, our typical estuary passed rapidly to maturity. The processes of bar and delta deposition came into balance with opposing erosive forces; current patterns were established; and the entire estuary became, on a human time scale, a more or less permanent feature of the coastline.

Two facts should be recognized at the beginning of any study on estuaries. First, the water level in the estuary cannot exceed the level of the adjacent ocean, except for short periods during floods. Second, the estuary shows a salt balance—it is not tending on the average to become either fresher or saltier. The estuary is, however, a dynamic system. Water flows in from the river; tides move in and out at the mouth; mixing of fresh water and seawater occurs in the basin. These considerations are the basis for assuming that the mature estuary exists in a dynamic steady state(*12*).

Under the steady-state assumption, certain conditions must be fulfilled. Since the estuary receives a net import of fresh water from the river, it must show a net export of fresh water to the sea. Also, since an influx of seawater occurs on the flood tide, there must be, on the ebb tide, an efflux equal in volume to tidal influx plus freshwater influx. Further, in order to maintain salinity as a conservative quantity, each cross section of the estuary must, on the ebb tide, show a net seaward freshwater flux equal to one river flow volume per tidal cycle.

The dissolved and suspended solids in the various waters are subject to essentially the same material balance considerations that apply to the solvent. Minor exceptions exist, relative to settling of suspended matter and uptake of micronutrients by the biota, but the total estuary mass balance during a tidal cycle is still quite simple:

$$M_E = M_F + M_R - M_L \tag{1}$$

where M_E is the mass exported to the sea during ebb tide, M_F is the mass imported from the sea during flood tide, M_R is the mass contributed by streams, watershed, etc., during a full tidal cycle, and M_L is the mass lost because of factors such as evaporation, seepage, sorption, and biological activity.

The tidal prism theory of estuarine flushing evolved from such considerations, but required the assumption of complete mixing in each segment of the estuary. Ketchum(*12,13*) developed a flushing theory that eliminated the need for this assumption. Using this approach, he obtained good agreement between computed and observed values for several bays. Since publication of his work, several modifications and refinements of the approach have been developed(*28,29*).

The original Ketchum technique has the advantage of simplicity. After an initial steady-state assumption, Ketchum divides his estuary into a number of volume segments such that the length of each segment is equal to the average transport distance of a particle of water during the flood tide. This distance, of course, varies as basin conformation varies in each segment. An analogous concept is to view the flood tide stage as a period of laminar flow, whereby incoming seawater in the seawardmost segment pushes an equivalent volume of water into the next upstream segment and this process is iterated in succeeding upstream segments.

Ketchum defines three parameters: R, the river flow volume during a full tidal cycle; V_i, the low tide volume of each volume segment; and P_i, the local intertidal water volume in each segment (i.e., the volume of water necessary to produce the observed rise in level at flood tide in each segment). Segment definition starts upstream at a segment where $P_i = R$. This segment, designated as segment 0, is regarded as the inner end of the estuary. Ketchum points out that there will be no net flux of water through this segment during flood tide, the entire river flow being employed to raise the level of the segment.

From these considerations, it is shown that the segments are interrelated as follows:

$$V_1 = V_0 + R \tag{2}$$

$$V_2 = V_1 + P_1 = V_0 + R + P_1 \tag{3}$$

$$V_3 = V_2 + P_2 = V_0 + R + P_1 + P_2, \text{ etc.} \tag{4}$$

and, in general,

$$V_i = V_0 + R + \sum_{k=1}^{i-1} P_k \tag{5}$$

In a further development, the exchange ratio (r_i), defining the proportion of fresh water (and accompanying conservative dissolved and suspended materials) removed from any particular volume segment on ebb tide, is shown to be

$$r_i = P_i/(P_i + V_i) \tag{6}$$

By extensions of the method, the proportion of fresh water from any previous tide still remaining in a volume segment can be computed. A summation of such expressions over a large number of tidal cycles gives the total proportion (Q_i) of river water accumulated in any volume segment. The latter is computed as

$$Q_i = R/r_i \tag{7}$$

These equations are based on the assumption that vertical mixing in the estuary is complete. If this is not the case, a series of top-to-bottom salinity checks will show the true depth of mixing. The exchange ratio expression is then modified to

$$r_i = \frac{P_i}{P_i + V_i} \times \frac{D}{H} \tag{8}$$

where D is the mean depth of volume segment i, and H is the depth of the mixed stratum. In the stratified case, the volume segments should be established only for the mixed layer, the unmixed lower stratum being regarded as a functional bottom.

Mixing and circulation in estuaries are complex phenomena depending on at least four influences: Coriolis force, tidal flow, river flow, and wind action.

Coriolis force, one of the prime movers of air and water masses, derives from gravitational and centrifugal forces related to the earth's rotation. The Coriolis force produces the great overall oceanic circulation patterns of the world, engendering clockwise rotation in the northern hemisphere and counterclockwise rotation in the southern hemisphere. On a planetary scale, the total Coriolis-related energies are of fantastic magnitude; however, in small, localized areas such influences, while detectable, are meager. In general, a reasonably broad northern hemisphere estuary with minimal turbulence from such factors as river flooding and wind action exhibits salinity isopleths sloping seaward from right to left, when viewed from the mouth of the estuary. The clockwise rotation tendency induces incoming seawater to flow toward the right side of the estuary and outflowing freshened water to move along the left side. Right and left bank salinities at equal distances from the estuary mouth may differ by 1‰ or more under proper conditions. Figure 2 gives a diagrammatic representation of this effect and an actual set of isopleths from Chesapeake Bay (*8,30,31*).

As a result of wind, tide, and river flow, the fresh and salt water in an estuary may, at any particular time, be completely mixed, partially mixed, or stratified (*15,32*). Actually, these categories are not isolated but show continual gradation from one to another.

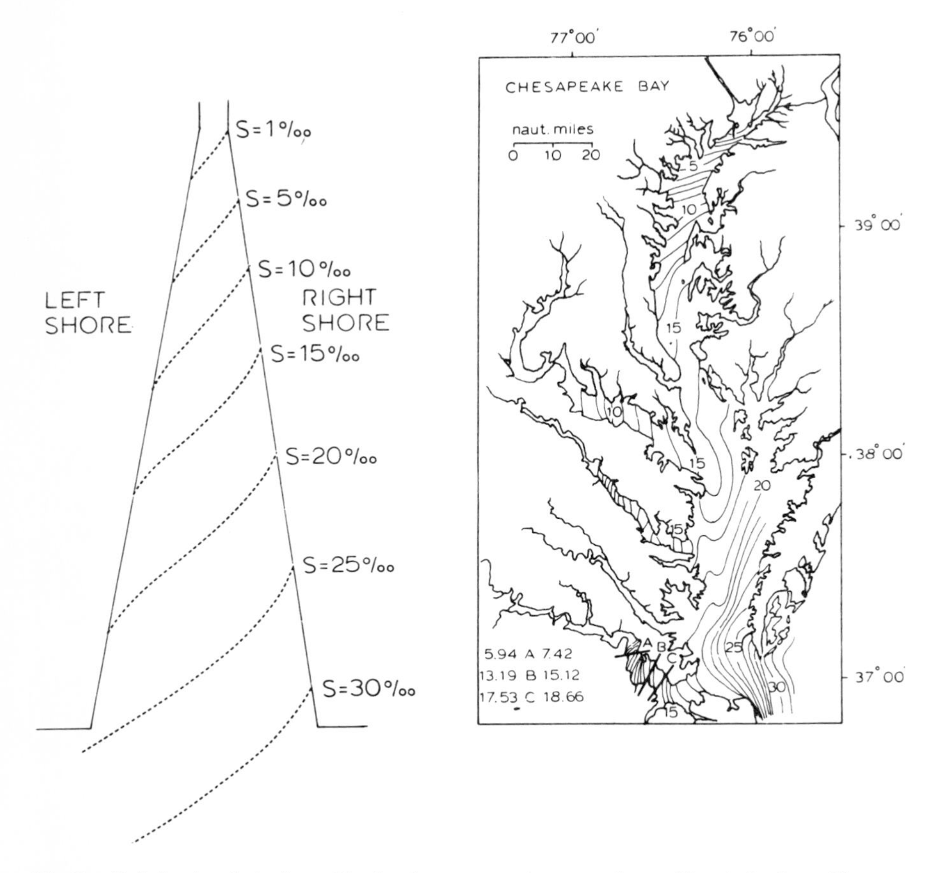

Fig. 2. Salinity isopleths in an idealized estuary and an actual set of isopleths from Chesapeake Bay. Data in the lower corner of the Chesapeake Bay map are left and right shore salinities along the indicated transects. [Redrawn, slightly modified, from Ref. *8*, p. 684, and Ref. *30*, p. 370, by courtesy of The Wildlife Management Institute, The Geological Society of America, and D. W. Pritchard.]

In extreme cases, stratification is so pronounced that a special situation called a "salt wedge" estuary exists. The condition is that of a deep, relatively narrow channel with a large freshwater flow. Incoming high density seawater tends to flow along the bottom of the channel, and the less dense fresh water of the river flows out across the top of the seawater layer. Owing to frictional forces at the interface, the seawater tends to be forced back by the fresh water, giving rise to an interface that is somewhat recurved. As a result of Coriolis force, the interface in northern hemisphere estuaries may slope upward toward the right side of the channel, looking upstream. So long as the relative velocity of the outflowing fresh water is below a critical value at the interface, essentially no mixing takes place between the layers. The salt wedge advances and retreats along the bottom as tides flood and ebb, and shows an internal circulation produced by upstream movement of water along the bottom of the wedge and seaward flow in the upper portions of the wedge [cf. (*15*)].

The Mississippi River is frequently offered as an example of a salt wedge estuary. Distance of penetration of the salt wedge into the river channel varies with discharge status of the river. During dry summer periods, when river flow is minimal, salt water may advance upriver beyond New Orleans, which utilizes the Mississippi as a municipal water supply. In such situations, the water intake system of the city must be adjusted to draw only from the upper strata of the river to avoid introducing salt into the drinking water system.

The usual estuary, however, does not show such pronounced stratification. When freshwater outflow rises during rainy periods, relative interfacial velocity reaches values sufficient to generate waves at the saltwater-freshwater boundary. Crests of these waves are torn off by the outflowing fresh water, entrapped and mixed with the outflow, and returned seaward. Usually, when interfacial turbulence becomes this extreme, some reciprocal exchange between the layers also takes place, so that the saltwater stratum becomes measurably diluted. When this occurs, the interface becomes more poorly defined as turbulence progresses or as one moves upstream. Diagrams of the various degrees of stratification and associated circulation patterns are given by Simmons (*32*), Bowden (*15*), and Emery and Stevenson (*8*).

During a flood condition on the watershed, the salt water may be driven out of an estuary completely. This condition developed in the usually hypersaline, relatively shallow Baffin and Alazan Bays of south Texas as a result of the floods which followed Hurricane Beulah in the fall of 1967 (*33*). Obviously, such a catastrophic disturbance is not regarded as a normal phase in estuarine hydrography. Immediate economic losses in terms of fisheries are extreme in cases such as this, but the freshening effect may be beneficial over the next several years.

In a broad, shallow estuary with tidal exchanges that are large relative to freshwater inflow, and with considerable wind action, mixing may be sufficient to eliminate or markedly reduce salinity stratification. The horizontal, Coriolis-induced salinity gradient occasionally persists, but more often it is eliminated. Many of the broad, shallow bays of the U.S. Gulf Coast exhibit various degrees of such mixing during most of the year. In these cases, wind is extremely influential. Although basic tidal oscillations range from 6 to 18 in. (15–45 cm), prevailing moderate winds from the ocean can generate tides of 3–4 ft (1.0–1.2 m), and hurricane-force winds can produce influxes of 10–15 ft (3.0–4.6 m) or more. In a small, protected inlet from Biloxi Bay, a well-mixed estuary, we found (*34*) a vertical mixing time of 8–10 min in a 30-cm depth. Equivalent mixing in an adjacent, wind-exposed area of the same depth required less than 1 min.

The effects of tides on flushing and embayment in estuaries have recently attracted considerable attention because use of estuaries as waste water-receiving bodies is increasing. In estuaries where total volumes are much greater than tidal prism volumes, particularly in estuaries protected from wind action, horizontal mixing and flushing by tides may proceed slowly. Especially in the upper reaches of the basin, water masses sometimes merely oscillate upstream and downstream as tides flood and ebb. Under such conditions, waste materials discharged into a water mass during flood tide may move upstream from the outfall until tidal turnover permits them to return to the point of discharge. If the wastes have reasonably conservative properties, restricted portions of the estuary can accumulate concentrations sufficient to generate deleterious conditions where none would have been expected on the basis of simple dilution computations.

Considerable flushing can occur, however, in shallow, mixed estuaries where tidal prism volume is large compared with total low tide volume. This was found to be the case with Bear Lake, a blind estuarine embayment in the lower reaches of the San Jacinto River of Texas (*35*). Tidal prism volume was approximately equal to low tide volume. With the mixing and flushing exhibited, cumulative concentration of sewage effluent from a proposed outfall could never exceed twice the daily plant output, even assuming total conservation of the waste.

In broad, shallow estuaries on low energy coasts, a few hours of stiff breeze from offshore can drive into the estuarine basin seawater sufficient to double or triple the normal volume of the estuary. Conversely, strong seaward winds can drive most of the water out of these estuaries, exposing broad mud flats which may represent more than half the basin area, and flushing the remaining deeper areas with freshened water from upstream (Fig. 3). This is not always beneficial. In many estuaries, the sanitary status of shellfish-harvesting areas deteriorates markedly with efflux of upstream waters containing high concentrations of human wastes.

An estuary may be regarded as an assemblage of transient, small, individual water masses whose aggregate statistical behavior generates the whole-system properties. These water masses are continually in motion in all directions, fusing, mixing, and separating. When operations

Fig. 3. Davis Bay, an arm of Biloxi Bay, Mississippi, during a brisk (10–15 knot) seaward wind (upper photo) and a moderate (5–7 knot) landward wind (lower photo). The outermost piling protrudes 3 m above the sediments. Tidal variation in this area is normally less than 0.4 m.

such as waste discharge or channel dredging are to be carried on in an estuary, information usually is needed on rates of dilution and dissipation of any substances introduced into these localized water masses. Two processes are involved. First, the overall translation of the water mass in three dimensions, related to general estuarine circulation, tends to remove dissolved or suspended material from the point under consideration. Second, eddy diffusion tends to spread the material in all directions, even in the absence of translational effects. These conditions can be summarized mathematically as follows:

$$\frac{\partial c}{\partial t}=\frac{\partial}{\partial x}\left(\frac{\alpha_x}{\rho}\frac{\partial c}{\partial x}\right)+\frac{\partial}{\partial y}\left(\frac{\alpha_y}{\rho}\frac{\partial c}{\partial y}\right)+\frac{\partial}{\partial z}\left(\frac{\alpha_z}{\rho}\frac{\partial c}{\partial z}\right)-\left(V_x\frac{\partial c}{\partial x}+V_y\frac{\partial c}{\partial y}+V_z\frac{\partial c}{\partial z}\right) \quad (9)$$

where c is the concentration at point xyz at time t; α_x, α_y, α_z are eddy diffusivities for mixing in directions x, y, z; ρ is the density of the water mass; and V_x, V_y, V_z are translational velocities in each direction.

Simplifying assumptions are usually made. For example, if the prevailing current is running in the x direction, eddy diffusivity along the x coordinate is neglected and current velocity in the y direction is assumed to be zero. Moon et al.(*36*), among others, have presented both a theoretical treatment of this problem and a practical method for measuring diffusivity in estuaries. Their work includes data from field studies on the Lydia Ann Channel near Port Aransas, Texas.

C. Gas Exchange

Of the many physical processes that occur in estuaries, gas exchange across the air-water interface is among the most dynamic. Theoretically, a body of water with an extensive air-water interface should come into equilibrium with the atmosphere with regard to dissolved gas content. In fact, however, gas solubility exhibits continual adjustment. Water temperature tends to reflect changes in air temperature quickly in a shallow estuary. Since any change in water temperature alters the equilibrium solubility of the various atmospheric gases, losses to or gains from the air must take place in order to regain equilibrium. The rates of transinterface gas exchange and attainment of equilibrium are profoundly influenced by surface turbulence.

Dissolved oxygen content is subject to influences of various chemical and biological processes in addition to the physical effects just discussed. Turbulence and activities of burrowing organisms continually expose highly reduced bottom sediments, which consume sizable amounts of dissolved oxygen. The respiration of the biota adds to this consumption. Opposing these effects, but during daylight hours only, photosynthesis liberates significant amounts of oxygen into the water. Theoretical treatments of the reoxygenation process have been published by Velz(*37*), Haney(*38*), O'Connor and Dobbins(*39*), and many others, while practical techniques for measuring exchange and reaeration have been developed by Odum and associates(*40–44*), Copeland and Duffer(*45*), Gameson and Barrett(*46*), and Tsivoglou et al.(*47*). Odum and Hoskin(*43*) report a number of reaeration constants determined by their field technique. Most of their values relate to shallow estuaries of the Texas coast and indicate reaerations in the range 0.1–1.5 g/m^2 per hour for a 100% oxygen saturation deficit, the exact value depending on water depth and wind velocity (Fig. 4).

D. Sedimentation

The suspended matter load in any estuary derives from a variety of sources. Freshwater influx contributes a portion, depending on rainfall, river flow conditions, and nature of the watershed. Similarly, strength of tidal movement, wind conditions, and composition of the ocean floor determine the amount of material carried into the mouth of the estuary from the sea. Within the main basin, currents and winds interact in stirring up loosely packed bottom sediments, while in situ growth of plankton communities adds measurably to suspended matter load in the water column. In warm climates, the plankton effect is often particularly important. Mulkana(*48*) has demonstrated that the mean dry weight of nannoplankton-sized material in Mississippi Sound is 36 g/m^2, not corrected for ash weight, on a total water column basis.

A number of forces tend to diminish the suspended matter load in the estuary basin. First, the net flushing action in a positive estuary carries a variable proportion of the suspended matter seaward during each ebb tide. Second, given reasonably calm conditions, colloidal and near colloidal particles tend to flocculate and settle, the flocculation being promoted somewhat by the high electrolyte content of the water as compared with fresh water(*49*). Third, in many cases particle-feeding organisms exist which tend to concentrate and remove suspended matter by filtration processes.

Information on sedimentation rates in estuaries, when available, may be of doubtful quality or lack general applicability. Since most estuaries are extremely heterogeneous, and most sedimentation measurements are performed in localized areas for specific purposes, such as engineering projects, the data gathered usually do not depict the average condition in the estuary. Often, they are not adequate even for their intended purpose.

Reliable average sedimentation data apparently are those of Nielsen [cited in Ref. (*50*)] for Danish estuaries, those of Rainwater(*51*) for Mississippi Sound, and those given by Gunter et al.(*52*) for Chesapeake Bay. The three sets of results are in surprisingly good agreement in spite of the diverse natures of the environments. Mean sedimentation rates,

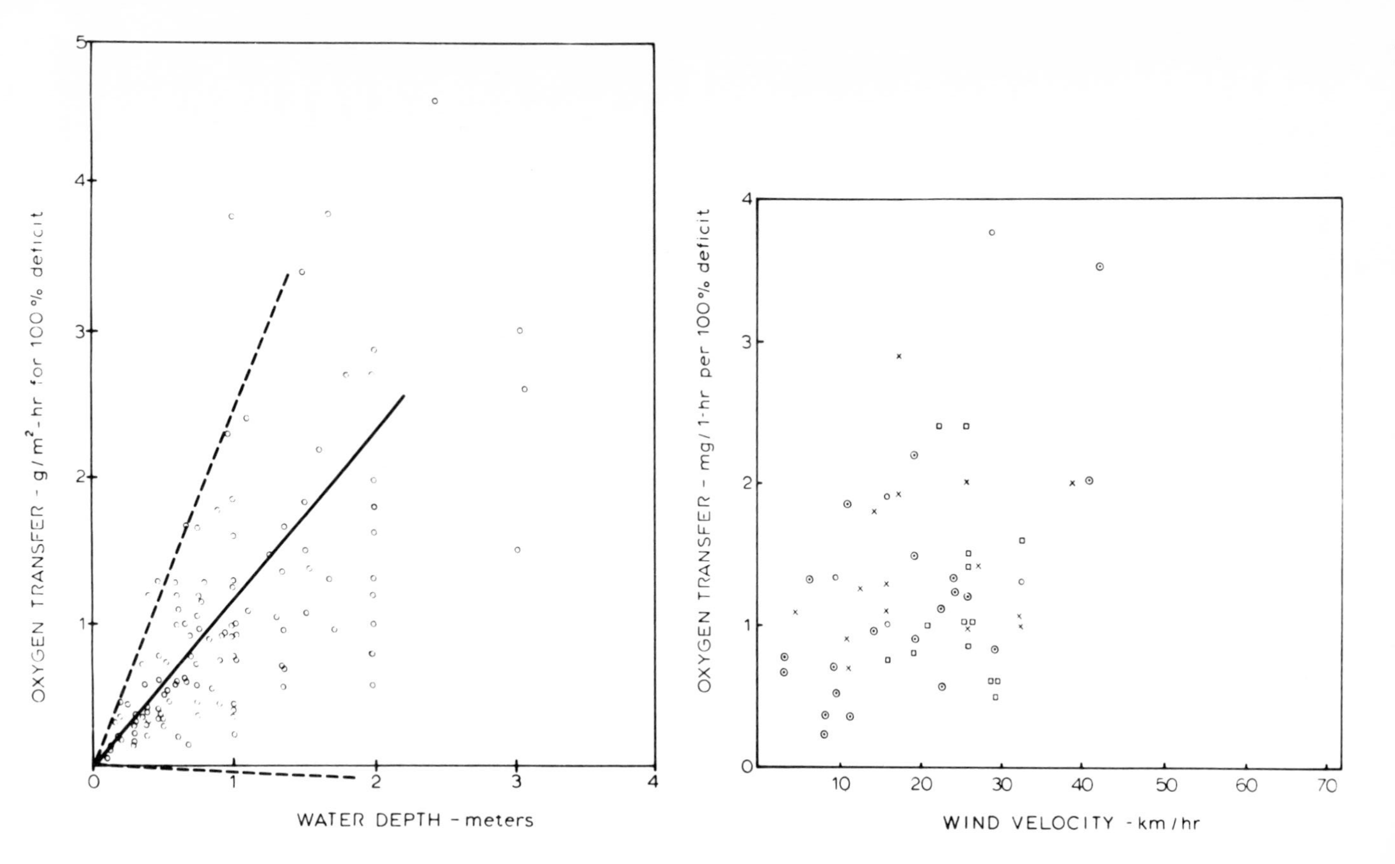

Fig. 4. Reaeration constants on an area basis (left) and on a volume basis (right). The volume-based graph illustrates the influence of wind. Dashed lines represent 95% confidence limits. [Modified from Ref. *44*, p. 28, by courtesy of H. T. Odum.]

after compacting, are 1–2 mm/year. This is less, by orders of magnitude, than many unrepresentative "spot" values which have been reported.

Numerous internal conditions in the estuary influence localized deposition or removal of sediments. Simmons(*32*), Ippen(*53*), Rusnak(*54*), and Emery and Stevenson(*55*) have reviewed sedimentation and scouring in estuaries but do not agree on all points. Simmons, for example, contends that before human intervention, most U.S. estuaries tended to be in equilibrium, and this view has been supported by Price(*56*), but Ippen points out that transport equilibrium cannot exist because of the time dependence of the currents involved.

Because of the circulation pattern in even moderately stratified estuaries, the impression has been that sediments deposited on the estuary bottom tend to be transported upstream(*53*). The mean rates of sedimentation already mentioned, however, indicate that the net result of this transport must be slight. In channels with high water velocity, sedimentation will be minimal, nonexistent, or even negative. Over shallow flats with low water velocities, sedimentation will be maximal. Postma(*57*, Fig. 1) shows graphically the relations between particle size and current velocity which determine whether an area is subject to scouring, is experiencing sedimentation, or is in equilibrium.

The biota of the estuary may influence channeling and sedimentation, especially if reef-building organisms are present. Although mussels and oysters do not necessarily form reefs, the reef does appear to be the preferred growth situation in large oyster-producing estuaries(*25*). Further, since oysters are particle feeders and gather their food by filtering suspended matter from the water, a reef located in a current gains a strong competitive advantage. Oyster reefs do tend to be so located, substrate permitting. Reefs in such positions slow the current passing over them, thereby decreasing the kinetic energy of the water and enhancing sedimentation. In addition, by filtering suspended matter and concentrating it in feces and pseudofeces, oysters deter resuspension of filtered materials. Lund(*58,59*) has reported quantitative laboratory studies on rates of turbidity clearance and self-silting by the American oyster, *Crassostrea virginica*. Figure 5 illustrates slowing of currents and removal of suspended matter (in this case, bacteria) by a small oyster reef in Matagorda Bay, Texas.

On high energy coasts, spontaneous closing of natural channels is seldom a problem. The channel in the mouth of the estuary is usually one of the deeper points in the basin because of intensive scouring by both incoming and outgoing tides. On medium energy coasts, with offshore barrier island formations, a similar deep channel or set of channels generally opens into the coastal lagoon. However, because of buffering

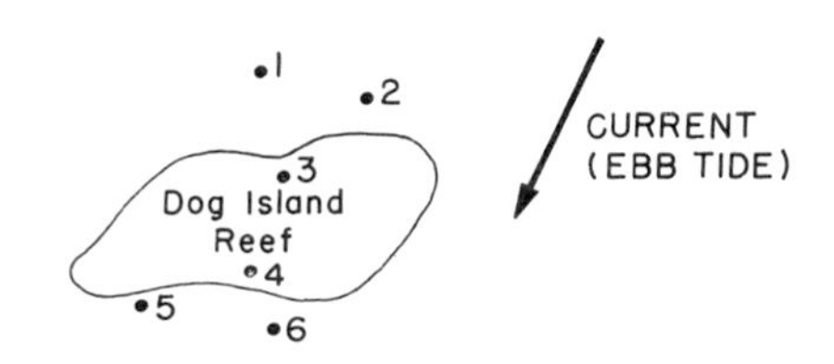

STATION NO.	PRESUMPTIVE COLIFORMS MPN/100 ml	CURRENT VELOCITY FT/MIN
1	13	25.2
2		34.2
3	7	12.7
4	5	16.3
5	0	23.4
6		21.1

Fig. 5. Influence of an oyster reef on current velocity and on coliform most probable number, Matagorda Bay, Texas. [Data from Ref. *195*.]

and slowing of tidal exchange by the lagoon, a secondary bar or spit may form across the mouth of the primary inshore estuary. On low energy coasts, channels between barrier islands often close because of sedimentation, especially if freshwater influx is small.

Hedgpeth (*25–27*) has discussed the high salinities and lack of flushing and exchange caused by channel closing in the Laguna Madre of Texas. Attempts have been made to open these channels by dredging, but they close again as soon as active dredging operations are halted. Unfortunately, these same channels open naturally during storm tides, disrupting the highway which connects Mustang Island with Corpus Christi.

In areas with lagoon-protected coasts, political pressures for dredging artificial channels to flush out the estuaries, diminish pollution, improve fishing, and so on, are frequent. Consequences may be far reaching and expensive. For example, in 1955 the State of Texas opened Rollover Pass, an artificial channel dredged across Bolivar Peninsula to help flush East Galveston Bay. Immediately afterward, extensive erosion began gobbling up the neighboring landscape. Cost of emergency repairs and bulkheading almost equaled the initial engineering and construction costs for the project (*60*). The pass is now maintained as a narrow, bulkheaded channel. This offers little in the way of flushing and general benefit to the bay as a whole, but it does provide a convenient shortcut through which various marine organisms can move from the bay to the Gulf of Mexico or vice versa.

The total energy normally available for channel scouring in an estuary is essentially the potential energy of the tidal prism and river flow. If an artificial channel is opened between the ocean and the estuary, two outcomes are possible. The new channel may soon close, probably after considerable lateral erosion accompanied by general shallowing of the cut. Alternatively, if the new channel remains open, whether naturally or artificially, diversion of some fraction of the tidal prism and associated scouring energy will lead to increased sedimentation in preexisting channels, the extent depending on the proportion of total water and energy diverted by the new channel.

Grain size of sediments might be expected to be largest at the mouth of an estuary, where sand grains and other heavy particles are deposited at the first slowing of the incoming seawater. Virgin sediments do tend to show an upstream gradation of diminishing size and density, but if the river also carries a load of large or dense particles, the trend may reverse at some point. In most cases, reworking of sediments by currents, burrowing organisms, and other processes begins immediately and quickly obscures the original pattern. In the Pascagoula River–Mississippi Sound system of the Gulf Coast, however, Siler(*61*) finds a persistent gradation from sand in the river channel to fine, silty sediments in mid-estuary to sand again in the vicinity of the barrier islands. Presumably, the innermost sandy area derives from sand beds through which the upstream portions of the river channel pass.

Core samples from estuarine sediments usually reveal distinct layering (*34*). Immediately underlying the water is a loosely agglomerated, yellow-brown, oxidized layer. Below this occurs a black, generally tacky, highly reduced layer, which releases strong hydrogen sulfide odors on exposure to air. At the sediment-water interface, the oxidized layer is flocculent and is easily stirred up by currents, vertical turbulence, or activities of burrowing and bottom-dwelling organisms. Beneath the flocculent portion, compaction of sediments ensues, with the rate of compaction depending upon the physical composition of the sediments. Thus, sediments composed largely of sand become firm at minimal depth, while those composed mostly of finely divided clays (mud) often tend to remain mushy to depths of 1 m or more. Sediment deposition is not uniform. Cores often show apparently random layering of sandy and muddy strata, and horizontal transects of the basin reveal interspersed sandy and muddy areas whose distribution often cannot be associated with distinct topographic causes.

Our review of available information indicates that no one has proposed an objective definition of the estuary bottom. Bottom is freely referred to in discussions of water column phenomena, biota, chemical exchanges, and other subjects, but the general implication is the indefinite one that bottom is the material underneath the water.

To resolve this, we withdrew cores from random points in Biloxi Bay. These cores were frozen and sliced into disks 2.5 mm thick, the uppermost disk being the segment immediately above the visually obvious sediment-water interface. Disks were weighed, dried to constant weight at 105°C, and reweighed to determine water content. From these data, rates of change of water content with depth were computed as change in proportion of total weight per millimeter. Figure 6 shows graphs of water content and rate of water content change versus depth for four cores. Cores 1 and 2 were from muddy areas, core 4 was from a sandy area, and core 3 was from a mixed area. The water content in the sandy area drops pronouncedly more rapidly with depth increase than does that in the muddy area, but qualitative patterns are similar. From these data, we propose that bottom be defined operationally as follows:

Bottom is that depth in the sediments at which mean rate of change of gravimetric water content (weight of water content per unit weight of sediment sample) with depth first becomes equal to or less than 0.01 mm^{-1}.

E. Temperature Effects

Since coastal plain estuaries are generally relatively shallow and unprotected from the wind, they respond rapidly to changes in air temperature. In warm, temperate areas, such as the U.S. Gulf Coast, this rapidity of response can generate biological problems, such as sizable fish kills. These kills are discussed in Sec. IV.D.

Temperature and, particularly, temperature differences influence the mixing of water masses in estuaries. Resistance to mixing in even partially stratified estuaries is proportional to density differences between the water masses. Under isothermal conditions, these differences are usually produced by salinity differences, the lower water mass being denser and more saline. If the freshened upper water mass is chilled without corresponding temperature change in the lower stratum, the density differential between the two layers can be materially reduced or eliminated, thereby enhancing mixing. Conversely, differential warming of the upper layer or chilling of the lower can lead to increased stratification and mixing resistance. Such considerations are increasingly significant as disposal of freshwater wastes into marine situations becomes more common (*62,63*) and as the need grows to discharge heated waters that have been used in cooling industrial plants, nuclear reactors, and desalination units.

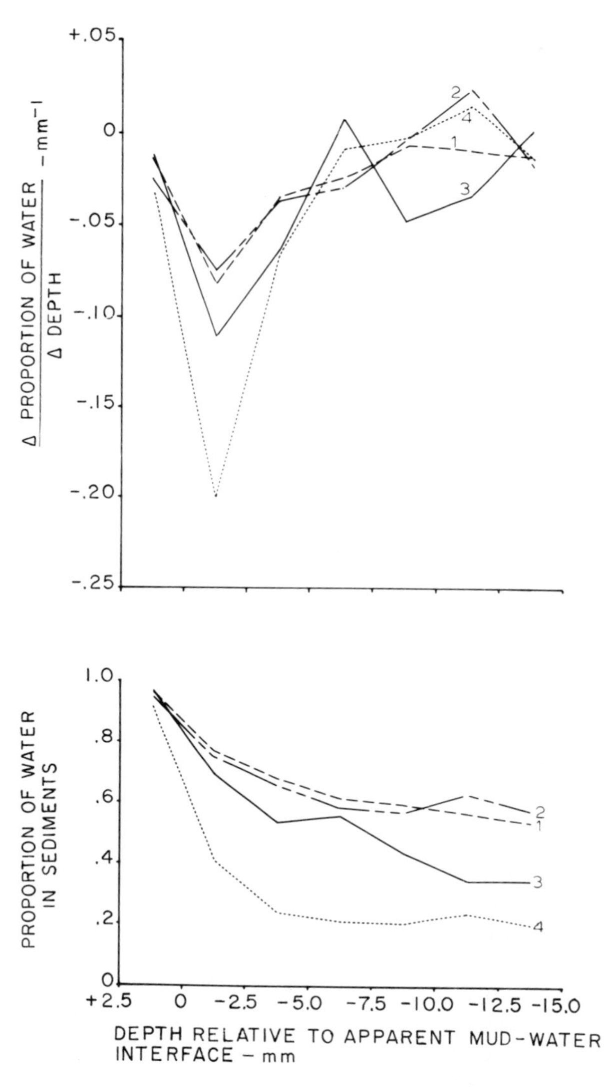

Fig. 6. Relative gravimetric water content of Biloxi Bay sediments (below) and rate of change of water content (above) as a function of depth below sediment-water interface. The numbers refer to cores: Cores 1 and 2 were from muddy areas, core 4 from a sandy area, and core 3 from a mixed (muddy and sandy) area.

III. Chemical Characteristics

A. Inorganic Aspects

The ocean is an aqueous solution containing most of the known, naturally occurring elements (*64*). This solute aggregation presumably is derived from megayears of weathering and leaching of the crust of the earth (*65*). Once dissolved, however, the various elements and ionic groups have interacted, recombined, precipitated, etc., so that the final intricate mixture constituting sea salt no longer reflects precisely the elemental ratios present in the primordial basaltic crust.

1. Salinity and Chlorinity

The salt mixture in the sea, while complex, shows a marked constancy of composition and concentration in pelagic ocean waters. For this reason, oceanographers have long treated the entire dissolved mass as a conservative, single-composition variable expressed in grams per kilogram (parts per thousand, $^{o}/_{oo}$) and designated as salinity (S) according to the following definition [cf. (*66,67*)]:

Salinity is the weight of solids which can be obtained from 1 kg of water when all organic matter has been oxidized, all bromide and iodide have been replaced by chloride, all carbonates have been converted to oxides, and the residue has been dried to constant weight at 480°C. Weighings are presumed to have been made in vacuo.

Since actual gravimetric analysis is not feasible for routine salinity determination, analytical methods are based on the extreme constancy of composition of oceanic waters. A sample of water is titrated argentimetrically by the Mohr procedure. The titration is carried out under rigorously defined empirical conditions, using a silver nitrate solution standardized against so-called normal water (*eau de mer normale*) prepared by the Laboratoire Hydrographique, Copenhagen. Worldwide comparability of data is assured by this approach. In the titration procedure, all chloride, bromide, and iodide are measured and are expressed as grams per kilogram chlorinity ($^{o}/_{oo}$ Cl). Chlorinity can then be converted to salinity by the relationship [(*68*) cited in (*67*)]:

$$^{o}/_{oo}S = 0.030 + 1.8050(^{o}/_{oo}\,\mathrm{Cl}) \qquad (10)$$

To maintain continuous comparability of data in the face of periodic revisions in atomic weights of the elements, chlorinity has also been defined in terms of the weight (in grams) of silver precipitated by 1 kg of sea water:

$$^{o}/_{oo}\,\text{Cl} = 0.3285234\,(\text{wt Ag}) \tag{11}$$

The ubiquity of the salinity concept in the oceanographic disciplines stems from its prominence in three more or less distinct areas.

The physical oceanographer uses salinity to compute specific gravity of a water mass. In practice, a defined quantity, σ_t, is obtained from tables cross-indexed by salinity and temperature. All specific gravity values are referred to the value 1.00000 for distilled water at 4°C, and conversion to σ_t is as follows:

$$\sigma_t = 1000\,(\text{specific gravity of sample} - 1.00000) \tag{12}$$

σ_t values are employed in various ways, such as tracing movements of currents and water masses and computing resistance to mixing in water columns.

To the chemical oceanographer, salinity is a measure of two milieu factors. First, as a conservative variable, salinity offers a convenient index of the strength of the buffer systems (carbonate and borate) in seawater. Since activities of these buffer systems regulate rates of precipitation and of redissolution of carbonates and other mineralogically important compounds, this aspect is also pertinent to geologists. Second, salinity represents the ionic strength of the aqueous medium in which specific chemical reactions are being studied. So-called "neutral salt effects" can significantly influence rates and courses of reactions that occur rapidly and in a straightforward fashion in distilled water or fresh water. In specific cases, reactions can be blocked completely, often by precipitation of one of the reactants. These influences are particularly notable in analytical reactions, especially colorimetric ones (Fig. 7).

Finally, to the biologist, salinity is an indication of the osmotic environment where organisms of interest live. Salinity represents the dehydrating influence which these organisms must resist, or to which they must accommodate, in order to exist. Thus, salinity may act as a barrier, either protecting an organism from predators and competitors or prohibiting its movement into an otherwise favorable habitat (*69*).

As mentioned at the beginning of this chapter, an estuary is an area where seawater is measurably diluted by fresh water. This immediately implies that the salinity range in an estuary is from that of pure seawater ($S \simeq 35^{o}/_{oo}$) to that of the fresh water feeding the estuary. In fact, extremes of range in the main body of the estuary are uncommon, unless flooding of the watershed occurs or storm-driven seawater intrudes into the basin. Under normal conditions, both salinity levels

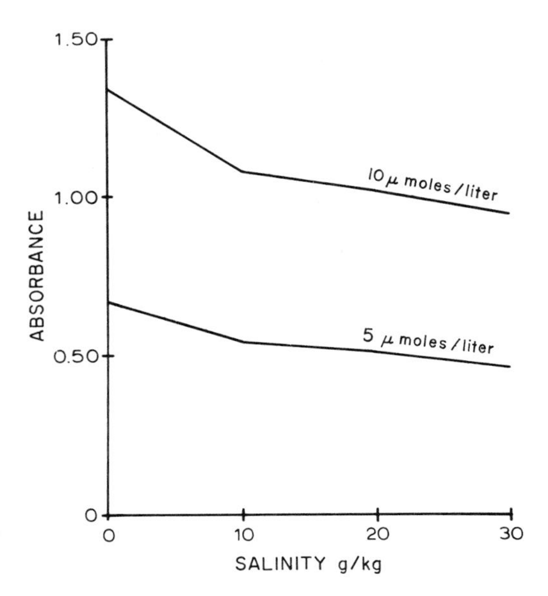

Fig. 7. Effect of salinity level on final color intensity in the phosphomolybdenum blue determination of orthophosphate. All values are corrected for reagent blanks at appropriate salinity. Plots are based on mean values of duplicate determinations.

and rates of change show regular variations associated with tidal cycles and with seasonal progressions. Generally, main basin salinities will range from 2–3 to 25–30‰. Figure 8 shows short-term changes in salinities of the bottom water and the mud at several points in the Pocasset River estuary of Massachusetts over several tidal cycles. Long-term variations in surface and bottom salinities at three stations in Mississippi Sound are shown in Fig. 9.

The relationship between salinity and actual total dissolved solids in estuarine waters has been debated for some years. In part, this involves the issue: When does water become brackish, i.e., when does a measurable amount of seawater appear in the water of an estuary? Several absolute answers, ranging from chlorinities of 0.01–0.5‰, have been proposed [cf. (*9,70–72*)]. Since the fresh water itself has a dissolved mineral content that varies widely from stream to stream and is not necessarily constant for any particular stream, attempting to set an absolute level seems futile.

However, the mineral composition of seawater differs from that of fresh water in several significant respects, and one or more of these might offer a useful index for measuring onset of brackishness. For example, Ca:Mg gravimetric ratios are about 0.2 in seawater versus about 6.0 in

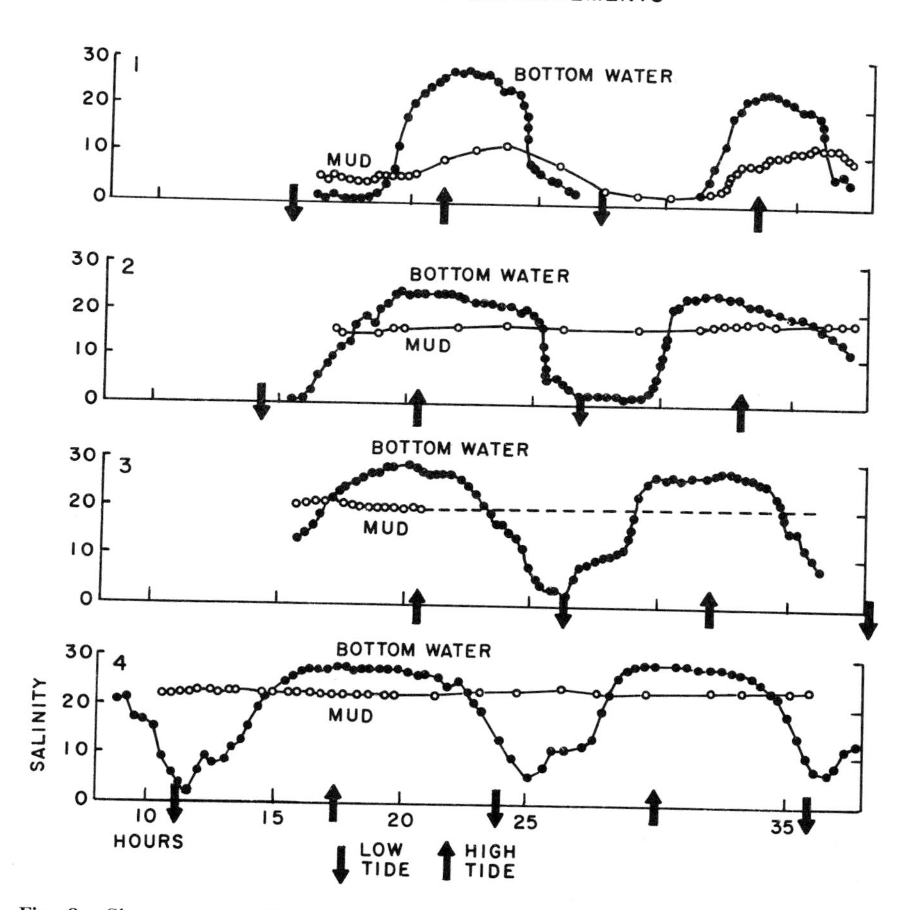

Fig. 8. Short-term salinity variations in the Pocasset River estuary, Massachusetts. [Reprinted from Ref. *75*, p. 77, by courtesy of American Association for the Advancement of Science and P. C. Mangelsdorf, Jr.]

fresh water, while corresponding Na:K values are 28 and 2.9, respectively(*64,73*). Similarly, Odum(*74*) has shown oceanic Sr:Ca atomic ratios to be 0.009 versus 0.005–0.0005 for fresh waters.

When these shifts in mineral composition occur, and whether they occur gradually or suddenly, is uncertain. Mangelsdorf(*75*) showed theoretical nonlinear changes that should occur in several salinity indices as seawater is progressively diluted with fresh water having a 200 ppm total dissolved solids content (Fig. 10). Recently we evaluated freezing

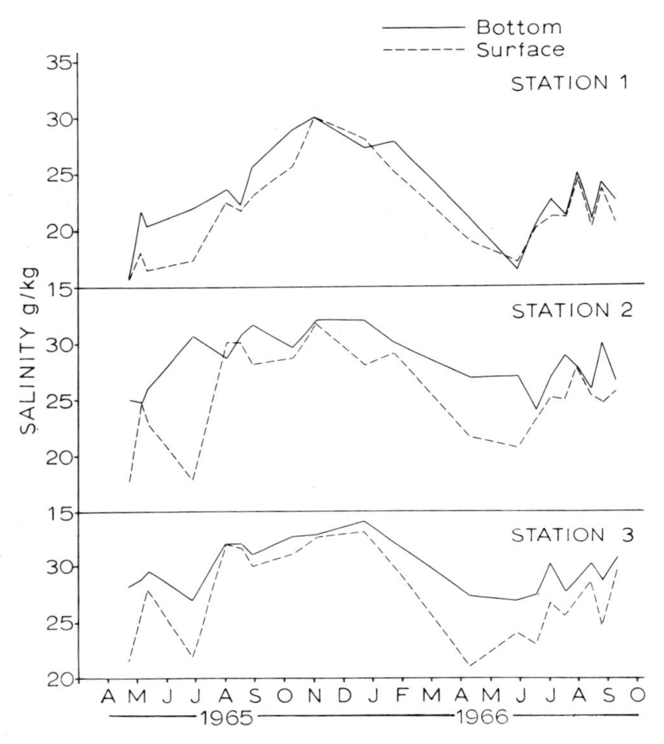

Fig. 9. Long-term salinity variations at three stations in Mississippi Sound. [Reprinted from Ref. *48*, p. 20.]

point osmometry as a method for precise salinity determination (*76*). We found that argentimetric chlorinity, or salinity computed from it, and osmolality, which presumably measures solute particles per unit volume, are tightly related linearly over the chlorinity range 0.010–15.46$^o/_{oo}$ in Mississippi coast estuaries (Fig. 11). Whether this relation holds for estuaries in general is not known as yet, but the results do suggest that the general concept of salinity may be useful at dilutions much greater than previously assumed. In this connection, Føyn (*77*) has pointed out that Ca : Mg ratios in estuarine waters as dilute as 1$^o/_{oo}$ S do not show any shift from the oceanic value.

2. Chemical Composition

Tables 1 and 2 show the elemental composition of average seawater and of some fresh waters, respectively. Ionic composition for waters of normal (positive) estuaries would be expected to fall between the levels

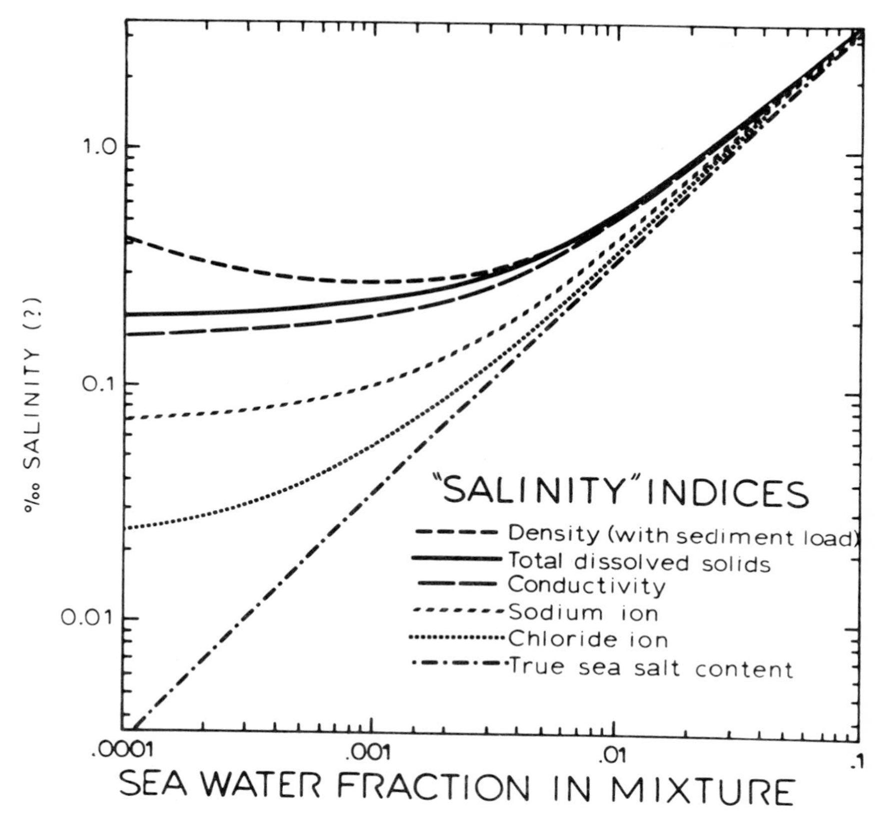

Fig. 10. Theoretical behavior of several salinity indices as seawater is progressively diluted with fresh water having 200 ppm total dissolved solids. [Reprinted from Ref. *75*, p. 78, by courtesy of American Association for the Advancement of Science and P. C. Mangelsdorf, Jr.]

given in the tables, except where significant pollution by specific substances is involved.

A convenient distinction can be made between substances present in quantities greater than 1 mg/kg, the macrochemical components, and those of lesser concentration, the microchemical components or trace elements. Roughly, the groupings correspond to those components that can be estimated directly by conventional analytical techniques and those that usually require special techniques, such as colorimetry, spectroscopy, activation analysis, or preconcentration. Goldberg(*64*), however, has defined a trace element in seawater as one that shows an atomic ratio to the chlorine content of less than 10^{-3}.

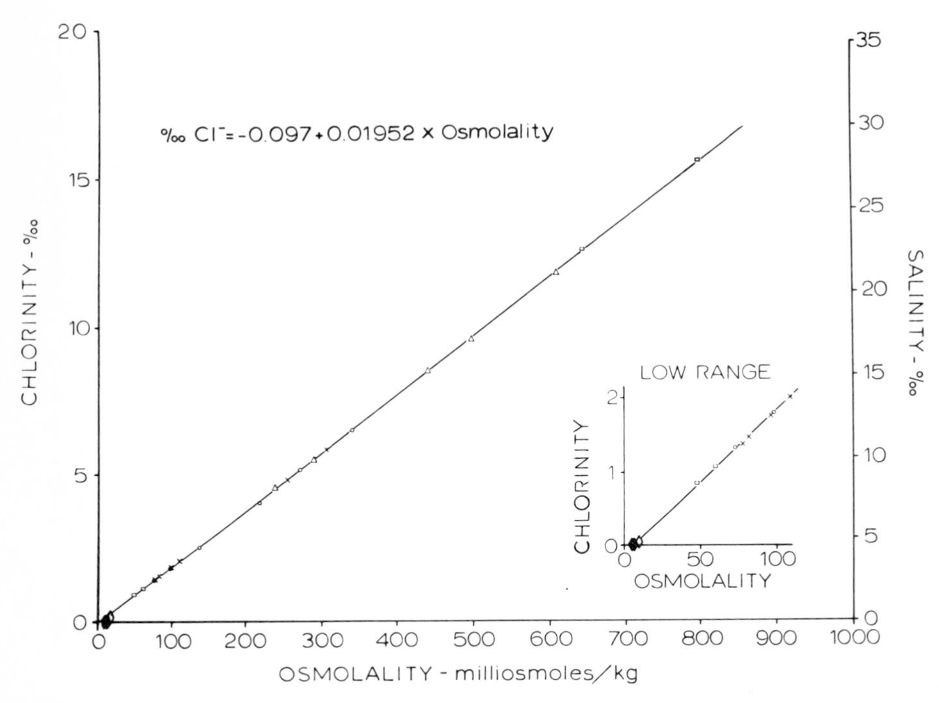

Fig. 11. The relationship between chlorinity and osmolality. [Reproduced from Ref. *76* by courtesy of Agustín Ayala-Castañares and Instituto de Biologia, Universidad Nacional Autonoma de México.]

a. Macrochemical Components

Only about 14 elements are included in the macrochemical group (> 1 mg/kg), the two most abundant being hydrogen and oxygen derived primarily from the water molecules. In the case of hydrogen, the water involvement accounts for essentially all of the tabulated amount. The element is of no further interest here except as it is involved in hydronium ion variations associated with pH changes. For oxygen, more than 99% is bound up in water. Of the remaining fraction, some part is involved in the ions of the buffer systems, carbonate and borate, while some is present in combination with sulfur, silicon, and other oxide-forming elements. A small amount, usually less than 10 mg/liter, is present as the dissolved oxygen of the water.

The buffer system of seawater assists in maintaining the chemostatic nature of the marine habitat. To a lesser extent, this is also true in estuaries. The oceanic buffer system is dual, consisting of a 10^{-3}–10^{-4} M borate buffer and a carbonate buffer of about 2×10^{-3} M. Since boron is

TABLE 1
Chemical Composition of Seawater[a]

A. Macrochemical components			
Element	mg/liter	Element	mg/liter
B	4.6–4.8	K	380
Br	65	Mg	1.8×10^3
C	28	Na	1.1×10^4
Ca	400	O	8.6×10^5
Cl	1.9×10^4	S	900
F	1.3	Si	0.2–4.0
H	1.1×10^5	Sr	8.0–13.0

B. Microchemical components			
Element	μg/liter	Element	μg/liter
He	5×10^{-2}	Nb	0.01–0.02
Li	100–200	Mo	0.7–12
Be	5×10^{-4}	Ag	0.1–0.3
N	6–700	Cd	0.11
Ne	0.3	In	20
Al	10–70	Sn	3
P	1–100	Sb	0.5
A	400–700	I	50–60
Sc	4×10^{-2}	Cs	0.5–2
Ti	0.02–1	Ba	6–54
V	0.3–2	La	0.3
Cr	0.05–0.25	Ce	0.4
Mn	0.4–10	Y	0.3
Fe	2–50	W	0.1
Co	0.1–0.5	Au	0.004–0.4
Ni	0.1–2	Hg	0.03–0.27
Cu	1–10	Tl	0.01
Zn	5–10	Pb	3–400
Ga	0.007–0.5	Bi	0.2
Ge	0.01–0.1	Rn	9×10^{-12}
As	3–38	Ra	3×10^{-8}
Se	4–6	Th	0.4–0.7
Rb	120–200	U	2–3

[a]Compiled from Refs. *64*, *73*, *74*, *78*.

normally a rarity in fresh water, the borate buffer is greatly diluted as mixing progresses in the estuary. Carbonate, however, does operate as the buffer mechanism in most fresh waters. Thus, because of supplementation by freshwater-derived carbonate, carbonate buffer dilution does not necessarily occur during absolute seawater dilution. As a result,

TABLE 2
Some Representative Data on Dissolved Inorganic Materials in Surface Fresh Waters

Water type	Units	Substance											References
		Na	K	Mg	Ca	CO_3	SO_4	Cl	NO_3	SiO_2	$(Fe,Al)_2O_3$	Total	
World average	%	5.79	2.12	3.41	20.39	35.15	12.14	5.68	0.90	11.67	2.75	100.00	(*73*)
Bicarbonate water	mg/liter	7.1	2.6	4.2	25.0	43.8	14.9	7.1	—	—	—	104.7	(*79*)
Bicarbonate water	mg/liter	13.9	5.0	8.3	49.4	86.6	29.3	13.9	—	—	—	206.4	(*79*)

estuarine pH is maintained as a conservative property except in unusual cases involving highly acidic bog drainage or discharge of strongly alkaline or acidic wastes. Generally, pH levels in well-mixed estuaries are in the range 7.5–8.1.

Two of the macrochemical constituents, sodium and chlorine, as sodium chloride, form the basis for localized salt industries, especially in areas where hypersaline estuaries and lagoons occur. Although no salt industry has been established on the Laguna Madre of Texas, large halite crystals attached to plants and other immersed objects have occasionally been found there. This occurs during dry periods when Laguna Madre salt concentration is two- to fourfold that of seawater [cf. (*25*)].

Calcium and magnesium, especially the former, are vital for reef formation by corals, for sedimentary processes, and for molluscan shell formation. Although the massive limestone areas of the world generally represent marine rather than estuarine deposits, oysters and other mollusks are prominent components of estuarine fisheries in many parts of the world today. Further, on the Atlantic and Gulf Coasts of the United States, a major dredging industry exploits fossil deposits of shells of *Crassostrea virginica*, the American oyster, and of *Rangia cuneata*, an abundant estuarine clam. Shell from the latter organism is generally used for surfacing parking lots, unpaved roadways, and driveways and for comparable purposes. Shell of the former, although sometimes used similarly, is a major source of crude lime for a number of industries, such as cement manufacture and magnesium production (Dow process).

Along the Gulf Coast, exploitation of "mud shell" by dredges has been surrounded by controversy for years. Conservation groups, real estate developers, and fishing interests have ranged themselves against dredging companies, industrial users of shell, and chambers of commerce. Voluminous transcriptions of regulatory hearings have accumulated. Presumably, the problem is self-limiting. Since exploitation rate appears to exceed biological replacement rate, soon there will be no more shell about which to argue.

The other dissolved macrochemical constituents are commonly abundant chemical elements. Dissolved silica may, under bloom conditions, be a limiting factor for diatom population growth, but blooms are atypical situations. The remaining macrochemical entities, while normal components of protoplasm, are present in estuaries in quantities far too great to engender nutritional limitations.

Figure 12 illustrates fluctuations over several tidal cycles in a number of macrochemical components for one point in Biloxi Bay, Mississippi. Pertinent changes in several physical parameters are also given.

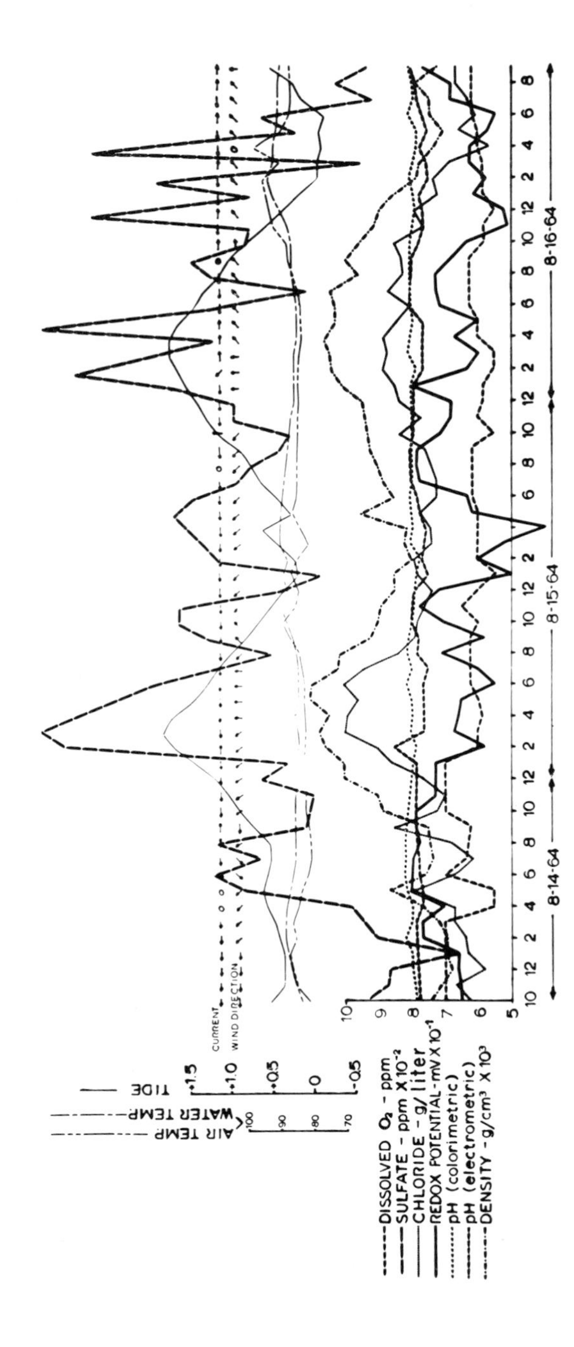

Fig. 12. Short-term variations in several physicochemical factors and dissolved macrochemical components in Biloxi Bay, August 14–16, 1964. [Courtesy of R. R. Priddy.]

b. Microchemical Components

Strong interactions between biosphere and environment exist in the microchemical realm. In 1840, Justus Liebig stated his "law of the minimum," and Blackman restated the principle in 1905 as his "law of limiting factors"(*80*). Under the terms of these concepts, growth and persistence of any biotic population are controlled and limited by that essential nutrient which is present in least amount in the environment. A feasible habitat is, of course, presumed. Modern usage tends to modify Liebig's principle to allow for interaction of vital factors, but the basic concept is unchanged.

In an estuary, the "law of the minimum" is involved in a situation wherein both biotic and abiotic systems compete for trace materials. The usual positive estuary receives these substances as imports from the watershed and tide and loses a portion of them to the sea either as dissolved materials or as particulate suspended matter, living or nonliving. In the estuary, however, processes occur which tend to retain the imports.

First, the estuary is generally much broader than the freshwater sources feeding it. Thus, water entering the estuary is dispersed over a wider channel, and flow velocity and turbulence diminish accordingly. Since these factors represent the kinetic energy holding particulate matter in suspension, settling ensues, with larger, heavier particles and associated sorbed materials tending to settle first.

Second, primary productivity in estuaries tends to be high(*1,42–44, 48*), and nutrient uptake and cycling rates are correspondingly great. We have recently shown(*81*) the great rapidity with which dissolved orthophosphate is taken up by an estuarine system and incorporated into sediments and suspended matter (Fig. 13). Parker(*82*) has used tracer methods to document the rapid turnover and diurnal cycling of cobalt, iron, manganese, and zinc in a bay ecosystem.

Levels of micronutrients such as phosphorus and nitrogen tend to be variable but high in estuaries compared with adjacent oceanic and freshwater bodies(*83–87*). This is largely caused by initial trapping in sediments followed by active sediment-water interchange. Berner(*88*), on the basis of sulfur studies in the Gulf of California, suggests that the upper sediments of an estuary exist in an open chemical steady state with respect to the water. It appears to us(*81*) that the sediments may function as "nutrient buffers." In shallow estuaries, turbulence almost continually stirs up the fine, silty sediments and moves them through the water column. This creates a general flux of sediment-associated nutrients. Additions to or withdrawals from this flux depend on what the various chemical and biochemical processes of the estuary require.

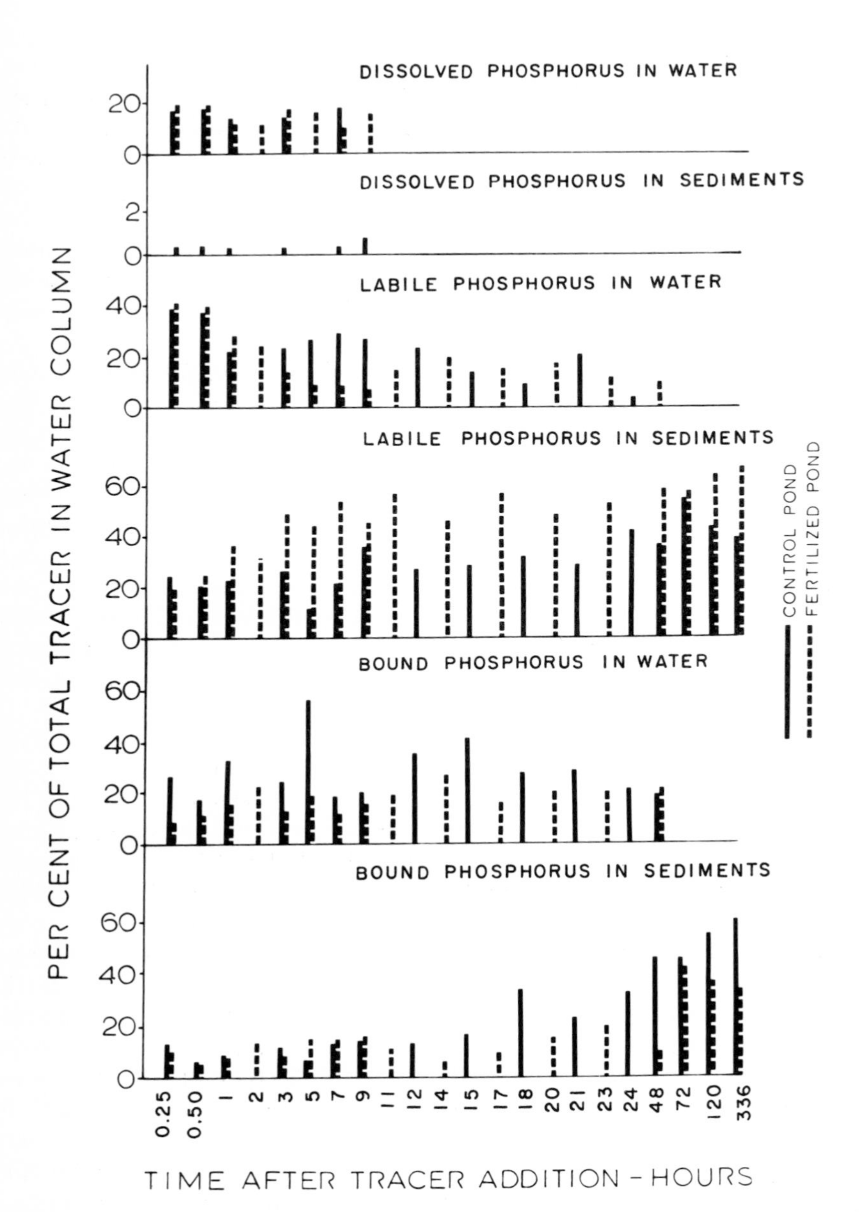

Fig. 13. Assimilation and distribution of $^{32}PO_4^{3-}$ by fertilized and unfertilized estuarine pond ecosystems. [Reproduced from Ref. *81* by courtesy of International Association on Water Pollution Research.]

Many marine organisms are able to concentrate trace materials from their environment. In some cases, presence of a material in the oceans was first inferred from its occurrence in organisms. Estuarine organisms also exhibit this concentration phenomenon.

Perhaps the best example of trace element concentration occurs among tunicates. Levine (*89*) showed that *Eudistoma ritteri*, an aplousobranch ascidian common along central California shores, contained 1512 ppm titanium, 471 ppm vanadium, and 144 ppm chromium on a dry weight, whole zooid basis. These elements are known in the ocean only as traces of a few parts per billion or less. Tunicates in general apparently concentrate vanadium and utilize it as the metallic portion of the respiratory compound in their blood. The blood also contains fair amounts of free sulfuric acid and does not exhibit the differential loading–unloading effect known for hemoglobin and other respiratory pigments. Since vanadium sulfate is widely employed for oxygen sorption in polarographic work, one is tempted to postulate a similar function here, but experimental data are lacking.

Winogradov (*90*) has tabulated the elemental composition of large numbers of marine organisms, including many estuarine forms, and should be consulted for further information on presence of trace elements in organisms. Goldberg (*64*, Table 2) gives enrichment factors for a number of elements concentrated from seawater by marine organisms.

B. Organic Aspects

Except for a thin upper layer, the bottom sediments of estuaries are highly reduced and anaerobic. They release a strong hydrogen sulfide odor when exposed to air and often exhibit in situ redox potentials in excess of −200 mV versus a saturated calomel electrode (*34,91*). This reduced situation derives from relatively high accumulations of organic matter. Emery and Stevenson (*55*), citing Trask (*92*), specify a minimum organic content of about 3% for sediments in positive estuaries, and this represents only the equilibrium differential between unknown, but probably high, rates of deposition and degradation. Biggs (*91*) gives data on Chesapeake Bay which seem to confirm Trask's value.

Darnell (*93,94*) reviewed the general organic detritus situation and concluded that this material "represents a major storage, transport, and buffer mechanism for the estuarine ecosystem."

Most of the organic material in the water occurs as particulate matter, either living plankton or fine debris. Some is generated endogenously, while some is carried in from freshwater sources. Although considerable export to the ocean does occur, eventually the nonexported fraction finds its way into the sediments, either as a result of direct settling or in fecal material following consumption by various organisms. The number of compounds involved is potentially vast. Bacterial action in the sediments begins immediately with release of methane and hydrogen sulfide and formation of organic acids and other breakdown products. As a result, pH of the mud may fall as low as 5.4 (*95*).

Not all the organic matter is decomposed. Some is eventually buried below the biologically active zone and thereby preserved. This happens, for example, with the protein matrix of mollusk shells. Peat formation by burial of fringing salt marshes is also common (*96*). Such preserved organic matter has been employed for carbon-14 dating of estuarine sediments (*10*).

Duursma (*97*) and Hood (*78*) reviewed existing information on dissolved organic carbon in the marine environment. Generally, reported values for carbon range from 0.4 to 6.0 mg/liter, with pelagic waters at the lower level and inshore waters at the upper. Wilson (*98*), however, studied some atypical environments, such as brine pools, mangrove swamp runoffs, and hypersaline lagoons, and found values in excess of 100 mg/liter. In his study, "dissolved organic matter" is defined as all material passing a 0.45-μm membrane filter.

Some data on specific types of organic compounds are available. Jeffrey (*99*) reviewed work on dissolved lipids in seawater and added new data on the Texas coast including values of 6.10 and 9.09 mg/liter for Aransas Pass and Bolivar Pass, respectively. Park et al. (*100*) reported a dissolved amino acid level of 0.3 μmole/liter (leucine equivalents) for a single water sample from Redfish Bay, Texas. This value is not surprising in view of the high level of the free amino acid pool in marine invertebrates (*101*). Hood (*78*) has summarized various reports on specific compounds and groups of compounds dissolved in seawater.

IV. Biological Characteristics

A. Introduction

In perusing literature on estuarine ecology, one finds a number of seemingly contradictory statements concerning the character of the biota. Some authors (*24,25*) state that estuaries are characterized by low species diversity and high population levels, whereas others (*102,103*) report considerable species diversity. In some cases (*43,104*) one gets the impression that estuaries are nutrient-rich, highly productive ecosystems devoid of macroorganisms.

Part of the problem lies in the fact that there is no such thing as a typical or standard estuary. Biotic complexes are often entirely different in neighboring estuaries due to contrasting physiography, nutrient levels, salinity characteristics, tide and current patterns, or other factors. According to the literature, temperate or northern estuaries frequently exhibit low species diversity and high population phenomena; whereas the opposite may be reported for tropical environments. Since generalizations have usually been made on the basis of a few phyla, a different picture may unfold when the entire biota is adequately enumerated.

The greatest obstacle to the development of workable standards for interestuarine biotic comparisons probably lies in the diversity of methodology, techniques, and interests of investigators or funding sources. In addition to the fact that there is no general agreement as to what should be sampled, efforts to standardize sampling methods have met with little success. Unless the new methods and instrumentation that are continually being developed yield results amenable to comparison with previous studies, they tend to compound the problem rather than aid in its solution.

Since major funding for estuarine studies has normally come from organizations interested in fisheries or pollution control, sampling parameters have been rather narrowly channeled toward solution of specific problems. The estuarine ecosystem is a rapidly fluctuating, integrated complex of chemical, physical, and biological processes. Restricted studies, evaluating only a portion of the biota, permit only general conclusions regarding biological characteristics of a particular estuary. Hedgpeth(*105*) has pointed out the fallacy of describing estuaries on the basis of gross photosynthesis, or any other single factor, since the values observed may be similar for entirely different biological conditions. Likewise, merely listing and enumerating the biota in one or more subenvironments within the estuary fails to indicate real or potential productivity levels.

Adequate evaluation of the estuarine biota will only result from long-term synoptic application of standardized methodology and techniques to all physical, chemical, and biological parameters. This will entail considerable expense and the concentrated cooperative efforts of specialists in each discipline, but the results should provide sufficient justification. Since industrialization and population pressures are rapidly encroaching upon the estuaries, the number of undisturbed environments is rapidly decreasing throughout the world. Carriker(*24*), Hedgpeth(*105*), and others have stressed the urgent need for immediate comprehensive estuarine studies. In view of our evaluation of the thermal pollution problem, such studies may be merely of historical interest by the year

2000, but at least they will provide an accurate record of one of the earth's more important biological environments.

Estuarine aquatic biota may be roughly grouped into fresh-brackish water, marine, and estuarine components. The latter consists of organisms which spend their entire lives within the estuary. These are usually polyhaline and are either adapted to rapid environmental fluctuations or even require such variations for survival. The fresh-brackish water component, usually represented by the smallest number of both individuals and species, is normally confined to the upper portions of the estuary. This faunal component is frequently absent or nearly so in a high salinity estuary, but may be seasonally abundant in tropical situations where there is a well-defined "rainy season." The marine component, including those forms utilizing the estuary as a nursery for juveniles and larvae, is probably the most diverse and widely distributed of the estuarine assemblages. Although most late juveniles and adults are restricted to moderate or high salinity waters, many early juveniles and larvae are euryhaline and range well into brackish waters. Adults of benthic marine forms frequently intrude into the estuary within the salt wedge beneath fresher surface waters. Although salinity apparently exerts a major influence, other factors, such as substrate type and stability, current velocities and wave action, temperature, and food supplies, all contribute to the distribution and abundance of estuarine biota.

Viewed from the standpoint of habitat, estuarine biotic groupings range from planktonic communities, through fouling associations, to benthic epifaunal and infaunal assemblages. Classical studies on benthic organisms have usually distinguished organisms living on the bottom (epifauna) from those living in the bottom (infauna), either in burrows or between substrate particles. Since these are not biologically discrete groups, differentiation has frequently been artificially based on the type of collecting apparatus employed. Application to benthic studies of our proposed definition of "bottom" (Sec. II.D) will doubtless show that many organisms previously considered as infauna are actually epifaunal in character.

The estuarine biota is a finely integrated complex tolerating rapid environmental variations, yet highly susceptible to major deviations from established local conditions. Any gross modification of the chemical or physical environment is sooner or later reflected by some biological change. Some aspects of the many types of estuarine organisms and their interrelationships with other organisms and with the environment are outlined in the following sections.

B. Estuarine Biota

1. Macrofauna

a. Mammals

A number of semiaquatic mammals, such as the raccoon (*Procyon*) and otter (*Lutra*), frequent the middle and upper reaches of south Atlantic and Gulf Coast estuaries. These carnivores are gradually being driven to the more remote and inaccessible regions by the encroachment of civilization. In many areas the secretive otter and even the ubiquitous raccoon have vanished completely from former ranges. The herbivorous nutria (*Myocastor*), originally imported from South America during the early 1930's (*106*), has flourished in the Louisiana marshes. It now supports a substantial fur industry and is simultaneously condemned for its depredations on inland sugarcane, corn, and rice crops.

Of perhaps more immediate interest are the marine mammals which either permanently inhabit or temporarily invade the estuaries. These are conveniently separated into the sirenians (manatees and dugongs), cetaceans (whales and porpoises), and pinnipeds (seals, walruses, and sea lions).

The sea cows (manatees and dugongs), as the name implies, are grazers on aquatic plants. Manatees (*Trichecus*) inhabit shallow tropical and subtropical marine, estuarine, and river environments on the west African coast from Senegal to Angola(*107*) and along western Atlantic coasts from the Carolinas south to the West Indies and the northeastern coast of South America(*108–110*). The dugong (*Dugong*), a related Indo-Pacific form ranging from the Red Sea to the Solomon Islands and northern Australia, apparently prefers more saline waters(*109*) and seldom ascends rivers to any great extent. Although nowhere abundant, and listed among the endangered and vanishing mammals, sea cows are individually the largest primary consumers in estuarine habitats. Their grazing efficiency is such that they are used in British Guiana to clear canals and other watercourses of weeds and algae(*109*).

Numerous species of dolphins (porpoises) are permanent or semi-permanent estuarine residents, and certain tropical forms are restricted to river or lacustrine environments. The Atlantic bottlenose dolphins (*Tursiops truncatus*) are widely distributed on both sides of the Atlantic (*111*) and are common in western Atlantic and Gulf Coast estuaries. They are most common in the mouths and lower reaches of the estuary, but will often, depending on the availability of food, range well into shallow, low salinity waters. Gunter(*108*) reported that mullet (*Mugil*) constitute a high percentage of their diet but that dolphins will feed upon most estuarine fishes. They occasionally ingest shrimp (*Penaeus*), but crustaceans apparently are not a preferred dietary item. On a local basis, the predations of *Tursiops* can be disastrous to schooling fishes. The dolphins may fish alone or in groups of several individuals. In the latter case, it is not unusual to see one or two continually circling the fish school, thereby keeping it relatively intact, while others make repeated successful strikes toward the center of the school. Gunter(*108*) suggested that historical records indicate a decline in the populations of dolphins in Gulf Coast estuaries.

Whales are not normal components of the estuarine biota, but certain species, especially the pilot whale or blackfish (*Globicephala macrorhyncha*), may enter the mouths and lower reaches of estuaries for short periods. The harbor seal (*Phoca*) also occurs in the lower reaches of middle and northern Atlantic estuaries but seldom in numbers.

Of considerable interest to both the biologist and the general public are the occasional groundings of marine mammals on estuarine beaches or shallows. In some species (e.g., pilot whale), this is a relatively frequent occurrence involving as many as 200 whales at one time(*112*). Grounding is apparently deliberate in this species, in that they seem to "follow the leader" blindly when one of their number is stranded. In other species, grounding may result from the live animal being trapped in shallows on a falling tide or, more frequently, from the carcass of a dead or dying creature drifting into the estuary. Caldwell and Golley(*110*) noted the grounding of 18 species of marine mammals from Georgia to Cape Hatteras, and many of these records are from well within estuaries.

Grounding and subsequent decomposition of a single *Tursiops* may not cause great concern, but the stranding of a single large whale or a school of pilot fish near a populated area presents an immediate and major disposal problem. A number of methods have been followed (burying, dousing with quicklime, burning) with varying degrees of success. Refloating and towing the carcass to sea is usually an expensive and temporary measure since tidal currents and onshore winds frequently combine to wash the beast ashore near its original location.

Marine mammals frequenting the estuaries are of but minor economic value since they are seldom exploited in a commercial fishery. Although they may, especially in the case of sea cows and manatees, be of local biological importance, their generally transitory nature and relatively low numbers minimize their significance in the estuarine biotope.

b. Fish

The diversity of the estuarine environment is reflected in its fishes. Freshwater forms usually inhabit the upper reaches of the estuary but may invade the lower and middle estuary at flood stages within an overlying, unmixed freshwater stratum. Euryhaline species, ranging throughout the salinity gradient from fresh to oceanic waters, usually constitute the bulk of estuarine fish populations. Marine or polyhaline species are generally found in the middle or lower estuary, but demersal forms may intrude the headwaters or upper estuary under drought or abnormal tidal conditions. To these we must add the anadromous forms—some salmons (*Salmo*, *Oncorhynchus*), sea sturgeons (*Acipenser*), and striped bass (*Roccus*)—which spend much of their lives in the open sea but return inland to spawn in fresh water; and the catadromous species. The latter, e.g., the common eel (*Anguilla*), develop in low salinity or fresh waters and return to the sea to spawn.

Although it is hazardous to categorize estuarine organisms, particularly the highly motile fishes, in relation to a single environmental factor, McHugh(*113*) has provided the following salinity-related general classification which may be convenient:

(1) Freshwater forms that occasionally enter brackish water.
(2) True estuarine species that are confined to the estuary.
(3) Anadromous and catadromous species.
(4) Marine species that seasonally enter estuaries, usually as adults.
(5) Marine species that utilize the estuary as a nursery.
(6) Occasional marine visitors with no apparent estuarine requirement.

All but the freshwater species and the last category are estuarine-dependent species in that they utilize the estuary at some stage in their life history. Most of the world's great fin-fisheries are based on these estuarine-dependent species.

There is little need to describe the morphology of the various species of fish inhabiting estuaries, but a generalized discussion of their biology and life history is in order.

Pelagic species inhabit the upper portions of the water column, whereas demersal forms live on or near the bottom. Most fishes are "actively swimming pelagic organisms" at some stage in their life history and may then be termed nektonic. Estuarine fishes are extremely varied in size and mode of life and range from gobies, which mature when they are less than 1 in. (2.5 cm) long, to large sharks and, in lower latitudes, manta rays. Most sharks and giant rays are casual intruders, but the bull shark (*Carcharhinus leucas*) and the sawfish (*Pristis*) frequently enter the upper estuary and have become well established in Lake Nicaragua(*114,115*). Most estuarine species, however, are small, averaging under 1 ft (30 cm) in length and less than 1 lb (454 g) in weight.

Juvenile and adult fishes may be broadly separated into filter feeders and predators. The few herbivorous species that occur are primarily freshwater forms. Filter feeders, such as menhaden (*Brevoortia*), sea herrings (*Clupea*), and shad and river herrings (*Alosa*), feed directly on plankton by straining large quantities of water through their sieve-like gill rakers. McHugh(*116*) estimated that an adult menhaden can filter about 3.9 gal/min (14.8 liters/min) of water. Using this as a base, he computed that one Chesapeake Bay menhaden could filter about 3826 m^3 of water containing some 9565 g of zooplankton in a period of 6 months. These primary carnivores support some of the world's largest fisheries, and most, if not all, are estuarine-dependent species. The demersal and pelagic predators feed on a wide variety of items ranging from minute polychaetes and other burrowing invertebrates to fishes as large as or larger than themselves. Some, such as the common sea catfish (*Galeichthys*), are scavengers and eat any type of available food, whereas others have more circumscribed feeding habits and exhibit specialized morphological feeding adaptations. The bottom-feeding mullet (*Mugil*) has a specialized stomach, or "gizzard," which serves to grind up ingested sediments and associated organic matter. The drum (*Pogonias*) has strong, band-like teeth suitable for crushing crustaceans and mollusks, while the voracious trout (*Cynoscion*) has sharp, pointed teeth for cutting or tearing its swimming prey.

Numerous modes of reproduction occur in estuarine fishes, but that followed by many commercially important species is characterized by the absence of parental care and usually by the relatively large numbers of eggs produced by the females – 50,000–100,000 in species of *Clupea* and *Alosa*. Mixed schools of sexually mature fish release eggs and sperm freely into the surrounding water, where fertilization occurs by chance union of the gametes. If, as in many species, the fertilized egg is buoyant, it becomes part of the plankton during incubation. Demersal eggs may lie free on the bottom or be attached thereto by adhesive processes. Larvae, hatched after varying periods of incubation, enter the plankton where they soon begin to feed on other planktonic organisms. With subsequent growth and development, the young fish eventually leave the plankton as early juveniles or postlarvae to enter a mode of life similar to that of the adult in respect to behavior and food preferences. Throughout this life cycle, the adults, eggs, larvae, and postlarvae are subject to the depredations of other species of vertebrates and invertebrates as well as all positive and negative environmental influences. Early larvae may, for example, be eaten in quantity by copepods and crustacean larvae, but those which survive may shortly be classed as predators on this same crustacean fauna.

Although the foregoing life history can be completely confined to the estuary, most species are estuarine dependent at some stage but are not restricted to the estuary throughout life. Many of the dominant estuarine-dependent species of the Gulf of Mexico, such as the croaker (*Micropogon*), spot (*Leiostomus*), menhaden (*Brevoortia*), and mullet (*Mugil*), exhibit a rhythmic, seasonally correlated, inshore-offshore migratory pattern(*2,117*). Adults move out of the estuaries in late summer and early winter and spawn in offshore waters of the Gulf. Postlarvae and juveniles appear in the upper estuaries in the early spring. As they grow, they tend to move toward the middle and lower estuary.

Although the seaward movement within estuaries follows a general trend from lower toward higher salinities, salinity per se has not been conclusively demonstrated to be the primary environmental factor influencing this migration. Unfortunately, fishery workers have attempted to relate distribution and abundance patterns of estuarine fishes directly to salinity and temperature, subordinating, or entirely overlooking, other physical, chemical, and biological factors that may be of equal or greater importance. The ease and rapidity with which salinity data may be obtained, as well as the many apparent correlations between salinity and species distribution(*113,118,119*), partly account for the overemphasis. Undoubtedly, estuarine fishes are restricted by certain salinity extremes and, in the absence of other influences, would inhabit areas that provide optimal salinity conditions concomitant with age and developmental stage. Such ideal conditions do not, however, occur in nature. Except at the extremes of the known salinity tolerance of a given species, we are not certain whether observed salinity correlations reflect a primary influence of salinity or whether salinity is interacting with other environmental factors which, in consort, influence observed distribution and development.

Temperature is one of the more important ecological factors directly or indirectly influencing the distribution of estuarine fishes. Since water density and oxygen content are temperature-related, temperature indirectly affects osmoregulation and respiration. It also has direct effects since oxygen requirements and food uptake of an individual fish may vary with temperature(*120*). Species may be adapted to temperatures of high or low latitudes, but they also respond to the amplitude of temperature variations. Stenothermal species may be influenced by ranges involving only a few tenths of a degree. Onset of gametogenesis, spawning and associated migrations, local vertical distribution in the water column, as well as geographic distribution, are temperature dependent in many species. Reaction to temperature range and rate of variation changes with age and development. Consequently, the optimum temperature for the young may be entirely different from that for the adult. Since temperature affects general growth and development, fishes of northerly (cooler) habitats generally take longer to reach maturity and often attain greater average size than fishes of the same species in warmer waters.

Although most estuarine species are eurythermal, they are profoundly affected by rapid temperature variations. Fishes that could normally adjust to a gradual 10 or 15 degree temperature reduction may be stunned or killed by a sudden cold spell(*121–123*). Similar results can also occur with an influx of abnormally warm water(*124*).

Despite rather extensive knowledge of the general effects of temperature on the physiology, growth, and behavior of fishes, particularly in relation to temperature extremes, little is known concerning developmental optima.

Type of substrate, turbidity and light penetration, availability of food, abundance of predators and parasites, current velocity, and other physical and chemical factors all exert their effects on estuarine fishes. In most cases, however, our knowledge of the complex interactions is incomplete. Studies on estuarine fish populations have usually been restricted to a single species and often to a limited segment of its life history, or to fishes taken with a single type of gear. These studies usually include salinity and temperature data (frequently only surface observations) but seldom pay more than cursory attention to correlative influences of the environment as a whole. Yet all environmental factors exert their effects simultaneously on the individual, whose ability to integrate and adjust to these influences determines the distribution, abundance, behavior, and well-being of the organism or even the species. McHugh(*113*) reported possible correlations between annual fluctuations in abundance of the blue crab (*Callinectes*) and menhaden (*Brevoortia*) in Chesapeake Bay and thereby called attention to the pronounced interphyletic relationships within an estuary. If McHugh's observations are projected through the several hundred plant and animal species present in the typical estuary, we have some concept of the complexity of biological interrelationships. Further, if these are considered in conjunction with the chemical and physical factors affecting each biological entity in a specific manner, they indicate that an infinitely complex system underlies the apparent distribution and abundance of estuarine organisms.

A surprising degree of similarity exists, however, in the types of fishes inhabiting widely separated estuaries, particularly in tropical and subtropical latitudes. Many of the numerically or commercially important families are represented in estuaries as geographically diverse as those of the North and South Atlantic, the Indian Ocean, and the western Pacific (Table 3). Most of the great fin-fisheries are based on estuarine or estuarine-dependent species, and the often-repeated reference to the estuaries as nursery grounds of the fisheries is certainly a truism.

TABLE 3
Representation of Certain Common Families of Estuarine Fishes in Several Geographic Regions

Family	Chesapeake Bay (*113,125*)	Gulf of Mexico (*2,117*)	Venezuela (*126*)	West Africa (*3*)	Indian Ocean (*4*)	Australia (*127,128*)
Atherinidae (silversides)	+	+	−	−	−	−
Bothidae (flounders)	+	+	+	+	+	+
Clupeidae (herrings)	+	+	+	+	+	+
Cynoglossidae (tonguesoles)	+	+	−	+	+	+
Engraulidae (anchovies)	+	+	+	+	+	+
Gobiidae (gobies)	+	+	+	+	+	+
Mugilidae (mullets)	+	+	+	+	+	+
Sciaenidae (croakers)	+	+	+	+	+	+
Stromateidae (butterfishes)	+	+	−	−	+	−
Soleidae (soles)	+	+	+	+	+	+

c. Crustaceans

Crustaceans are a conspicuous segment of the estuarine fauna. Shrimp and crabs support extensive fisheries in many areas, while other crustaceans constitute an economic menace. Boring isopods, such as *Limnoria*, do untold amounts of damage to wooden piers, pilings, and other installations in favorable habitats. Barnacles that foul piers, boats, and other structures cause significant economic loss, particularly in the more saline middle and lower estuary. These forms, however, make up only a fraction of the crustacean population. The majority of crustaceans are small, inconspicuous, burrowing, planktonic, or benthic species having little direct economic importance.

Crustaceans occupy all segments of the estuarine environment from the bottom sediments, where species such as minute cumaceans and the mudshrimps (*Callianassa*) burrow, to the shore, where one encounters intertidal crabs (e.g., *Uca*, *Sesarma*) and, in the tropics and subtropics, land crabs (*Cardisoma*). They occur throughout the salinity gradient, with river and grass shrimp (*Macrobrachium* and *Palaemonetes*) in the upper estuary, euryhaline species such as the blue crab (*Callinectes*) throughout the area, and high salinity species such as rock shrimp (*Sicyonia*) in the lower estuary. Planktonic crustaceans include abundant holoplanktonic copepods and innumerable meroplanktonic larvae and postlarvae of benthic species. The benthic fauna is extremely diverse in most estuaries, and species population levels may be high. Although this is most apparent in commercially exploited species, other forms may be equally abundant. Recognition of 69 macrocrustacean species associated with oyster beds in a North Carolina estuary(*129*) and of some 87 species, excluding amphipods, copepods, and similar forms, in a south Florida estuary(*130*) gives some idea of species diversity.

Life histories, modes of reproduction, growth rate, feeding habits, and migratory patterns are extremely varied in the Crustacea. Although a few types resemble the adult when first hatched, most undergo a somewhat complicated development through several larval or postlarval stages that are completely different from the adult form. The commercial shrimp (*Penaeus*) may pass through a dozen or more stages between the egg and its final juvenile form, and the blue crab has four or five zoeal stages and a megalops stage before attaining the typical adult form. Some crustaceans may have several generations in a single season, but most of the larger forms require one or more seasons for completion of their life cycle.

Since they are more or less imprisoned within a semirigid calcareous exoskeleton, all typical crustaceans grow by a series of steps. Each growth period is completed in a relatively short time between the shedding of one exoskeleton and the hardening or calcification of the next, which is already developed in the underlying dermal layers. This shedding process (ecdysis) may take less than a minute in small forms or, as in the case of the blue crab, may require a day or more to complete. Growth occurs during the soft stage by a swelling of the body, due to the intake of water, before the new exoskeleton hardens. Frequency of shedding and amount of water intake are hormone-regulated(*131*), and hormonal activity in turn is apparently influenced by the environment. Shedding, and consequently growth, may be inhibited by unfavorable conditions(*132*) but is frequently rapid under optimum food and physicochemical conditions. Allen(*133*) reports an average increase of 8–30% in body size at each ecdysis in juvenile decapod crustaceans, a group in which commercially important species complete their overall growth within 1–3 years.

Many of the burrowing crustaceans are filter feeders; holoplanktonic herbivorous crustaceans frequently dominate the zooplankton; larger crustaceans are most frequently omnivorous scavengers. The blue crab ingests a considerable quantity of estuarine plant materials (*134,135*), as well as fresh or decaying fish, shellfish, and other animal matter. Darnell (*136*) considers the species to be a "detritivore, bottom predator and general scavenger." Commercial shrimp (*Penaeus*) have similar feeding habits. Darnell (*135*), in reviewing the feeding of the white shrimp (*Penaeus setiferus*), found this organism to be omnivorous but to feed primarily on the organic detritus of estuarine bottoms. Shrimp, crabs, and probably most of the carnivorous crustaceans are cannibalistic and will attack their soft or recently shed brethren with abandon. In aquarium studies, shrimp first immobilize their victim by destroying the eyes and then devour their now defenseless prey.

Aside from being extremely variable, life histories of estuarine crustaceans are, in many cases, poorly understood. Larvae of most benthic species have at least a short planktonic existence before taking up their final mode of life. This planktonic phase allows for widespread distribution throughout the ecosystem and partly accounts for the continual restocking of areas depleted by natural mortalities or fishing pressure.

The general life histories of common commercial shrimps and crabs of the South Atlantic and Gulf coasts are similar in many respects. The larvae are hatched in relatively high salinity or oceanic waters and later appear in low salinity upper estuary habitats as late larvae or early juveniles. With subsequent growth and development, shrimp move out of the estuaries to spawn in offshore waters. Conversely, crabs mate within the estuaries. Shortly thereafter, the mature females migrate to saltier waters of the lower estuary or into open, shallow coastal waters, at which time the mature crab population may be almost completely separated into males in reduced salinity environments and females in high salinity waters. Darnell (*136*) and Van Engel (*132*) have reviewed the life cycle of the blue crab, and shrimp life histories have been discussed by Gunter (*2, 137*) and Lindner and Anderson (*138*).

Crustacean growth rates, behavior, and spawning are directly influenced by temperature. Darnell (*136*) suggests that estuarine blue crabs may hibernate during winter, and Gunter (*2*) considers that the annual temperature cycle is chiefly responsible for seasonal migratory patterns of shrimp and crabs. Allen (*133*) includes both food relationships and temperature as the important determinants of decapod migratory patterns.

Dissolved oxygen levels appear to be critical for a number of crustacean species. Anderson and Reish (*139*) reported interspecific differences in oxygen requirements of the isopod *Limnoria* and found daily egestion rates to be significantly reduced at oxygen concentrations below 3.0 mg/liter. Mass mortalities from oxygen deficiencies in estuarine blue crabs and other crustaceans have been discussed by Brongersma-Sanders(*124*), Carpenter and Cargo(*140*), and Gunter(*122*).

Some crustaceans range throughout the salinity gradient from fresh to hypersaline environments, but most are restricted to a narrower span of salinities. Osmoregulation in crustaceans, reviewed by Kinne(*141,142*), has been shown to vary seasonally(*143*). Ballard(*72*) has shown the seasonal effect to persist in blue crabs under constant temperature conditions. Her work also demonstrated that female blue crabs have higher blood osmoconcentrations at environmental extremes over the observed 1.0–70.8‰ range. Sex-linked variations in osmoregulatory ability apparently contribute to the observed separation of mature male and female crabs along the estuarine salinity gradient. These and other factors, such as substrate type and consistency, current velocities, and turbidity, strongly influence the distribution of benthic estuarine crustaceans.

A number of the same or closely related crustacean genera are represented in widely separated estuaries, suggesting a duplication of the environmental limitations governing local abundance and distribution. Important, or potentially important, commercial populations of shrimp (*Penaeus*, *Metapenaeus*, etc.) occur in North and South American estuaries as well as in comparable Indo–Pacific habitats. In low temperate and tropic latitudes, portunid crabs (*Callinectes*, *Portunus*), river shrimp (*Macrobrachium*), mudshrimp (*Callianassa*, *Upogebia*), and other forms are conspicuous elements of the biota in estuarine environments.

Some crustaceans, such as estuarine spider crabs (*Libinia*), are characterized by low levels of abundance, whereas groups such as barnacles, shrimp, crabs, and amphipods occur in great numbers. Although annual and seasonal population levels may fluctuate widely, data on selected estuarine forms (Table 4) provide some indication of typical abundance

TABLE 4

Representative Maximum Counts of Selected Estuarine Crustaceans

Crustacean	Locale	Collection method	Maximum counts (estimated)	Reference
Neomysis americana	Delaware	Plankton net	3300/m^3	*145*
Balanoids	England	Petersen grab	7200/m^2	*146*
Balanus improvisus	Maryland	Fouling collector	4000/m^2	*147*
Corophium lacustrae	Maryland	Fouling collector	> 100,000/m^2	*148*
Rhithropanopeus harrisi	Maryland	Fouling collector	7800/m^2	*148*
Callinectes sapidus	Texas	Drop net	26/m^2	*149*
Penaeus duorarum	Texas	Drop net	208/m^2	*149*
Palaemonetes pugio	Texas	Drop net	828/m^2	*149*

levels. Crustaceans are subject to the depredations of other crustaceans and also constitute a significant part of the diet of many fish species. Darnell(*135*) reported a crustacean component in the diet of 29 of 30 examined species of filter feeders, omnivores, and predators. Gunter(*144*) found blue crabs and mudshrimp to account for about 90% of the food of large catfish (*Galeichthys felis*) in Texas waters. Although various parasitic barnacles (*Rhizocephala*), bopyrid isopods, nemerteans, and other organisms attack estuarine macrocrustacea, these appear to have only a minor role in determining population levels except in unusual cases.

d. Mollusks

The molluscan fauna probably makes the greatest mass contribution to the biota of temperate and tropical estuaries. Extensive intertidal oyster reefs and mussel beds, paving large portions of many shallow estuaries, are familiar to even the casual observer. In addition, varied and frequently abundant populations of less obvious forms are distributed throughout the estuary. Gastropods, such as *Littorina*, may occur on marsh grass (e.g., *Spartina*) well above normal tide levels, whereas the mud snail (*Nassarius*) crawls just beneath the substrate with only its siphon extended into the overlying waters. Mussels and oysters may line the intertidal zone, where they survive several hours of daily exposure with concomitant cessation of feeding and rapid temperature changes. Other pelecypods, such as the mud-dwelling *Barnea*, may live as much as 2 ft (0.6 m) below the substrate surface and extend their retractile siphons upward to the overlying waters. Other specialized forms, such as the so-called shipworms (*Teredo*, *Bankia*), boring clams (*Diplothyra*, *Martesia*), squid, octopus, various opisthobranchs, and scaphopods, may inhabit restricted habitats. Estuarine mollusks occur as planktonic, nektonic, burrowing, or benthic organisms, the latter existing on almost all types of substrate.

Some indication of the diversity of estuarine molluscan populations is shown by Wells' faunal study of North Carolina oyster beds(*129*). Mollusks accounted for 99 of 303 collected species, and these included 39 gastropods, 21 opisthobranchs, and 39 pelecypods. Moore(*150*) listed 163 mollusks from the Mississippi Sound, and similar numbers of molluscan species occur in many temperate and tropical estuarine environments.

(*1*) *Salinity effects*. Many estuarine mollusks are polyhaline or euryhaline and successfully accommodate to relatively great variations in salinity. Other less adaptable forms cannot tolerate salinity extremes or short-term variations of any magnitude.

Lologinid squid, octopi, scaphopods, chitons, and such forms as *Chione* and *Oliva* are usually found in the high salinity waters of the lower or middle estuary. Squids, such as *Lolliguncula brevis*, frequent these waters but are also found in salinities as low as $17^{o}/_{oo}$(*137,151*). Although *L. brevis* is collected over a broad salinity range, the species is most abundant at salinities above $25^{o}/_{oo}$. This is a polyhaline-marine organism rather than euryhaline, as described by Dragovich and Kelly(*152*).

Certain clams (e.g., *Rangia*, *Polymesoda*) and gastropods (e.g., *Neritina*) inhabit the upper estuary and are generally restricted to low salinity or freshwater habitats. Although tolerance levels vary considerably, the majority of estuarine species occur over a rather wide salinity range. Wells(*129*), in a series of 5-day tests using diluted seawater, found the salinity death point in eight common estuarine gastropods to range from 9 to $21^{o}/_{oo}$.

Similar data for five pelecypods varied from 0 to $21^{o}/_{oo}$, with $7^{o}/_{oo}$ being the observed death point for the common oyster, *Crassostrea virginica*. Although growth, shell movement, and water transport rates for natural populations of oysters in low salinity waters are abnormal, oysters have been observed to feed at salinities of $5^{o}/_{oo}$. Butler(*153*), Dawson(*154*), and others have reported natural populations of Gulf Coast oysters in areas with mean salinities of about $10^{o}/_{oo}$. In general, spatfall in such environments is low, but first-year growth may be rapid. Studies at Crystal River, Florida, showed that 3.5-in. (9-cm) oysters could be produced from natural set within 12 months in localities where salinity ranged from 7.7 to $15.5^{o}/_{oo}$ and had a mean weekly variation of $1.2^{o}/_{oo}$(*154*). Individual oysters may thrive at average salinities near $30^{o}/_{oo}$, but in high salinity environments mortality rates from various predators and diseases can approach 90%. For example, Copeland and Hoese(*155*) reported massive summer mortalities of Texas oysters at salinities of $40^{o}/_{oo}$ when temperatures rose above 37°C. Oyster populations acclimated to one salinity range cannot be moved to a significantly different mean salinity ($\pm 10^{o}/_{oo}$) without subsequent depression of growth rate and survival potential. In cases where such salinity change is not immediately lethal, recovery and acclimation occur, but high mortality rates may follow. Ideal salinity conditions for the American oyster appear to be in the $15–20^{o}/_{oo}$ range, with lowest salinities occurring in late summer and fall.

Salinity appears to have similar effects on the activities and distribution of estuarine gastropods. Laboratory experiments indicate that the crown conch (*Melongena*) survives for limited periods at $8–9^{o}/_{oo}$(*156*). "Normal activity," however, was noted only above $12.8^{o}/_{oo}$, and long-term survival required salinities of $21.5^{o}/_{oo}$ or above. Hathaway and Woodburn(*156*) found natural populations where daily variations approximated $12^{o}/_{oo}$, but specimens from low salinity environments ($\simeq 12^{o}/_{oo}$) did not exhibit normal reproductive activity. Carriker(*157*) showed that low-salinity tolerance of the oyster drill (*Urosalpinx*) varied with temperature. With summer temperatures, mortality was high under conditions of low salinity (12–17‰), but these salinities could be survived for considerable periods at low water temperatures. A number of apparent relationships between salinity and other factors, e.g., size, shell configuration, and thickness, in estuarine mollusks have been reviewed by Pearse and Gunter(*158*). These authors concluded that such morphological variations are not necessarily correlated with salinity alone.

(2) *Temperature effects.* Oysters survive water temperatures of about 1–38°C and internal shell temperatures of 45–49°C, the latter occurring during short-term exposure on tidal flats(*159*). Although temperature continually regulates molluscan metabolic rates, these effects are most apparent near the upper and lower temperature tolerance levels. Maximum ciliary activity and, consequently, maximum water transport in the oyster occur at 25–26°C. Above 32°C, ciliary motion is reduced, and feeding stops near 6–7°C(*159*). Temperature effects on overall growth rate have been amply demonstrated. Three to four years are required to produce marketable oysters in northern waters, whereas 12–18 months are sufficient in Florida waters. Many mollusks spawn only within a limited temperature range, and spawning is sometimes initiated by a definite sequence of temperatures. Although oysters can spawn at temperatures from 15 to 34°C, mass spawning of the population usually occurs when temperatures reach 22–23°C(*159*). Spawning of the scallop (*Aequipecten*) seems to occur only in conjunction with decrease in temperature following summer maxima(*160*).

Carriker(*157*), reviewing the literature on the drill (*Urosalpinx*), found it to be eurythermal over the range −3 to 30°C. These gastropods hibernate at temperatures approaching 2°C, feed over an 8–28°C range, and spawn from about 12 to 25°C.

Similar temperature-correlated effects have been found in the wood-boring *Bankia*. Observed growth rates averaged about 50 mm/month below 10°C and reached 120 mm/month at about 13°C(*161*). Accelerated growth, however, is not solely related to direct temperature effect on the individual but results from increased metabolic activity, greater availability of food (plankton flowering), and other factors.

(3) *Osmoregulation.* Although osmoregulation in marine and estuarine mollusks is poorly understood, some evidence suggests that the excretory system has a regulatory function(*159,162*). Of a few species studied, most were found to concentrate potassium, and several, including the European oyster (*Ostrea*), concentrated magnesium. Knight(*163*) demonstrated an increase in total free amino acids with increase in salinity from 5 to 25‰ in *Crassostrea*. Decrease in ninhydrin-reactive substances between 25 and 30‰ was attributed to osmotic stress.

(*4*) *Food relationships.* Food relationships of adult estuarine mollusks are varied and often highly selective. Oysters, clams (*Mya*), scallops (*Pecten*), and most other pelecypods ingest small phytoplankton or nannoplankton that are entrapped on mucus of the gills and mantle as water is continually transported through the organisms by ciliary action. Although water transport rates vary with size of the individual and with ecological conditions, rates as high as 457 liters/day in the American oyster(*159*) and 25.4 liters/hr in a scallop(*164*) have been observed. Many gastropods are herbivores or detritus feeders. Others, such as the oyster drill, are predatory omnivores or carnivores.

While some species accept a wide variety of food items, others exhibit considerable selectivity, apparently based on particle size or shape. The herbivorous sea hares (*Aplysia*) show considerable food selectivity. Carefoot(*165*) found *Aplysia punctata* to prefer *Enteromorpha* over seven other algae. Growth rates on unialgal diets generally decreased with decreasing order of food preference. Wood fibers, which the boring *Teredo* and *Bankia* shred by rotary movement of their shells, are apparently hydrolyzed by cellulolytic enzymes of their specialized digestive systems(*166*).

(*5*) *Life histories.* From the molluscan life histories, two general types are outlined.

One type is illustrated by the American oysters, which release their gametes directly into the water after mutual biochemical stimulation of the sexes. Fertilization occurs by chance union of egg and sperm. The egg develops into a planktonic trochophore larva within a few hours. Shortly thereafter (24–48 hr), this matures into the early veliger stage. After 2 or 3 weeks of planktonic life, during which the early veligers are distributed throughout the estuary and undergo further development, the late veligers settle to the bottom and seek suitable substrates for attachment. If a hard, relatively clean, stable surface is found, the larval oyster undergoes rapid metamorphosis to the spat or juvenile form. Unless the substrate moves, the oyster is now permanently fixed in one location for the remainder of its life. Veligers that settle on such substrates as silt and fine sand normally die within a short time.

Although hermaphrodites occur infrequently, the American oyster is considered to have separate but unstable sexes. Males usually outnumber females in first-year populations, but the species is protandric and sex change may result in equal sex ratios in later years. In most cases, however, males continue to outnumber females.

The number of eggs produced by a single oyster varies with size of the individual and ranges from 15×10^6 to 114×10^6 in 9- to 13-cm specimens (*159*). Recent studies on the larviparous European oyster (*Ostrea edulis*), which produces between 9×10^4 and 1×10^6 eggs, show that total larvae production may exceed $1 \times 10^6/yd^2$ ($1.2 \times 10^6/m^2$) of bottom(*167*).

Another type of life history is illustrated by gastropods. These mollusks frequently provide some degree of protection for developing young in the form of egg cases or brood pouches, and the total young produced per individual is usually much less than in the pelecypods. The drill (*Urosalpinx*) has separate sexes, fertilization is by copulation, and the eggs are deposited in leathery, oval egg cases. The number of eggs per egg case ranges from none to about 30, but is usually 9 or 10. An average of 45 egg cases is produced per season(*157*). Egg cases are attached to any hard substrate. The incubation period, depending on temperature conditions, ranges from about 18 to 78 days. The young drills then emerge from the egg cases as 1- to 2-mm replicas of the adult and immediately begin lifelong depredations on small oysters, mussels, barnacles, and similar prey.

(6) *Population levels*. Molluscan population levels are controlled, in part, by a large number of predators, parasites, and diseases. Besides being predators on other mollusks, gastropods are frequently cannibalistic. Fishes such as the drum (*Pogonias*) consume considerable quantities of small gastropods, bivalves, and other estuarine animals, such as the blue crab (*Callinectes*). The flatworm (*Stylochus*) occasionally attacks oyster and clam beds. Various trematode larvae can infect pelecypods and gastropods, and mortalities associated with the fungus *Dermocystidium* or the sporozoan *Haplosporidium* are occasionally disastrous to commercial oyster beds(*159,168*). Various other organisms, such as the blister worm (*Polydora*), the boring sponge (*Cliona*), and the boring clam (*Diplothyra*), contribute to mortalities of estuarine mollusks. Significant mortalities of oysters can accompany heavy fouling growths of mussels, sponges, barnacles, and even dense sets of oyster spat. The latter, for all practical purposes, can "smother" the preexisting oysters.

Oysters sink and die in soft substrates. They are generally absent from sand, where they can be covered by the shifting bottoms. The growth pattern of some oyster bars can in itself result in death or debility of oysters. For example, successive sets of oysters, one on top of the other, can elevate the bar until its upper portions are exposed for long periods during the tidal cycle. The uppermost oysters, subjected to temperature extremes and shortened feeding periods, usually exhibit poor growth and survival patterns, whereas those on the lower bar may be of marketable quality. When development of the bar reduces or redirects water currents, unfortunate consequences follow. Reduced current velocities result in increased siltation of the bar. Simultaneously, oysters are buried by the accumulation of pseudofeces when not adequately swept by water currents. With massive oyster populations, reduced current velocities in overlying waters can result in local depletion of planktonic food organisms. Starvation, together with excessive sedimentation, has destroyed many previously productive oyster beds.

In addition to the familiar oysters, clams, and scallops, other estuarine mollusks, such as mussels and squids, may support commercial fisheries. At times, individual molluscan populations can become large enough to be of major economic importance. For example, massive shell deposits are exploited as sources of lime or of fill and construction materials. Such large molluscan populations, with their varied and complex community interrelationships, can exert widespread influences on local estuarine ecology, both physically (e.g., by altering current and sedimentation patterns) and biologically.

e. Other Macrofauna

The estuaries support considerable populations of miscellaneous biota that make major contributions to the ecosystem. Benthic forms living on (epifauna) or beneath (infauna) the bottom, including larvae, postlarvae, and adults of both vertebrates and invertebrates, are frequently minute, inconspicuous, and present in massive numbers. Wieser(*169*), in a study of the meiofauna (small metazoans) of Buzzards Bay, Massachusetts, reported maximum counts of $1.86 \times 10^6/m^2$ and dry weights as high as 600 mg/m^2. Infaunal and epifaunal components consist mainly of small invertebrates, such as foraminifera, ostracods, nematodes, polychaetes, copepods, turbellarians, echinoderms, small crabs, mollusks, and even some burrowing fishes. Detailed ecological and biological relationships of benthic estuarine invertebrates have recently been reviewed by Carriker (*24*).

Cronan(*170*) found mollusks to constitute 86.7% of the animal food of the greater scaup (*Aythya marila*) in Connecticut waters. Cleaver and Franett (*171*) found an average of 11,945 herring eggs in the stomachs of 15 scaups and scoters (*Melanitta deglandi*) in Puget Sound. Although the total effect of predatory birds on estuarine populations is unknown, their depredations must be great in many areas. Birds, aside from their role in the estuarine food chain as scavengers, predators, and herbivores, are significant contributors of organic matter. The number of bird species that make continual or temporary use of estuaries is apparently large. Oberholser(*172*) listed more than 93 species recorded from Louisiana coastal and marsh areas, and many of these species occurred in large numbers. Bird wastes from commercial fowl-rearing operations have been shown to contribute to the development of local plankton blooms(*173,174*) and estuarine hypereutrophication.

The amphibian and reptile populations are of minor importance in tropical and low temperate estuaries. Various snakes inhabit the marsh and protected waterways but generally are not abundant and usually are most common in the middle or upper estuary. Sea turtles are occasionally found, and freshwater turtles often occur in low salinity areas. Caimans, alligators, and crocodiles are occasionally abundant and hazardous in tropical and subtropical environments. The economic contribution of the estuarine herpetofauna is small when compared with the remainder of the fisheries but may be of some local importance.

Practically any form of animal life from protozoans to mammals may inhabit some portion of the estuarine ecosystem. Many of these are of direct economic importance to man, but all contribute in one way or another to the biological balance of the estuarine environment.

2. Macroflora

Upland plants, salt marsh and marine spermatophytes, and various types of large algae all contribute to the estuarine ecosystem. The role of upland flora primarily involves the deposition of detrital material. Chiefly this is in the form of leaves, seeds, and pollen, but in heavily wooded areas large quantities of branches and tree trunks enter the estuaries as flotsam after upstream floods or storm tides. Much of this detritus remains in the estuary and is eventually broken down by mechanical, chemical, and biological agencies. While released nutrients become available to the biota, the relatively stable detrital fractions, such as cellulose, contribute to sedimentation of the estuary. Waterlogged tree trunks and branches frequently support significant, locally unique populations of fouling organisms. In muddy estuaries, large detritus sometimes constitutes the only available hard substrate suitable for the attachment of such species as oysters and bryozoans and the only habitat for boring organisms such as *Bankia* and *Teredo*.

The spermatophytes and algae are normally the dominant macrophytes of temperate and tropical estuaries. The marsh grasses (e.g., *Spartina*, *Salicornia*, *Distichlis*, *Juncus*) fringe protected muddy and sandy shores and generally become more abundant upstream from the mouth of the estuary. Along the salinity gradient there appears a general succession of facies from more-seaward *Spartina* toward *Juncus* in the upper estuary. A similar succession occurs across the intertidal marsh, which is generally characterized by *Spartina* on the lower slope, *Spartina–Salicornia–Distichlis* on the middle slope, and such species as *Juncus* and *Panicum* at maximum tidal levels. Exact generic sequence of succession exhibits considerable geographic variation but generally agrees with the foregoing.

The lower limit of marsh grass intrusion on the tidal slope depends on the mechanical action of currents and waves, the composition of the substrate (*25*), and the frequency of tidal inundation(*175*). Some insight into the extent of salt marsh is provided by Thorne's summary (*176*) for the South Atlantic and Gulf of Mexico. Of some 5.6×10^6 acres (2.3×10^6 ha) of marsh tabulated, more than half of the total U.S. acreage was in Louisiana.

Marine phanerogams, such as *Zostera*, *Halodule*, *Ruppia*, and *Thalassia*, frequently develop as extensive meadows over shallow bay and marsh bottoms. These can extend from the intertidal zone to depths of 50 m in the case of *Zostera*, and to about 30 m for *Thalassia*(*177*). Although such meadows withstand a wide range of salinity and physical conditions, *Zostera*, at least, has been ravaged from time to time by an epidemic fungal disease, probably caused by *Labyrinthula*(*177*). These sea grasses exhibit seasonal trends in abundance but can develop extremely dense growths. Moore(*178*) reports growths of *Zostera* providing dry weights as high as 960 g/m^2.

Blue-green (Cyanophyta) and green (Chlorophyta) algae often develop as mat-like coverings between low tide level and the marsh edge. On a gently sloping, protected shore, these algal mats are occasionally extensive and aid in the initial stabilization of fringing bottoms.

The larger forms of green, brown (Phaeophyta), and red (Rhodophyta) algae frequently develop a varied flora in estuaries. In areas of relatively low turbidity and high temperature, they occasionally form dense and extensive growths over much of the protected middle and lower estuary bottom. Such floras exhibit seasonal succession and variation in both abundance and distribution. Frequently, they support significant epiphytic populations of smaller blue-green and other algae. The large algae are grazed upon by many invertebrates, especially gastropods and opisthobranchs, and are also ingested by certain omnivorous fishes.

Various mangroves fringe many tropical and subtropical estuaries, and these have been shown by Davis(*179*) and others to be effective "land builders." Succession of mangrove species frequently occurs in a newly colonized area, wherein the softer substrate is first pioneered by species of *Rhizophora*, characterized by its extensive system of prop roots, and later by communities of *Avicennia* and *Conocarpus*. Succession in mangrove communities has been discussed for west African estuaries by Lawson(*180*), for Australian estuaries by Macnae(*181*), and for the Gulf of Mexico by Thorne(*176*).

In addition to the stabilizing action of algal mats and the land-building tendency of root and rhizome systems of sea grasses and mangroves, the estuarine macrophytes contribute to the ecosystem as sources of organic materials and as habitats for numerous vertebrate and invertebrate species. Odum and de la Cruz (*182*) have shown that organic detrital material, 95% from decaying *Spartina*, is a major link between primary productivity in Georgia salt marshes and secondary productivity in the associated estuaries. Similar situations likely will be encountered in analyses of the detrital contributions of mangroves, sea grasses such as *Zostera* and *Thalassia*, and larger algae. In many instances, specific plant-animal associations have developed. This aspect of the mangrove community has been discussed by Hedgpeth (*25*), Lawson (*180*), Macnae (*181*), and Walsh (*183*). Various faunal relationships with the salt marsh, sea grasses, and macroscopic algae have been treated by Kilby (*184*), Springer and Woodburn (*185*), Hoese and Jones (*149*), Carriker (*24*), and Day (*103*).

3. Plankton

Plankton can be defined as all drifting and floating organisms whose existence is at least temporarily independent of the bottom. The broad category "plankton" includes three types of organisms. True plankton (holoplankton) are free-floating throughout their life history. Meroplankton, organisms such as larvae of decapod crustaceans, pelecypods, and gastropods, have a limited planktonic period. Tychoplankters, including normally benthic organisms, are occasionally carried into the water column by forces such as strong currents.

Although large organisms, such as *Sargassum*, with its complex faunal associations, and the common jellyfish (*Aurelia*), are planktonic, these constitute only a small part of the biomass and are relatively unimportant in the estuarine ecosystem. The majority of planktonic organisms are small, usually microscopic, and can conveniently be categorized as net plankton and nannoplankton. Net plankton are those organisms normally retained by the finest-mesh plankton net; nannoplankton, usually less than 50 μ in maximum dimension, are those small, photosynthetic organisms which will pass through such a net. Plankton are further differentiated into unicellular photosynthetic plants, such as diatoms and flagellates, termed phytoplankton, and animals, or zooplankton.

Although some primary productivity is carried out by bacteria and macrophytes, phytoplankton are the main source of autotrophic production. Provided with adequate insolation, with such nutrients as nitrates and phosphates, and with acceptable temperature, the phytoplankters photosynthesize, grow, and reproduce rapidly by binary fission. Under optimum conditions, this capacity for geometric reproduction may result in a massive population explosion or "bloom," wherein a species normally present in small numbers can reach levels in excess of 10^6 cells/

liter. Although such blooms are often detrimental to local fauna and may occur with some frequency in certain areas (*124*), these are abnormal conditions not representative of seasonal abundance patterns in estuaries.

a. Seasonal Cycles

As with other phenomena of the estuarine environment, net plankton populations show seasonal cycles varying with latitude and even between neighboring estuaries. Relative magnitude of cyclic production modes is depressed or at least less apparent in tropic and subtropic estuaries as a result of more stable radiation and temperature levels or, perhaps, of a scarcity of nutrients. Despite such variations, well-defined bimodal seasonal cycles are evident in many estuarine net phytoplankton populations. Nannoplankton, however, do not show any pronounced seasonally correlated cycles (*48,186*).

Minimum net phytoplankton populations are usually observed in late autumn or early winter and coincide with decreasing water temperature, shortening days, increased turbidity resulting from winter storms, and minimal concentrations of essential nutrients. During these winter months, estuarine nutrient levels are replenished through a combination of bacterial decomposition of organic matter, runoff, and tidal flushing.

A resurgence of the planktonic population occurs in late winter or early spring. This spring "flowering," which is usually the major mode in the annual phytoplankton cycle, is apparently triggered by vernal increases in radiation and water temperature. Increased insolation is considered to be the prime initiating factor (*187,188*). The population rises to a peak, then declines rather rapidly and maintains a relatively low and stable level through the summer months. The primary cause of this decline has often been regarded as the grazing effect of an expanding zooplankton population during late spring and summer, with reduction in nutrient levels suggested as a secondary factor. However, Riley (*187*) considers that reduced levels of nutrients, primarily nitrates, terminate the spring flowering, while Raymont's observations (*189*) suggest that light saturation may serve to limit photosynthesis and reproduction.

Coinciding with slight increases in available nutrients, stability of the water column, and shorter days, a secondary phytoplankton bloom often occurs in late summer or early autumn. This is usually short-lived and is apparently terminated by reduction of insolation below compensation levels, by increasing turbidity, and by dropping temperature.

The described seasonal phytoplankton cycle changes from year to year with annual variations in radiation, rates of nutrient recycling or replenishment, temperature, and other factors. Riley (*187*) shows that under certain conditions the autumn mode may equal or exceed the spring flowering.

Relationships between nutrients and phytoplankton have been greatly simplified in this discussion. The general interactions are, in fact, highly complex. The importance of trace elements, such as iron, manganese, and zinc, and organic compounds, such as cobalamine, in developing phytoplankton populations has been indicated by Provasoli(*190*), Raymont (*189*), and Riley(*187*). Patrick(*188*) shows that relative nutrient levels, particularly for nitrate, phosphate, and vitamin B_{12}, may greatly affect species composition and species dominance in estuarine diatom populations.

Seasonal cycles of zooplankton abundance show considerable annual and geographic variation, with modal periods that do not necessarily parallel those of the phytoplankton. It is not unusual, however, to find zooplankton maxima occurring a few weeks or a month later than the spring phytoplankton flowering. Major factors influencing zooplankton abundance appear to be salinity, temperature, interspecific competition, and predation. Since estuarine phytoplankton populations normally exceed grazing requirements of herbivorous zooplankton, available food supplies probably are never a limiting factor. Indeed, massive phytoplankton blooms can possibly limit zooplankton populations in local situations. Meroplankton, representing several phyla, ingest quantities of phytoplankton but are most frequently omnivorous or carnivorous. Many will ingest a wide variety of food organisms, whereas others exhibit considerable selectivity in respect to size and type of food. Food relationships of carnivorous plankters are complex but generally follow a typical pattern of smaller organisms being devoured by those which are larger or more voracious.

b. Species Diversity and Abundance

Estuarine phytoplankton are characterized by high species diversity and, except under abnormal conditions, relatively low abundance of each species. In an 8-year study of Long Island Sound, Riley(*187*) reported the recognition of 150 species. Of these, however, single species which accounted for at least 5% of the population during any month for 4 of the 8 years numbered only 13. Patrick(*188*) found estuarine diatom populations to consist of freshwater or low salinity forms together with euryhaline and marine components.

Estuarine phytoplankton provide an abundant food source for herbivorous zooplankton which are, in turn, preyed upon by holoplanktonic, meroplanktonic, and nektonic carnivores as well as by benthic organisms. Jeffries(*191*) described estuarine holoplankton as "a monotonous assortment of calanoid copepods" and showed that holoplanktonic copepods usually account for 60% of the spring-through-fall zooplankton populations. These copepod populations are also characterized by having a large biomass but a relatively small number of dominant species. Riley (*187*) and Jeffries (*191*) found that *Acartia*, *Eurytemora*, and *Oithona* were the dominant genera of estuarine copepods and that three or four species accounted for more than 80% of the holoplankton. Similar observations by Cuzon du Rest (*192*) in Louisiana showed *Acartia tonsa* to be as much as 40 times more abundant than other copepods and to exceed the total of all other zooplankters. Jeffries (*191*) considered knowledge of the biology of dominant holoplanktonic genera essential to an understanding of estuarine productivity.

Nannoplankton, mainly small diatoms, flagellates, and coccolithophores, are an important component of the phytoplankton, especially with respect to primary productivity. Although net plankton may predominate in northern waters, nannoplankton frequently account for the greater portion of the standing crop of phytoplankton in tropical and subtropical estuaries. Nannoplankton have been found to make up 97% of the phytoplankton population in Brazilian lagoons (*193,194*), and Mulkana (*48*) found the mean standing biomass of nannoplankton to be 72 times that of net plankton in the Mississippi Sound. Nannoplankton as small as about 3 μ in mean diameter are eaten by certain herbivorous zooplankters and are also ingested by many of the benthic filter feeders.

4. Bacteria

Until recent years, estuarine bacteriology has been oriented almost exclusively toward sanitary aspects. Thus, the literature abounds with methods for determining bacterial counts and coliform most probable number in saline waters (*196,197*) and studies on survival of enteric organisms in estuaries (*198,199*). Primary emphasis has been on classification of bathing beaches and on standards for waters in which shellfish may be grown and harvested (*200,201*).

Of the relatively few academically oriented microbiologists working to elucidate the bacteriology of saline waters, most have dealt primarily with fully oceanic environments (*202–205*).

In general, those scientists who have studied estuarine environments have shown that the majority of bacteria in estuaries are sediment-associated (10^5–10^8 cells/ml) rather than planktonic (10^4 cells/ml or less) (*206*). The sediment-dwelling forms include both aerobic and anaerobic organisms and are further classifiable as autotrophic (mainly chemautotrophic) and heterotrophic assemblages. These organisms are intimately involved in estuarine food and energy webs, wherein they function as consumers of carbonaceous materials and act to remineralize essential nutrients.

Studies have been made of rates of sulfide formation, of phosphorus release, and of decomposition of organic matter in near shore environments(*204,207*). Results indicate that decomposition proceeds rapidly (sediment pH change from 7.6 to 6.3 in 5 days) and that sulfide production exceeds sulfate reduction, the differential amount of sulfur being derived from organic materials. This suggests that the generally high levels of productivity in estuaries result from rapidity of bacteria-associated turnover of nutrients such as phosphorus.

Wood(*206,208*) has provided reviews of the general status of estuarine microbiology. He considers(*208*) that the estuarine environment permits the maximum number of microbial reactions because of pH and redox potential conditions. Further, he concludes that among the most important of these reactions are those that restore phosphorus to the ecosystem. Basically, these are the reactions that remineralize organically bound phosphorus and dissolve calcium phosphate, the sulfur cycle reactions that alter the chemical environment, and secondary reactions that restore the phosphorus to the general estuarine chemical flux.

Some of the most pronounced chemical changes in estuarine waters and sediments result from normal metabolic processes of microorganisms. Microbial activity sometimes consumes oxygen more rapidly than photosynthesis and diffusion can replace it in the environment, thereby giving rise to anaerobic conditions. The redox transition associated with the development of the anoxic situation permits insoluble compounds, such as iron hydroxide, to dissolve. The iron is then freed to migrate and combine with sulfides produced by anaerobic sulfate-reducing bacteria(*209*). This leads to zones rich in reduced iron sulfide. When these zones are again oxidized, either in distinct layers in the sediments or around the burrows of various organisms, brownish-red layers of ferric hydroxide result. If the iron hydroxide is reduced and subsequently dissolved in the water column rather than in the sediments, adsorbed inorganic and organic materials are released into the water.

Production of carbon dioxide or hydrogen sulfide in waters and sediments leads to a lowering of pH, with resultant chemical changes such as shift in carbonate equilibrium and increase in solubility of calcium carbonate. Uptake of carbon dioxide during photosynthesis causes the water to become more alkaline, while metabolism and concomitant carbon dioxide release at night lower the pH. This results in the sequence: precipitation of carbonate, solution of silicate, and redissolution of carbonate.

Microorganisms can actively concentrate trace elements to levels several thousandfold greater than those of seawater(*210*). This may produce areas within the environment where individual elements predominate. An example of this is the concentration of manganese by *Metallogenium*(*211*).

Physicochemical processes in the environment may be enhanced or altered by microbial processes. Production of organic or inorganic acids and alkalis by microorganisms may influence weathering of minerals. Thixotropic properties of sediments can be affected directly by production or consumption of organic materials that allow sedimentary particles to manifest colloidal properties. When bacteria at the water–air or water–sediment interfaces produce materials that give rise to foams, aerosols, sapropel, etc., other colloidal effects become evident. Those colloidal properties affected by bacteria may be regulators of diagenetic processes during sediment compaction.

The role of bacteria in mineralization and cycling of nitrogen, sulfur, carbon, and other elements is outlined in most general microbiology textbooks. However, many subtle cycling effects are also mediated by microorganisms. For example, microorganisms can selectively concentrate lighter isotopes from a mixture during membrane transport, leaving the environment enriched in heavier isotopes(*212*). Isotope effects are known for elements as heavy as sulfur, for which the mass difference among isotopes is sufficient to permit detection of enrichment by present-day techniques. Specific crystal formation may be altered by the sorption of microorganisms or their metabolic products. This can result in variation in the formation of minerals found in nature. It is thought by many scientists that sedimentary mineral deposits may be affected by microbes. Iron deposits, for example, are known which show evidence of iron bacteria(*213*).

C. Productivity of Estuaries

1. Theory and Methodology

The primary productivity of any environment is the gross rate of photosynthetic fixation of inorganic carbon. Thus, all measures of primary productivity must ultimately measure components of the photosynthetic scheme. Rabinowitch(*214,215*), Vishniac(*216*), Forti(*217*), and many others have reviewed the state of knowledge of the photosynthetic mechanism, and Bassham and Calvin(*218*) have provided an eminently readable account of the transit of carbon through the biochemical pathways involved. The scheme proposed by Holzer(*219*) seems to provide a valid and useful photosynthetic model for ecological purposes and reveals the various inputs and outputs through which measures of photosynthetic rate may be attempted under field conditions:

H_2O ↘ —— $h\nu$ —— ↗ $2[H] + 1/2O_2$

$NADP$ ↙ —————— ↘ $NADPH$

ATP ↙ —————— ↘ $ADP + PO_4^{3-}$

$CO_2 + H_2O$ ↙ —————— ↘ glyceraldehyde-3-phosphate

$+$ 3-phosphoglyceric acid ↓ carbohydrate

Most of the materials shown are metabolic intermediates not easily measured, but carbon dioxide, water, and phosphate are distinct inputs, and oxygen and carbohydrate are definite outputs. Water is too ubiquitous to offer a convenient index, but the other components have all been employed at various times for estimates of productivity. Thus, Steeman Nielsen (*220*) has advocated measurement of ^{14}C uptake from labeled carbon dioxide or carbonate to determine productivity, and the technique has been used widely. Alternatively, Park et al. (*221*) and Beyers and co-workers (*222–226*) have used pH changes calibrated against carbon dioxide content of water as an index of carbon fixation. The validity of their approach was questioned by Verduin (*227,228*), and Lyman (*229*) suggested that direct measurement of carbon dioxide stripped from the water was the only reliable method for such determinations.

Phosphate assimilation [cf. (*230*)] is attractive as a reaction rate index. Bruce and Hood (*86*) were able to correlate phosphate diurnal cycles in several Texas bays with primary productivity estimates derived from other techniques. They pointed out, however, that communities could maintain primary production in the face of phosphate deficiencies, apparently by employing organic phosphorus compounds as a phosphorus substrate. In this connection, Levin and Shapiro (*231*) demonstrated the ability of some microorganisms to store excess phosphate during periods of abundance. Further, variability in phosphate turnover rates by organisms and sediment-associated fluxing of phosphorus through the environment as a result of wind and current actions can influence observed daily cycles. Thus, this method may lead to erroneous results.

De Saussure (*232*) showed that green plants gained weight during photosynthesis and that the weight gained plus weight of oxygen evolved exceeded the amount of carbon dioxide consumed. He correctly ascribed the difference to water uptake. These data represent the first measurement of net carbohydrate formation in plants. In more modern times, increases in weight per unit area of leaf and similar techniques have been employed to estimate carbohydrate formation and, hence, productivity, but the methods are imprecise and may yield deceptive results.

Measurement of oxygen production has probably been the most popular technique for productivity estimation in aquatic environments. This popularity results largely from the ease with which dissolved oxygen is determined either by the classical Winkler technique or by various instrumental approaches. The general stoichiometry of the photosynthetic reaction indicates production of one molecule of oxygen per molecule of carbon dioxide assimilated, which makes the index convenient. Most of the available estuarine primary productivity data are based on oxygen production and consumption computed over a 24-hr period.

Regardless of the index selected, primary production measurements in aquatic ecosystems are plagued by several problems. One difficulty is biochemically based. The compounds consumed and produced during photosynthesis are the same as those involved in the carbohydrate respiratory metabolisms of both the primary producer plants and the animal components of the system. Thus, the rate of plant respiration during photosynthesis has not yet been established unequivocally [cf. (*233*)]. Holm-Hansen(*234*) points out that entry of newly fixed carbon into the Krebs cycle is somewhat suppressed during illumination. Presumably, this results from the reduction of thioctic acid, which in oxidized form is the cofactor for pyruvate decarboxylation [cf. (*235*)]. An absolute blockage seems unlikely, however, and ^{14}C assimilation data must, therefore, be regarded as providing productivity values intermediate between gross and net primary productivity. Carbohydrate accumulation data necessarily are subject to the same considerations, while oxygen production data generally reflect, in addition, the respiratory demands of the ecosystem in general. From the same general considerations, plus known variations in phosphorus turnover times in various organisms and environments, phosphorus cycles also presumably reflect some intermediate productivity value resulting from the differences in anabolic and catabolic phosphorus demands.

The second difficulty in productivity estimation relates to the measurement technique itself. Customarily, to measure oxygen production or ^{14}C assimilation two aliquots of water are bottled. Under conditions specified for the particular study, these aliquots are incubated, one in the dark and the other in light. Various absolute and differential changes observed in the two bottles are then interpreted as net productivity, gross productivity, and respiration. Incubation periods vary greatly in length, and incubation may be carried out by suspending the bottles in the environment at some depth, such as depth of sampling; by incubating on the deck of the research vessel, usually in a flowing water incubator under natural illumination; or by incubation under artificial illumination. The exact method seems to depend upon the whim and convenience of the investigator.

The light–dark bottle procedure appeals on the basis of its great convenience, but even if exact procedure were standardized, the method would still have several drawbacks. First, turbulence reduction concomitant with enclosure probably affects metabolic rates. Second, enclosure halts nutrient replenishment by currents and upwellings. Third, introduction of a surface, the container wall, can induce growth of atypical biotic assemblages.

The alternative procedure, measurement of diurnal change in oxygen, phosphate, pH, or some other index in an open body of water, has a strong theoretical appeal [cf. (*40*)] but poses its own problems, especially in estuaries. Examples of these problems are the uncertainty of estimates of exchange across the air-water interface if a gas is being measured; the difficulty of following a specific water mass, especially at night, in a rapidly circulating, probably wind-stirred, definitely tidal environment; and the logistic problems involved in obtaining broad-scale, representative data over prolonged time periods.

Raymont (*189*) reviewed productivity information on marine plankton and critically examined many of the methods employed for obtaining productivity data. Cassie (*236*) evaluated the general statistical sampling problem associated with field measurements of primary production.

2. Field Studies

a. Productivity and Standing Crops

Many estimates of estuarine primary productivity have been obtained by various techniques. Recently, Teal (*237*) reported values from the Woods Hole, Massachusetts, area determined by stripping gases from the water and measuring the CO_2 tension, the method advocated by Lyman (*229*). Teal's results are similar in magnitude to those obtained by other methods.

A good comparative summary of productivity values obtained by bottle methods versus open water diurnal curve methods is given by Odum and Hoskin (*43*). Odum and Wilson (*44*) have reported large amounts of diurnal curve-derived productivity data for Texas bays.

In general, estuarine organic matter production rate seems to range upward considerably from Ryther's(*238*) marine gross productivity estimates of 0.74–2.1 g/m^2 per day. Estuary primary productivity values as high as 30 g/m^2 per day, as oxygen, have been reported(*44*). Such data reinforce the concept of estuaries as extremely productive environments.

Of more immediate concern than productivity is information on standing crops, but available data on this subject are of variable utility. Odum and de la Cruz(*182*) have recently emphasized the role of detritus in linking primary productivity of fringing marshes to estuarine productivity as a whole. The overall contribution from this source, however, must be relatively small when compared with in situ production by the usually large plankton community in an estuary of any reasonable size. Mulkana (*48*), for example, reported average standing crop data for Mississippi Sound as 36.7 g/m^2 dry weight. Of this, material smaller than 75 μ comprised 36.2 g/m^2 and was presumed to be nannoplankton plus detritus. Microscopic examination revealed, however, that the detrital component resulted mostly from physical disruption of macroplankters, not from marsh vegetation. The standing crop values are deceptive because silt particles are inextricably mixed with the nannoplankton. Thus, a better index of available forage is provided by Mulkana's data on specific nutrients: 2.1 g/m^2 protein, 1.9 g/m^2 carbohydrate, and 0.9 g/m^2 lipid. The lipid datum comes from samples from a single cruise. Comparable data from other estuaries are not available. Riley(*239*) has estimated average zooplankton standing crop in Long Island Sound as 20 g/m^2, wet weight, representing 2 g/m^2 organic matter. The value seems high to us relative to Mulkana's data and to data on fish standing crops.

Most catch data for large estuarine organisms are reported as catch per unit effort values for specific types of gear used under specified conditions. Comparability among such reports usually is nonexistent. Hellier (*240*) has reported standing animal crops from Laguna Madre of Texas, based on the drop-net technique(*241*). Although this technique suffers from several objections, such as possibly attracting specific organisms to the sampling area, it does lead to catch figures expressible as weight per unit area. Hellier found wet weight standing crops of fish and larger invertebrates to range from 2.0 g/m^2 in winter to 37.8 g/m^2 in summer. Annual fish production was estimated as 15.4 g/m^2 (dry weight), and annual gross plant productivity in the area was measured as 4177 g/m^2. As with Mulkana's(*48*) nutrient composition data for plankton, comparable information on fish standing crops in other estuaries is not available.

The general status of primary productivity studies and of problems related to this area of ecology was the topic of a symposium held under International Biological Programme auspices in 1965(*242*).

b. Fisheries Statistics

Data to determine the total contribution of estuarine or estuarine-dependent biota to world fisheries production are not available. Published statistical summaries list catches under broad headings, e.g., herrings, catfishes, scallops, without indicating whether these are estuarine or estuarine-dependent species. In fact, whether a particular species is estuarine dependent or not is frequently problematical because little is known of its life history.

A partial estimate of estuarine-dependent fisheries has been made here by the selection of a few groups (Table 5) in which most forms utilize estuaries at some period of their lives. Although a number of herrings are

TABLE 5
Reported 1964 World Landings for Some Estuarine-Dependent Species[a]

Product	Metric tons
Diadromous fishes	750,000
Croakers and drum	340,000
Mullets	98,000
Herrings, etc.	18,570,000
Blue crabs	70,000
Common shrimp	597,000
Oysters	795,000
Total	21,220,000

[a]Data compiled from Ref. *243*.

not within this category, the majority certainly qualify. Conversely, some pertinent species have been omitted from the table to offset the inadvertent inclusion of some marine forms. These data, representing a minimum estimate, indicate that estuarine-dependent species probably accounted for over 40% of the 1964 total world fishery production of 51.5×10^6 metric tons (*243*).

Although more detailed data are available for U.S. fisheries, the figures are again considered to represent minimum estimates. Aside from species that could be included as estuarine dependent if their life histories were known, many sport and commercial fisheries catches do not enter the statistical reports.

Data on 15 estuarine-dependent species or species groups have been summarized for the U.S. Atlantic and Gulf Coasts (Table 6). Total reported U.S. Atlantic and Gulf coastal fisheries production in 1965 amounted to 1,577,179 metric tons, with a value of slightly more than

TABLE 6
Reported 1965 Landings and Value of Selected Estuarine-Dependent Fisheries of the Atlantic and Gulf Coasts of the United States[a]

Product	Metric tons (approximate)	Value (thousands of dollars)
Menhaden	784,594	27,974
Alewives	29,432	886
Mullets	18,804	2,585
Shad	3,496	1,095
Striped bass	3,506	1,457
Sea trout	5,401	1,999
Spot	2,330	625
Drum	1,964	562
Croakers	1,575	284
Sturgeon	103	38
Eels	711	271
Blue crabs	77,395	12,849
Shrimp	100,649	81,067
Oysters	20,692	25,638
Bay scallops	1,051	1,574
Total	1,051,703	158,004

[a]Data compiled from Ref. *244*.

280 million dollars (*244*). The species listed in Table 6 account for about two-thirds of the total production and 56% of its value. These data compare favorably with McHugh's analysis (*245*) of 1963 statistics. McHugh calculated the value of 39 estuarine-dependent species reported for the U.S. Atlantic and Gulf Coasts at 63.7% of the total production. A similar selection of Alaskan and West Coast estuarine-dependent species accounts for more than 153,000 metric tons, valued at about 67.4 million dollars (*244*). Total U.S. fisheries production in 1965, including freshwater fisheries, whaling, and other specialized operations, amounted to 2,167,000 metric tons, valued at about 445.7 million dollars. The minimum estimated contribution to this by estuarine-dependent species is about 55.6% by weight and 50.6% by value.

These estimates do not include any indirect estuarine contribution to the marine fisheries. Young and adults of many estuarine-dependent species probably constitute a significant portion of the diet of many coastal marine forms. When considered in this context, the estuarine contribution to total fisheries production is much greater than that indicated by available statistical records.

D. Stresses in the Estuarine Environment

The plants and animals of the estuaries are seldom exposed to optimum conditions. In the rapidly fluctuating environment characteristic of temperate and tropical estuaries, the biota must either adapt immediately to changes or have protective mechanisms against unfavorable conditions. All deviations from the various chemical, physical, and biological optima for a particular species or population result in some degree of stress. Inability to accommodate to a single unfavorable condition, or to the summation of a number of stress factors, results in death.

Successful estuarine forms are manifestly able to withstand rather broad variations in the environment, but each has definite maximum and minimum levels of accommodation. Such levels are not static. They usually vary up or down with other factors. Maximum and minimum salinity tolerance levels, for example, may change with temperature. Likewise, oxygen requirements change with temperature as well as with the general activity of the organism.

Little information is available on stress induced by cyclic phenomena, such as tidal currents and diurnal variations of temperature or light. Also lacking are useful data on cumulative effects of a number of individually sublethal stresses. Our knowledge in these areas derives from experimental procedures wherein one factor is varied in a controlled environment, or from after the fact observations of mass mortalities. Since stress occurs as a result of deviations from largely unknown optima, further research in this area is essential to a more complete understanding estuarine faunal relationships. Meanwhile, a brief outline of some stress conditions is pertinent.

1. Osmotic Stress

Pearse and Gunter(*158*) and Kinne(*141,142*) have reviewed osmoregulation and ionic regulation of marine organisms, and these authors should be consulted for details of regulatory mechanisms and systems. Some animals survive unfavorable osmotic environments by escape or by activating a mechanical defense, such as shell closure in the case of pelecypod mollusks. The majority of the biota, however, must counteract detrimental salinity variations by mechanisms such as volume regulation, ionic regulation, or osmoregulation. Capacity to regulate and the method employed may vary within different species of the same genus and even with age, sex, and previous environmental history of a single species or organism.

The lancelet (*Branchiostoma nigeriense*) normally inhabits a salinity range of about 13–36‰ in Lagos Lagoon, and adults survive experimental salinities as high as 58.9‰ (*246*). Threshold salinity (the lowest to which the animal can accommodate itself) is at or above 13.0‰ in larvae, but the adult threshold is between 12.5 and 13.0‰, and adults can accommodate themselves rapidly to wide salinity variations. A critical period was found some 14–20 hr after transfer to lower salinity, when death followed return to high salinity. Lancelets that were returned earlier to high salinity survived, as did those remaining at the lower salinity or those returned to high salinity after 20 hr. This species appears to accommodate itself to reduced salinity but to be incapable of surviving the stress of repeated osmotic shock without an intervening recovery period. Ciliary activity ceased with salinity change, and duration of this stoppage varied with magnitude of the salinity change and between adults and larvae. Relatively slow growth of lancelets in areas with large salinity fluctuations is considered to result from the salinity-activated interruption of the feeding mechanism. Following annual spring flooding, the lancelet population in Lagos Lagoon suffers a large mortality. Subsequent repopulation is from the open sea (*247*).

Despite their regulatory capabilities, many estuarine forms are unable to cope with osmotic stress imposed by periods of abnormally high or low salinities. Mass mortality of a lancelet (*B. caribbaeum*) has been observed in connection with rainstorms in Mississippi Sound (*248*). Gunter (*249,250*) described massive oyster mortalities in Louisiana bays following flooding by the Mississippi River, and other salinity-related catastrophes have been noted by Brongersma-Sanders (*124*). Although mass mortalities are obvious and immediate results of inability to cope with osmotic stress, the outcome is less clear in other situations. Long-term cyclic patterns in precipitation can result in gradual changes in estuarine population structures because of differentials in abilities of various species to osmoregulate at relatively higher or lower salinity levels. Gunter and Hildebrand (*251*), Kutkuhn (*252*), and others have discussed apparent correlations between abundance of white shrimp and extended drought.

2. Temperature Stress

Most estuarine organisms have little thermal regulatory ability and must conform to the temperature of the environment. Although metabolic rates usually increase with increasing temperature, there are temperature optima concomitant with best growth and reproduction potential for all species. Optimum temperature for a northern population may be entirely different from that of a conspecific group in more southerly waters.

Kinne(*141,142*), in reviewing literature on salinity–temperature relationships, recognizes several nonexclusive general categories of organisms. Some organisms exhibit their maximum tolerance for extremes (high or low) of temperature at their optimum salinity, or for extremes of salinity at their optimum temperature. Other organisms show their greatest tolerance for low salinity near their lower limit of temperature tolerance, or for low temperature near their lower limit of salinity tolerance. Still other organisms tolerate high salinities best near their higher limits of temperature tolerance or tolerate high temperatures best near their higher limits of salinity tolerance. These various combinations may be effective in relieving stress accompanying short-term deviations from optimum conditions, but they cease to operate at extremely high or low temperatures.

Gunter(*253*) states that tropical species normally live in environments near their thermal death point, whereas the environment of temperate or northern species is usually 10–15°C below the lethal limit. Tropical or subtropical forms may suffer more frequent thermal mortalities, but they are also better suited to tolerate continuously high temperature levels. Many South Atlantic and Gulf Coast organisms regularly tolerate summer water temperatures of 35°C or more and, in some cases, regular exposure on tidal flats where temperatures of 40°C or more are not uncommon. Although thermal death points for most estuarine organisms appear to be somewhat below 50°C(*253*), high-temperature mortalities may result from a combination of factors.

As Gunter(*122*) and Simmons(*254*) have pointed out, fish in the Laguna Madre of Texas are under stress to maintain salt and water balance in the face of an osmotic differential that can be as large as 30–45 atm in the usually warm, hypersaline water. If temperature of the water rises, or if wind action, associated both with cooling and with gas exchange, ceases, the fish soon rise to the surface gasping. A fish kill may ensue. Conversely, Simmons notes that if low temperatures accompany high salinities, species of fish not normally tolerant of hypersaline conditions begin to appear in the Laguna Madre. From this, the fish would appear to be operating at or near their maximum metabolic capability under the usual Laguna Madre conditions. Any additional stress associated with oxygen depletion or extra temperature-associated metabolic work is potentially lethal.

The second cause of temperature-associated Gulf Coast fish kills seems to be a simple chilling process. During the winter, the Gulf Coast is subject to sudden cold spells known as "northers"(*255*), which may drop temperatures from the mid-70's to the mid-30's (Fahrenheit) or lower in a few hours. The strong north winds associated with the northers drive much of the water out of the shallow estuaries and rapidly chill the remainder to near the temperature of the air. If the ambient temperature is sufficiently low, a cold kill may follow(*122,124,253*). The cold weather usually persists for only a few days, and the water does not reach freezing level. If, however, near freezing temperatures are attained and the cold persists, kills may reach massive proportions.

3. Other Stresses

Although estuarine dissolved oxygen levels are normally near saturation, the occasional combination of a high summer temperature, a stable water column resulting from reduced wind velocities, and other factors can result in low oxygen levels. Estuarine species accommodate themselves with varying degrees of success to this form of stress, but mass mortalities can occur(*122,140*). Although initial mortalities may be slight, the depressed oxygen levels are further reduced by decomposing organisms so that total mortality of the biota can eventually ensue. These essentially anaerobic conditions frequently continue until storms or other phenomena mix and reaerate the water column. Abbott(*35*) has shown that localized nocturnal oxygen depletion can result in migration of nektonic organisms from oxygen-depleted water at night and reimmigration the following day as dissolved oxygen levels increase.

Estuarine waters are usually well buffered, but variations in pH can occur in areas where circulation is reduced or where there is significant runoff from bogs, heavily forested areas, or industrial effluents. Calabrese and Davis(*256*) have shown that pH variation beyond narrow limits imposes significant stress on oyster (*Crassostrea*) and clam (*Mercenaria*) populations. The pH range for normal larval development was found to be 6.75–8.75 for oysters and 7.00–8.75 for clams. Growth rates in both cases diminished rapidly below 6.75 and above 9.00, and neither species could successfully reproduce when pH remained above 9.00.

Other physical factors, such as turbidity, sedimentation, and substrate type and stability, can impose some degree of stress on estuarine species, but little is known concerning their nonlethal effects.

Wohlschlag and Cameron(*257*) have shown that low level, presumably generalized, stress induced by pollutants results in depressed respiratory levels in the pinfish (*Lagodon*), with highest oxygen consumption at low temperature (10°C). Of the actively swimming fish tested, the larger individuals showed the more marked respiratory depression in polluted waters. This suggested that larger fish would have little reserve energy to cope with further abrupt temperature declines, a possible explanation for the apparent predominance of large fishes stunned or killed during sudden cold waves (see Chap. 6).

Every component of the estuarine biotope is subjected to stress as a result of the continuing competition for living space, food, oxygen, and other biological requirements. The ability of any type of organism to accommodate itself to the summation of instantaneous and long-term stresses is reflected by its survival and population characteristics within each estuary.

V. Pollution

Since a large part of the water in the world hydrologic cycle must eventually pass through an estuary in order to return to the sea, estuaries are the ultimate sewers of the terrestrial environment. Further, since the water must carry along any reasonably conservative dissolved or colloidal materials added to it, any water pollution problem is potentially an estuarine pollution problem.

Obtaining an objective definition of pollution is difficult because of personal biases which creep into proposed definitions. For example, a recent issue of the *SFI Bulletin* (*258*) offered the following:

> . . . water pollution is the specific impairment of water quality by agricultural, domestic, or industrial wastes (including thermal and atomic wastes), to a degree that has an *adverse effect upon any beneficial use* of water yet that does not necessarily create an an actual hazard to the public health.

The italics are ours. This is a semantically loaded definition whose meaning depends on what the reader personally believes to constitute an "adverse effect upon any beneficial use." While the multiple-use concept for water bodies is currently in vogue, the presumption that all possible uses must be feasible in all estuaries seems illogical [cf. (*259*)]. For example, is a waste discharge situation that tends to eliminate fishing or water-skiing in a heavily used estuarine ship channel necessarily bad? Traffic conditions would preclude such recreational activities in any case.

Without pretending personal immunity from biases, we propose a generalized operational definition as follows:

> Water pollution is the addition to a natural body of water of any material which diminishes the optimal economic use of the water body by the population which it serves.

This definition is subject to immediate attack on grounds that it ignores social and esthetic considerations. The statement has, however, the merit of being objectively definable in cost-accounting terms. For example, what is the annual value of a sport and/or commercial fishery, and what will be the capital and operating costs for waste treatment facilities to protect this fishery? Socio-esthetic aspects cannot be entered into a cost-accounting system such as this, except as arbitrary values for re-creation units. We do not deny that they have merit. We simply contend that no rigorous scientific treatment is possible. Another basis must be found for decisions on nonutilitarian uses.

A waste is dumped into a body of water because a city, industry, or individual wishes to eliminate a useless and somewhat noxious mess from its environment. Unless marine disposal is carried out with forethought, however, the waste not only may not leave the environment, but may actually accumulate from day to day. Two problems are generally involved. First, since most wastes are dilute solutions in fresh water, they tend to be less dense than the saline water into which they are discharged. Therefore, resistance to mixing of waste with receiving water may be considerable unless discharge is carried out through a carefully designed diffuser. Second, current patterns in the receiving body will dictate whether the discharged material is carried seaward or returned to shore to create a nuisance. Thus, a reliable hydrographic study of a potential discharge area is necessary for any competent outfall design. Numerous sources are available for detailed discussions of these problems. Perhaps the most convenient is the *Proceedings of the 1st International Conference on Waste Disposal in the Marine Environment* (62), especially the chapters by Bolomey, Wiegel and Johnson, Brooks, and Pomeroy.

Pollutants can be categorized as four types—nontoxic, toxic, thermal, and radioactive. Overlap between these broad categories is, of course, possible. However, although a hot, toxic, radioactive pollutant is not difficult to envision, the manifestations generally behave independently.

A. Nontoxic Pollution

This is definable as addition to a receiving body of substances which are not, intrinsically, metabolic poisons, but which act to alter the environment, thereby altering or eliminating the biological assemblages customarily inhabiting the receiving body.

Generally, the ultimate manifestation of nontoxic pollution is depletion of the dissolved oxygen in the water. Sludges and liquid wastes which cause hypereutrophication and have high oxygen demands are major contributors to this depletion.

The materials with large oxygen demands act in a direct fashion. Dissolved oxygen in the water is consumed chemically, and the aerobic biota is asphyxiated or driven away. The area may become azoic, or an anaerobic microbial-fungal community may develop. In essence, a large septic tank is created. To correct this situation, and to eliminate the unpleasant sights and smells associated with it, legislation usually is developed to require treatment of wastes before discharge. Facilities are constructed, and the septic situation is relieved.

Then a second problem develops. The treated waste effluents no longer contain vast quantities of reduced organic matter to create septic conditions. However, they do contain, especially in the case of domestic sewages, large amounts of various essential nutrients, which stimulate growth of the plankton algae population beyond the point at which control is possible by grazing pressure from the fauna. This leads to accrual of a large amount of secondarily derived reduced organic matter in the form of accumulated dead algae. Oxygen demand from this, coupled with nocturnal respiratory demands of the vastly expanded living algal population, can then lead to depletion of dissolved oxygen at night, asphyxiation of fauna, and creation of a generally septic condition. Numerous articles [(*42, 260–263*) and others] have appeared discussing this phenomenon and should be consulted for details.

Similar effects can result from runoff from fertilized agricultural areas. The only apparent method for eliminating this problem is total removal of nutrient-loaded wastes from the watershed in question. This, however, presumably only transfers the problem to another watershed.

Riley(*187*) suggests that nitrate depletion may be the ultimate growth limiter for diatom populations in northern bays and sounds on the Atlantic Coast of the United States. Along these lines, the hypereutrophication of large areas of Long Island Sound by nitrogen- and phosphorus-rich wastes from commercial duck-rearing operations has been documented (*173,174*). In the usual warm temperate or tropical estuary, however, turbidity is relatively high. Thus, in spite of naturally nutrient-rich conditions, lack of light tends to hold photosynthesis and concomitant phytoplankton population growth in check(*43,264*). In experimental fertilization studies on subtropical estuarine ponds(*81*), we found that massive doses of nitrate and phosphate do not engender "bloom" conditions. Patrick(*265*) has encountered a similar lack of response to nitrogen and phosphorus enrichment in studies on a moderately eutrophic creek in Pennsylvania, but comparability of the two situations is uncertain.

Copeland(*69*) has presented convincing arguments that diversion of freshwater streams feeding Gulf Coast estuaries will have a damaging effect on the estuaries. First, the turbidity flux through the estuary will decrease, as will general flushing. Second, salinity increase will promote flocculation and settling of suspended matter(*49*). The resulting clarification should lead to a large increase in photosynthesis in a naturally nutrient-rich situation, and may result in hypereutrophic conditions in the estuaries involved.

With regard to general pollution of estuaries, probably no estuary has been so thoroughly studied as the Thames [cf. (*46*)]. By drawing on this work plus historic data on New York Harbor, Torpey (*266*) has developed a strong thesis that exogenous biochemical oxygen demand loadings as small as 15 lb/acre per day (17 kg/ha per day; 0.2 ppm/day) may produce an oxygen deficit of 50% in an estuary, but that the deficit then does not increase further until loading increases manyfold. This is consistent with the notion that estuaries may function as environmental buffers for their biota [cf. (*267*)].

B. Toxic Pollution

Toxic pollution may be defined as addition to a receiving body of substances which are active metabolic poisons to some or all of the biota of the receiving body.

A plethora of substances may be toxic to aquatic life. These comprise various heavy metals, strong acids and alkalis, and a multitude of organic compounds (*268,269*). Perhaps the commonest toxins now entering estuaries are the various pesticides and herbicides widely employed in agriculture. Since these are, variously, metabolic analogs, protein denaturants, or enzyme inhibitors, they tend to be degraded slowly or not at all when broadcast into the environment. DDT, one of the earliest developed and most widely used of these compounds, is so ubiquitous after 20–25 years of use (*270,271*) that it may now be impossible to find a DDT-free area anywhere on the planet. Risebrough *et al.* (*272*) have recently documented this wide distribution by reporting DDT residues in a number of oceanic birds.

Sizable localized fish kills have resulted from sudden discharge or runoff of pesticides into streams [cf. (*271*)]. However, we have encountered no report of a pesticide-linked full-scale kill in a major estuary, and Ferguson and co-workers (*273–275*) have shown that various fish and bird populations can develop pesticide resistance after chronic exposure to low levels of the various compounds. Great concern, nevertheless, surrounds the uptake of pesticides by common estuarine organisms such as oysters and shrimps, since human consumption is involved. For the past several years, the U.S. Bureau of Commercial Fisheries has maintained an extensive program for monitoring pesticide residues in fisheries products from all the estuarine areas of the United States (see Chaps. 23, 28, and 30). Bureau personnel have also carried on considerable experimental and investigative work on pesticide effects in estuaries (*276–278*).

One serious facet of toxic pollution often escapes notice. Although low concentrations of various toxic wastes may not damage the standing crop of fish, they can seriously injure or destroy the photosynthetic organisms upon which the estuarine food chains depend. Thus, great public excitement will attend the destruction of a few fish, which are replaced almost immediately by a new generation or by recruitment from adjoining populations, while no notice may be taken, until too late, of a situation that is actually destroying the estuary from a biological standpoint (also see Chap. 6).

C. Thermal Pollution

Thermal pollution consists of addition to a body of water of heat derived from a source not normal for the watershed. This results in elevation of water temperature, increase in evaporation rate, and general speeding up of chemical reactions, including metabolic reactions, taking place in the water.

At present, thermal pollution seems to be a major problem only in certain northern estuaries and streams inhabited by commercially important, cold-adapted organisms. For example, the temperature of the Columbia River has been increased considerably by impoundment and by various cooling water discharges. This temperature increase has resulted in considerable damage to economically important salmon and steelhead trout populations of the river, primarily by allowing a cold-limited disease to flourish *(258)*.

Warm climate estuaries, where summer water temperatures often exceed 95°F (35°C), do not exhibit major thermal pollution problems at present. In these estuaries, the biota tends to be heat tolerant, even though it exists near its thermal death point. It is, in fact, relatively susceptible to death from cold *(124,253)*. This state of affairs, however, may not continue for long since thermal pollution of the environment appears to be one of the most serious threats facing the human race.

The main source of waste heat is cooling water discharge from industries such as steam-powered electrical generating plants, which are increasing in number on a worldwide basis. Naylor *(279)*, citing Ross *(280)*, reports that such plants, fired by fossil fuels, must dispose of approximately 5 Btu/hr (1.3 kcal/hr) of waste heat per watt of generating capacity. This represents a generating efficiency of 35–40%.

In the United States, demand for electrical energy has been doubling every 6–10 years *(258,279)*, and the trend seems to be continuing. Concurrently, pressures related to air pollution created by conventionally fueled plants are leading to emphasis on nuclear-powered generating systems. A great deal of controversy is accompanying this shift.

Apparently, nuclear-fueled generating plants must have capacities of

500 MW or more to be economically competitive with plants employing fossil fuels; however, construction of larger and larger capacity generating installations has been the general rule for some time, regardless of fuel source. These large plants have vast cooling water requirements, which often cannot be met by freshwater sources. Thus, paralleling increasing plant size is a trend to locate such units in coastal areas, especially along estuaries, in order to obtain sufficient water. Concern about thermal pollution of estuaries has arisen from this.

Inquiries by Senator E. S. Muskie and the Senate Public Works Subcommittee on Air and Water Pollution into cooling water-related thermal pollution of coastal areas have been reported in some detail by Nelson (*281*). The report implies that only nuclear power plants create heat problems, but whether this represents the view of the author of the article or of the Senate Subcommittee is unclear. In any case, Hetrick and Seale (*282*) offered effective rebuttal and pointed out the advantages of nuclear fuels relative to air pollution problems. During the same period, however, an editorial in *Oceanology International* (*283*) stated:

> Nuclear power plants produce far more thermal pollution than conventional coal-, oil-, and gas-fueled generating facilities. For safety reasons, nuclear installations must operate with lower steam pressure. As a result, nuclear plants are less efficient than other plants and they discharge about 50% more waste heat through their condenser cooling systems.

For evaluation purposes, we requested and obtained from Dr. D. L. Hetrick information on thermal efficiencies and related matters for various types of generating plants that exist or are under construction at the present time (*284*). Table 7 shows generating efficiencies computed as [watts of electricity/(watts of electricity + watts of waste heat)] × 100. Dr. Hetrick makes two other pertinent statements:

> (1) Future fast-neutron reactors using sodium cooling will exceed 40% efficiency.
>
> (2) Note that water-cooled reactors today are less efficient than coal-fired plants. This is dictated by economy and *not* by safety. Future technology will probably improve efficiencies for all types of plants, but the lack of atmospheric pollution clearly favors nuclear power.

Cairns (*285*) reported projections that by 1970 generating plants in the United States will be discharging 1.25×10^{12} Btu/hr (3.15×10^{11} kcal/hr) of waste heat; moreover, if the trend continues, heat discharge by the year 2000 will be sufficient to raise the total U.S. minimum runoff (220,000 ft^3/sec; 6200 m^3/sec) by 50°C. The runoff value seems extremely small (*286*), but total heat content can be recomputed from the data and applied to any desired runoff figure. Naylor (*279*) has reviewed heat effects on marine and estuarine organisms in detail (also see Chap. 6).

TABLE 7
Electrical Generating Plant Efficiencies Computed as [watts of electricity/(watts of electricity + watts of waste heat)] × 100[a]

Type of plant	% Efficiency
Coal-fired	38
Natural gas-fired	34
Boiling and pressurized water reactors	32
High temperature gas-cooled reactors	39

[a]Data provided by Ref. *284*.

No long-term solution to this problem is apparent. Alternative discharge of this heat into the atmosphere would merely introduce a time lag before the planetary heat cycle carried an appropriate portion of the heat into the oceanic environment.

D. Radioactive Pollution

As the name implies, radioactive pollution is the discharge into a receiving body of a radioactive waste material. Generally, such discharges are subject to strict governmental regulation so that, barring accident, total levels of discharged radioactivity are quite low.

Mauchline and Templeton(*287*) have reviewed effects of natural and artificial radioisotopes in marine environments. Under estuarine and marine conditions, many organisms tend to concentrate trace materials from the environment [cf. (*64*), Table 2]. This tendency may lead to localized accumulation of sizable amounts of radionuclides(*288*). A good, but nonestuarine, example of this is the general radioactivation of the biota at Eniwetok Atoll by biological reconcentration of radioactive trace elements generated during nuclear weapon tests(*289*). The reef organisms, in particular, tend to hold the nuclides rather than rerelease them. As a result, the organisms can generate their own autoradiographs [cf. (*290*)]. Templeton(*291*) mentions similar reconcentrations of dissolved radionuclides on the surface of fish eggs.

There seems to be general agreement [cf. (*292*)] that discharges at permissible levels are not harming the environment. However, care must be exercised to avoid reconcentration effects that could lead to potentially dangerous radiation levels in fisheries products. Recently a major symposium in Stockholm explored in detail the general problem of ecological concentration of radionuclides(*293*).

VI. Hypersaline Environments

Landlocked or semilandlocked bodies of water having total dissolved solids content considerably greater than that of seawater occur in various areas of the world. Most of these bodies seem, on the basis of ionic ratios, to be dried-up remnants of large lakes (*25, 73*). A few, however, are obviously ocean-related and may still have limited or periodic exchanges of water with the sea. Such embayments occur either in areas where average evaporation exceeds rainfall, or in areas where rainfall is spottily distributed. In the latter case, much fresh water is lost to the sea by stratified flow without significant dilution of the estuary or lagoon.

The two best known examples of hypersaline lagoons are the Laguna Madre system of Texas and Mexico and the Sivash at the western edge of the Sea of Azov.

The Sivash, also called the Putrid Sea, is described by Caspers (*294*, p. 854) as ". . . an extensive area of shallow bordering swamps, the Sivash, of 2630 sq km." Zenkevitch (*295*) discusses this area in some detail, as does Hedgpeth (*25*). The only connection to the Sea of Azov is a strait 120 m wide and 2–3 m deep located at the northern end of the basin. The main basin is somewhat dissected by islands and bars and varies in average depth from 0.63 m in the northern end to 0.86 m in the southern portion. Salinities in the blind southern end have been reported as high as 166‰. Citing from Zernov (*296*), Hedgpeth (*25*) gives a table relating salinity to standing crops of phytobenthos and zoobenthos. In general, as salinity increases, zoobenthos decreases, while phytobenthos first decreases and then increases to very high levels. Maximum standing crop of phytobenthos is 833 g/m^2 in the salinity range 126–144‰. Zenkevitch (*295*) points out, however, that suffocation of the benthos is a common occurrence during calm summer weather. This he attributes to massive organic decay in a shallow body at elevated temperatures. Presumably, a cessation of atmospheric reaeration is involved.

The Laguna Madre system has been investigated from many standpoints. Gunter (*122*) reports salinities in past years as high as 110‰ or more and mentions summer fish kills under calm conditions similar to the suffocation effects in the Sivash. Apparently, opening of the Intracoastal Waterway from Corpus Christi to Brownsville in 1948 increased circulation in the Laguna Madre so that maximum salinities now are in the 70–80‰ range (*254*).

Hedgpeth (*25–27*) reports fish yields from Laguna Madre of Texas as high as 2×10^6 lb/year (9×10^5 kg/year) and points out that in some years this area has been the source of 50–60% of the total fin fish catch in Texas coastal waters. Some ambiguity exists, however, in reports on primary productivity in Laguna Madre of Texas. Odum et al.(*297*) have reported rather low chlorophyll *a* values of 0.05–0.3 g/m^2 from various points in the basin during various seasons, but Odum and Hoskin(*43*) suggest rather high gross productivity. Odum and Wilson(*44*, p. 53) conclude that "maximum efficiencies, productivities, and sustained respiration were found in natural, high salinity, grass flat systems . . ." Copeland and Jones(*298*), however, surmise that (p. 189) ". . . a study of the Mexican Laguna Madre could yield results comparable to those expected from the Texas Laguna Madre before industrialization and civilization . . ." and conclude that (p. 203) "photosynthetic productivity, community respiration and *P*/*R* ratios were lower than in comparable bays with lower salinities." All the reports just mentioned are based on short-term studies at a few stations. A prolonged, systematic investigation seems necessary in order to resolve these contradictions.

The general problems related to hypersaline environments have been treated in a recent symposium(*299*), which should be consulted for detailed information.

VII. Glossary of Specialized Terms and Symbols

‰	parts per mille; parts per thousand; g/kg; the unit in which salinity and chlorinity values are expressed
P/*R*	the ratio of photosynthetic production to respiration for an organism or an ecosystem
ADP	adenosine diphosphate
ATP	adenosine triphosphate
Cl	standard abbreviation for chlorinity
NADP, NADPH	oxidized and reduced forms of nicotinamide adenine dinucleotide phosphate
Anadromous	normally dwelling in the ocean but entering fresh water for a specialized period during the life cycle
Autotrophic	not requiring preformed organic substrate for existence; capable of photosynthesis or of chemosynthesis from inorganic substrates
Benthos	organisms living on or in the bottoms of bodies of water
Biomass	the total weight of living organic matter occupying a specified area or volume of the environment; may relate to individual species or to entire ecosystems
Catadromous	normally dwelling in fresh water but entering the sea for a specialized period during the life cycle
Chlorinity	the mass of chlorine equivalent to the bromide, chloride, and iodide in 1 kg of seawater
Demersal	inhabiting the lower part of a water column
Diagenesis	a geologic process, physicochemically based, wherein loosely deposited materials are forced into compact, mineralized masses by the weight

	of overlying water and sediment; chemical recombination may accompany the physical processes
Diadromous	either anadromous or catadromous
Ecdysis	the molting process in an organism; specifically, in arthropods, the periodic growth-associated shedding of the chitinous exoskeleton and all related processes
Euryhaline	tolerating a broad range of environmental salinity levels
Eurythermal	tolerating a broad range of environmental temperatures
Facies	a particular geological or biological assemblage which is recognizable and distinguishable from adjacent assemblages
Heterotrophic	requiring a preformed organic substrate for existence
Holoplanktonic	having a planktonic existence throughout life
Isopleth	a line on a map or chart connecting points at which a specified variable has a constant value
Lacustrine	inhabiting lakes or ponds
Meroplanktonic	having a planktonic existence for only part of a life cycle
Nekton	free-swimming aquatic organisms
Pelagic	inhabiting the open reaches of a body of water as opposed to inhabiting near shore areas
Plankton	aquatic organisms, usually minute, which drift with a water mass and have little or no swimming ability
Polyhaline	a poorly defined term which generally refers to organisms capable of tolerating a broad but fairly high range of salinities
Salinity	the weight in vacuo of solids which can be obtained from 1 kg of water when all organic matter has been oxidized, all bromide and iodide have been replaced by chloride, all carbonates have been converted to oxides, and the residue has been dried to constant weight at 480°C, with appropriate correction for loss of volatile halides during drying
Spat	the final larval stage of a sessile mollusk, particularly an oyster, which attaches to a substrate and initiates the attached stage of the life cycle
Standing crop	a term derived from agricultural usage and often employed as a synonym for biomass, but also a term referring to readily harvestable portions of the biomass as opposed to roots, holdfasts, and other inaccessible structures
Stenohaline	tolerating only a narrow range of environmental salinity levels
Stenothermal	tolerating only a narrow range of environmental temperatures
Tidal prism	the volume of water involved in raising and lowering the water level in an estuary during a tidal cycle

REFERENCES*

1. G. Gunter, *J. Mississippi Acad. Sci.*, **9**, 286 (1963).
2. G. Gunter, *Publ. Am. Assoc. Advan. Sci.*, **83**, 621 (1967).
3. T. V. R. Pillay, *Publ. Am. Assoc. Advan. Sci.*, **83**, 639 (1967).
4. T. V. R. Pillay, *Publ. Am. Assoc. Advan. Sci.*, **83**, 647 (1967).
5. P. De Falco, Jr., *Publ. Am. Assoc. Advan. Sci.*, **83**, 701 (1967).

*Based on a literature survey concluded in January, 1968.

6. G. H. Lauff, ed., *Estuaries, Publ. No. 83*, American Association for the Advancement of Science, Washington, D.C., 1967.
7. A. T. Ippen, ed., *Estuary and Coastline Hydrodynamics*, McGraw-Hill, New York, 1966.
8. K. O. Emery and R. E. Stevenson, *Geol. Soc. Am. Mem.*, **67**(1), 673 (1957).
9. D. W. Pritchard, *Publ. Am. Assoc. Advan. Sci.*, **83**, 3 (1967).
10. R. J. Russell, *Publ. Am. Assoc. Advan. Sci.*, **83**, 93 (1967).
11. J. H. Day, *Trans. Roy. Soc. S. Africa*, **33**, 53 (1951).
12. B. H. Ketchum, *J. Marine Res.*, **10**, 18 (1951).
13. B. H. Ketchum, *Sewage Ind. Wastes*, **23**, 198 (1951).
14. D. W. Pritchard, *Proc. Hydraulics Div. Am. Soc. Civil Engrs.*, **81**, Separate 717 (1955).
15. K. F. Bowden, *Publ. Am. Assoc. Advan. Sci.*, **83**, 15 (1967).
16. H. Caspers, *Publ. Am. Assoc. Advan. Sci.*, **83**, 6 (1967).
17. K. M. F. Scott, A. D. Harrison, and W. Macnae, *Trans. Roy. Soc. S. Africa*, **33**, 283 (1952).
18. E. W. Behrens, *Publ. Inst. Marine Sci. Univ. Texas*, **9**, 7 (1963).
19. J. P. Morgan, *Publ. Am. Assoc. Advan. Sci.*, **83**, 115 (1967).
20. J. W. Hedgpeth, *Publ. Am. Assoc. Advan. Sci.*, **83**, 707 (1967).
21. W. L. Siler, personal communication, 1968.
22. J. W. Hedgpeth, *Publ. Inst. Marine Sci. Univ. Texas*, **3**, 107 (1953).
23. D. S. Gorsline, *Publ. Am. Assoc. Advan. Sci.*, **83**, 219 (1967).
24. M. R. Carriker, *Publ. Am. Assoc. Advan. Sci.*, **83**, 442 (1967).
25. J. W. Hedgpeth, *Geol. Soc. Am. Mem.*, **67**(1), 693 (1957).
26. J. W. Hedgpeth, *Trans. N. Am. Wildlife Conf., 12th, San Antonio, 1947*, p. 364.
27. J. W. Hedgpeth, *Publ. Am. Assoc. Advan. Sci.*, **83**, 408 (1967).
28. A. B. Arons and H. Stommel, *Trans. Am. Geophys. Union*, **32**, 419 (1951).
29. D. R. F. Harleman, in *Estuary and Coastline Hydrodynamics* (A. T. Ippen, ed.), McGraw-Hill, New York, 1966, Chap. 14.
30. D. W. Pritchard, *Trans. N. Am. Wildlife Conf., 16th, Milwaukee, 1951*, p. 368.
31. D. W. Pritchard, *J. Marine Res.*, **11**, 106 (1952).
32. H. B. Simmons, in *Estuary and Coastline Hydrodynamics* (A. T. Ippen, ed.), McGraw-Hill, New York, 1966, Chap. 16.
33. H. H. Hildebrand, personal communication, 1967.
34. W. Abbott, *J. Water Pollution Control Federation*, **38**, 258 (1966).
35. W. P. Abbott, technical report filed with Texas Water Pollution Control Board, November, 1963.
36. F. W. Moon, Jr., C. L. Bretschneider, and D. W. Hood, *Publ. Inst. Marine Sci. Univ. Texas*, **4**(2), 14 (1957).
37. C. J. Velz, *Trans. Am. Soc. Civil Engrs.*, **104**, 560 (1939).
38. P. D. Haney, *J. Am. Water Works Assoc.*, **46**, 353 (1954).
39. D. J. O'Connor and W. E. Dobbins, *Proc. Am. Soc. Civil Engrs., SA6, Paper No. 1115*, 1956.
40. H. T. Odum, *Limnol. Oceanog.*, **1**, 102 (1956).
41. H. T. Odum, *Ecol. Monographs*, **27**, 55 (1957).
42. H. T. Odum, *Proc. Intern. Conf. Waste Disposal Marine Environ., 1st, Berkeley, Calif., 1959*, p. 547.
43. H. T. Odum and C. M. Hoskin, *Publ. Inst. Marine Sci. Univ. Texas*, **5**, 16 (1958).
44. H. T. Odum and R. F. Wilson, *Publ. Inst. Marine Sci. Univ. Texas*, **8**, 23 (1962).
45. B. J. Copeland and W. R. Duffer, *Limnol. Oceanog.*, **9**, 494 (1964).
46. A. L. H. Gameson and M. J. Barrett, in *Robert A. Taft Sanit. Eng. Center Tech.*

Rept. W58-2, U.S. Dept. Health, Education, and Welfare, Public Health Service, Cincinnati, 1958.
47. E. C. Tsivoglu, R. L. O'Connell, C. M. Walter, P. J. Godsil, and G. S. Logsdon, *J. Water Pollution Control Federation*, **37**, 1343 (1965).
48. M. S. Mulkana, Ph.D. Thesis, Mississippi State Univ., State College, Miss., 1968.
49. U. G. Whitehouse, L. M. Jeffrey, and J. D. Debrecht, *Proc. Natl. Conf. Clays, Clay Minerals, 7th, 1960*, p. 1.
50. A. Schou, *Publ. Am. Assoc. Advan. Sci.*, **83**, 129 (1967).
51. E. H. Rainwater, *Mississippi State Geol. Surv. Bull.*, **102**, 32 (1964).
52. G. Gunter, J. G. Mackin, and R. M. Ingle, *A Report to the District Engineer on the Effect of the Disposal of Spoil from the Inland Waterway, Chesapeake and Delaware Canal, in Upper Chesapeake Bay*, technical report available from District Engineer, U.S. Army Corps of Engineers, Philadelphia, 1964.
53. A. T. Ippen, in *Estuary and Coastline Hydrodynamics* (A. T. Ippen, ed.), McGraw-Hill, New York, 1966, Chap. 15.
54. G. A. Rusnak, *Publ. Am. Assoc. Advan. Sci.*, **83**, 180 (1967).
55. K. O. Emery and R. E. Stevenson, *Geol. Soc. Am. Mem.*, **67**(1), 729 (1957).
56. W. A. Price, cited by A. Collier and J. W. Hedgpeth, *Publ. Inst. Marine Sci. Univ. Texas*, **1**(2), 125 (1950).
57. H. Postma, *Publ. Am. Assoc. Advan. Sci.*, **83**, 158 (1967).
58. E. J. Lund, *Publ. Inst. Marine Sci. Univ. Texas*, **4**(2), 296 (1957).
59. E. J. Lund, *Publ. Inst. Marine Sci. Univ. Texas*, **4**(2), 313 (1957).
60. *Texas Game and Fish*, **13**(6), 4 (1955).
61. W. L. Siler, unpublished work, 1968.
62. E. A. Pearson, ed., *Proc. Intern. Conf. Waste Disposal Marine Environ., 1st, Berkeley, Calif., 1959*, Pergamon, New York, 1960.
63. J. E. Foxworthy, R. B. Tibby, and G. M. Barsom, *J. Water Pollution Control Federation*, **38**, 1170 (1966).
64. E. D. Goldberg, *Geol. Soc. Am. Mem.*, **67**(1), 345 (1957).
65. W. W. Rubey, *Geol. Soc. Am. Bull.*, **62**, 1111 (1951).
66. H. Barnes, *Apparatus and Methods of Oceanography, Part One: Chemical*, George Allen & Unwin, London, 1959.
67. H. W. Harvey, *The Chemistry and Fertility of Sea Waters*, University Press, Cambridge, Mass., 1955.
68. M. Knudsen, *Kgl. Danske Videnskab. Selskab Mat. Fys. Medd.*, **12**, 1 (1902).
69. B. J. Copeland, *J. Water Pollution Control Federation*, **38**, 1831 (1966).
70. *Simp. Classificazione Acque Salmastre, Arch. Oceanog. Limnol.*, **11**, Suppl. (1959).
71. J. B. Price and G. Gunter, *Intern. Rev. Ges. Hydrobiol.*, **49**, 629 (1964).
72. B. S. Ballard, Ph.D. Thesis, Mississippi State Univ., State College, Miss., 1967.
73. F. W. Clarke, *The Data of Geochemistry, U.S. Geol. Surv. Bull. 770*, Washington, D.C., 1924.
74. H. T. Odum, *Publ. Inst. Marine Sci. Univ. Texas*, **4**(2), 23 (1957).
75. P. C. Mangelsdorf, Jr., *Publ. Am. Assoc. Advan. Sci.*, **83**, 71 (1967).
76. W. Abbott, *Simp. Intern. Lagunas Costeras, México, November*, 1967.
77. E. Føyn, *Simp. Intern. Lagunas Costeras, México, November, 1967.*
78. D. W. Hood, *Oceanog. Marine Biol. Ann. Rev.*, **1**, 129 (1963).
79. G. E. Hutchinson, *A Treatise on Limnology*, Vol. 1, Wiley, New York, 1957.
80. W. C. Allee, A. E. Emerson, O. Park, T. Park, and K. P. Schmidt, *Principles of Animal Ecology*, W. B. Saunders, Philadelphia, 1949.
81. W. Abbott, *Proc. Intern. Conf. Water Pollution Res., 4th, Prague, April, 1969*, Pergamon, London, in preparation.

82. P. L. Parker, *Publ. Inst. Marine Sci. Univ. Texas*, **11**, 102 (1966).
83. W. Abbott, *Ecology*, **38**, 152 (1957).
84. G. A. Riley, *Bull. Bingham Oceanog. Collection*, **13**, Article 1 (1951).
85. C. L. Newcombe and H. F. Brust, *J. Marine Res.*, **3**, 76 (1940).
86. H. E. Bruce and D. W. Hood, *Publ. Inst. Marine Sci. Univ. Texas*, **6**, 133 (1959).
87. K. M. Mackenthun, *Nitrogen and Phosphorus in Water*, U.S. Dept. Health, Education, and Welfare, Public Health Service, Washington, D.C., 1965.
88. R. A. Berner, *Marine Geol.*, **1**, 117 (1964).
89. E. P. Levine, *Science*, **133**, 1352 (1961).
90. A. P. Winogradov, *The Elementary Chemical Composition of Marine Organisms*, Yale Univ. Press, New Haven, 1953.
91. R. B. Biggs, *Publ. Am. Assoc. Advan. Sci.*, **83**, 239 (1967).
92. P. D. Trask, assisted by H. E. Hammar and C. C. Wu, *Origin and Environment of Source Sediments of Petroleum*, Gulf Publ. Co., Houston, 1932.
93. R. M. Darnell, *Publ. Am. Assoc. Advan. Sci.*, **83**, 374 (1967).
94. R. M. Darnell, *Publ. Am. Assoc. Advan. Sci.*, **83**, 376 (1967).
95. C. H. Oppenheimer and L. S. Kornicker, *Publ. Inst. Marine Sci. Univ. Texas*, **5**, 5 (1958).
96. A. C. Redfield, *Publ. Am. Assoc. Advan. Sci.*, **83**, 108 (1967).
97. E. K. Duursma, *Neth. J. Sea Res.*, **1**, 1 (1961).
98. R. F. Wilson, *Publ. Inst. Marine Sci. Univ. Texas*, **9**, 64 (1963).
99. L. M. Jeffrey, *J. Am. Oil Chemists' Soc.*, **43**, 211 (1966).
100. K. Park, W. T. Williams, J. M. Prescott, and D. W. Hood, *Publ. Inst. Marine Sci. Univ. Texas*, **9**, 59 (1963).
101. J. Awapara, in *Amino Acid Pools* (J. T. Holden, ed.), Elsevier, Amsterdam, 1962, pp. 158–175.
102. J. B. Lackey, *Publ. Am. Assoc. Advan. Sci.*, **83**, 291 (1967).
103. J. H. Day, *Publ. Am. Assoc. Advan. Sci.*, **83**, 397 (1967).
104. H. T. Odum, *Publ. Inst. Marine Sci. Univ. Texas*, **9**, 48 (1963).
105. J. W. Hedgpeth, *Am. Fisheries Soc. Spec. Publ. 3*, 1966, p. 3.
106. E. Waldo, *Louisiana Wildlife and Fisheries Commission Seventh Biennial Report 1956–1957*, Baton Rouge, 1958, pp. 45–50.
107. G. M. Allen, *Am. Comm. Intern. Wildlife Protection Spec. Publ. 11*, New York, 1942.
108. G. Gunter, *U.S. Fish Wildlife Serv. Fishery Bull.*, **89**, 543 (1954).
109. E. P. Walker, F. Warnick, K. I. Lange, H. E. Uible, S. E. Hamlet, M. A. Davis, and P. A. Wright, *Mammals of the World*, Vol. 2, Johns Hopkins Press, Baltimore, 1964.
110. D. K. Caldwell and F. B. Golley, *J. Elisha Mitchell Sci. Soc.*, **81**, 24 (1965).
111. V. B. Scheffer and D. W. Rice, *U.S. Fish Wildlife Serv. Spec. Sci. Rept. Fisheries No. 431*, 1963, p. 1.
112. J. C. Moore, *Am. Midland Naturalist*, **49**, 117 (1953).
113. J. L. McHugh, *Publ. Am. Assoc. Advan. Sci.*, **83**, 581 (1967).
114. H. B. Bigelow and W. C. Schroeder, *Sears Found. Marine Res. Mem.* **1**(2), 1 (1953).
115. T. B. Thorson, D. E. Watson, and C. M. Cowan, *Copeia*, **1966**, 385.
116. J. L. McHugh, *Virginia J. Sci.*, **13**, 144 (1962).
117. G. Gunter, *Ecol. Monographs*, **8**, 313 (1938).
118. G. Gunter, *Ecology*, **37**, 616 (1956).
119. C. E. Dawson, *Am. Midland Naturalist*, **76**, 379 (1966).
120. G. V. Nikolsky, *The Ecology of Fishes*, Academic, New York, 1966.
121. G. Gunter, *J. Wildlife Management*, **16**, 63 (1952).
122. G. Gunter, *Contrib. Marine Sci.*, **12**, 230 (1967).
123. H. W. Wells, M. J. Wells, and I. E. Gray, *Ecology*, **42**, 217 (1961).

124. M. Brongersma-Sanders, *Geol. Soc. Am. Mem.*, **67**(1), 941 (1957).
125. S. F. Hildebrand and W. C. Schroeder, *Bull. U.S. Fish. Bur.*, **43**(1), 1 (1928).
126. F. M. Leccia, *Bull. Marine Sci.*, **15**, 274 (1965).
127. I. G. Innes, *Australian Fisheries*, Halstead Press, Sydney, 1950.
128. T. C. Roughley, *Fish and Fisheries of Australia*, Angus & Robertson, Sydney, 1966.
129. H. W. Wells, *Ecol. Monographs*, **31**, 239 (1961).
130. D. C. Tabb and R. B. Manning, *Bull. Marine Sci. Gulf Caribbean*, **11**, 552 (1961).
131. F. G. W. Knowles and D. B. Carlisle, *Biol. Rev. Cambridge Phil. Soc.*, **31**, 396 (1956).
132. W. A. Van Engel, *Commer. Fisheries Rev.*, **20**(6), 6 (1958).
133. J. A. Allen, *Oceanog. Marine Biol. Ann. Rev.*, **4**, 247 (1966).
134. R. V. Truitt, *Contrib. Chesapeake Biol. Lab.*, **27**, 1 (1939).
135. R. M. Darnell, *Publ. Inst. Marine Sci. Univ. Texas*, **5**, 353 (1958).
136. R. M. Darnell, *Trans. Am. Fisheries Soc.*, **88**, 294 (1959).
137. G. Gunter, *Publ. Inst. Marine Sci. Univ. Texas*, **1**(2), 7 (1950).
138. M. J. Lindner and W. W. Anderson, *U. S. Fish Wildlife Serv. Fish. Bull.*, **56**(106), 555 (1956).
139. J. W. Anderson and D. J. Reish, *Marine Biol.*, **1**, 56 (1967).
140. J. H. Carpenter and D. G. Cargo, *Chesapeake Bay Inst. Johns Hopkins Univ. Tech. Rept. 13*, 1957, p. 1.
141. O. Kinne, *Oceanog. Marine Biol. Ann. Rev.*, **2**, 281 (1964).
142. O. Kinne, *Publ. Am. Assoc. Advan. Sci.*, **83**, 525 (1967).
143. P. A. Dehnel, *Biol. Bull.*, **122**, 208 (1962).
144. G. Gunter, *Publ. Inst. Marine Sci. Univ. Texas*, **1**(1), 1 (1945).
145. T. L. Hopkins, *Chesapeake Sci.*, **6**, 86 (1965).
146. M. M. Mistakidis, *Gt. Brit. Fishery Invest. Ser. II*, **17**(6), 1 (1951).
147. W. N. Shaw, *Chesapeake Sci.*, **8**, 228 (1967).
148. R. L. Cory, *Chesapeake Sci.*, **8**, 71 (1967).
149. H. D. Hoese and R. S. Jones, *Publ. Inst. Marine Sci. Univ. Texas*, **9**, 37 (1963).
150. D. R. Moore, *Gulf Res. Rept. 1*, 1961, p. 1.
151. A. Dragovich and J. A. Kelly, Jr., *Proc. Gulf Caribbean Fisheries Inst.*, **15**, 87 (1963).
152. A. Dragovich and J. A. Kelly, Jr., *Bull. Marine Sci. Univ. Miami*, **17**, 840 (1967).
153. P. A. Butler, *U.S. Fish Wildlife Serv. Fishery Bull.*, **89**, 479 (1954).
154. C. E. Dawson, *Publ. Inst. Marine Sci. Univ. Texas*, **4**(1), 280 (1955).
155. B. J. Copeland and H. D. Hoese, *Publ. Inst. Marine Sci. Univ. Texas*, **11**, 149 (1966).
156. R. R. Hathaway and K. D. Woodburn, *Bull. Marine Sci. Gulf Caribbean*, **11**, 45 (1961).
157. M. R. Carriker, *U.S. Fish Wildlife Serv. Spec. Sci. Rept. Fisheries No. 148*, 1955, p. 1.
158. A. S. Pearse and G. Gunter, *Geol. Soc. Am. Mem.*, **67**(1), 129 (1957).
159. P. S. Galtsoff, *U.S. Fish Wildlife Serv. Fishery Bull.*, **64**, 1 (1964).
160. A. N. Sastry, *Biol. Bull.*, **125**, 146 (1963).
161. D. B. Quale, in *Marine Boring and Fouling Organisms* (D. L. Ray, ed.), Univ. Washington Press, Seattle, 1959, pp. 175–183.
162. W. T. W. Potts and G. Parry, *Osmotic and Ionic Regulation in Animals*, Macmillan, New York, 1964.
163. J. M. Knight, *J. Mississippi Acad. Sci.*, **12**, 208 (1966).
164. W. A. Chipman and J. G. Hopkins, *Biol. Bull.*, **107**, 80 (1954).
165. T. H. Carefoot, *J. Marine Biol. Assoc. U.K.*, **47**, 565 (1967).
166. C. E. Lane, in *Marine Boring and Fouling Organisms* (D. L. Ray, ed.), Univ. Washington Press, Seattle, 1959, pp. 137–156.
167. P. R. Walne, *J. Marine Biol. Assoc. U.K.*, **44**, 293 (1964).
168. J. G. Mackin, *Publ. Inst. Marine Sci. Univ. Texas*, **7**, 132 (1961).

169. W. Wieser, *Limnol. Oceanog.*, **5**, 121 (1960).
170. J. M. Cronan, Jr., *Auk*, **74**, 459 (1957).
171. F. C. Cleaver and D. M. Franett, *Washington State Dept. Fisheries Biol. Rept. 46B*, 1946, p. 1.
172. H. C. Oberholser, *Bull. Louisiana Dept. Conserv.*, **28**, 1 (1938).
173. T. C. Nelson, in *Public Health Serv. Tech. Rept. W60-3*, U.S. Dept. Health, Education, and Welfare, 1960, pp. 203–211.
174. P. S. Galtsoff, in *Public Health Serv. Tech. Rept. W60-3*, U.S. Dept. Health, Education, and Welfare, 1960, pp. 128–134.
175. V. J. Chapman, *Coastal Vegetation,* Pergamon, London, 1964.
176. R. F. Thorne, *U.S. Fish Wildlife Serv. Fishery Bull.*, **89**, 193 (1954).
177. E. Y. Dawson, *Marine Botany—An Introduction*, Holt, Rinehart and Winston, New York, 1966.
178. H. B. Moore, *Marine Ecology*, Wiley, New York, 1958.
179. J. H. Davis, Jr., *Carnegie Inst. Washington Papers Tortugas Lab.*, **32**, 303 (1940).
180. G. W. Lawson, *Oceanog. Marine Biol. Ann. Rev.*, **4**, 405 (1966).
181. W. Macnae, *Publ. Am. Assoc. Advan. Sci.*, **83**, 432 (1967).
182. E. P. Odum and A. A. de la Cruz, *Publ. Am. Assoc. Advan. Sci.*, **83**, 383 (1967).
183. G. E. Walsh, *Publ. Am. Assoc. Advan. Sci.*, **83**, 420 (1967).
184. J. D. Kilby, *Tulane Studies Zool.*, **2**(8), 176 (1949).
185. V. G. Springer and K. D. Woodburn, *Florida State Board Conserv. Profess. Paper*, **1**, 1 (1960).
186. C. S. Yentsch and J. H. Ryther, *J. Conseil Conseil Perm. Intern. Exploration Mer*, **24**, 232 (1959).
187. G. A. Riley, *Publ. Am. Assoc. Advan. Sci.*, **83**, 316 (1967).
188. R. Patrick, *Publ. Am. Assoc. Advan. Sci.*, **83**, 311 (1967) .
189. J. E. G. Raymont, *Advan. Ecol. Res.*, **3**, 117 (1966).
190. L. Provasoli, in *The Sea* (M. N. Hill, ed.), Vol. 2, Wiley (Interscience), New York, 1963, pp. 165–219.
191. H. P. Jeffries, *Publ. Am. Assoc. Advan. Sci.*, **83**, 500 (1967).
192. R. P. Cuzon du Rest, *Publ. Inst. Marine Sci. Univ. Texas*, **9**, 132 (1963).
193. C. Teixeira, *Bol. Inst. Oceanog. Univ. Sao Paulo*, **13**, 53 (1963).
194. C. Teixeira and M. B. Kutner, *Bol. Inst. Oceanog. Univ. Sao Paulo*, **12**, 101 (1963).
195. W. P. Abbott, *Pollution Studies on Galveston Bay*, technical report filed with Texas Parks and Wildlife Commission, October, 1963.
196. *Recommended Procedures for the Bacteriological Examination of Sea Water and Shellfish*, Am. Public Health Assoc., New York, 1962.
197. G. D. Floodgate, *J. Marine Biol. Assoc. U.K.*, **44**, 365 (1964).
198. A. E. Greenberg, *Public Health Rept. (U.S.)*, **71**, 77 (1956).
199. H. H. Carter, J. H. Carpenter, and R. C. Whaley, *J. Water Pollution Control Federation*, **39**, 1184 (1967).
200. W. J. Beck, *Proc. Natl. Shellfish Sanit. Workshop, 5th, Washington, D.C., 1964*, p. 143.
201. E. T. Jensen, ed., *Sanitation of Shellfish Growing Areas, Publ. No. 33*, U.S. Dept. Health, Education, and Welfare, Public Health Service, Washington, D.C., 1962, Part 1.
202. C. E. ZoBell, *Marine Microbiology*, Chronica Botanica, Waltham, Mass., 1946.
203. C. E. ZoBell, *Geol. Soc. Am. Mem.*, **67**(1), 1035 (1957).
204. C. H. Oppenheimer, *Symposium on Marine Microbiology*, Thomas, Springfield, Ill., 1963.

205. A. E. Kriss, *Marine Microbiology (Deep Sea)*, Oliver & Boyd, Edinburgh, 1963.
206. E. J. F. Wood, *Microbiology of Oceans and Estuaries*, Elsevier, Amsterdam, 1967.
207. C. H. Oppenheimer, ed., *Marine Biology IV*, New York Acad. Sci., 1968.
208. E. J. F. Wood, *Oceanog. Marine Biol. Ann. Rev.*, **1**, 197 (1963).
209. C. H. Oppenheimer, *Geochim. Cosmochim. Acta*, **19**, 2244 (1960).
210. C. H. Oppenheimer, *Z. Allgem. Mikrobiol.*, **5**, 284 (1965).
211. B. V. Perfiliev and D. R. Gabe, *Capillary Methods of Studying Microorganisms*, Academy of Sciences USSR, Moscow, 1961 (in Russian).
212. H. G. Thode, A. G. Harrison, and J. Monster, *Bull. Am. Assoc. Petrol. Geologists*, **44**, 1809 (1960).
213. S. I. Kuznetzov, M. V. Ivanov, and N. N. Lyalikova, *Introduction to Geological Microbiology* (C. H. Oppenheimer, ed.), McGraw-Hill, New York, 1963.
214. E. I. Rabinowitch, *Photosynthesis*, Vol. I, Wiley (Interscience), New York, 1945.
215. E. I. Rabinowitch, *Ann. Rev. Plant Physiol.*, **3**, 229 (1952).
216. W. Vishniac, *Ann. Rev. Plant Physiol.*, **6**, 115 (1955).
217. G. Forti, in *Primary Productivity in Aquatic Environments* (C. R. Goldman, ed.), Univ. California Press, Berkeley, 1966, pp. 17–35.
218. J. A. Bassham and M. Calvin, *The Path of Carbon in Photosynthesis*, Prentice-Hall, Englewood Cliffs, N.J., 1957.
219. H. Holzer, *Z. Naturforsch.*, **6B**, 424 (1951).
220. E. Steeman Nielsen, *J. Conseil Conseil Perm. Intern. Exploration Mer*, **18**, 117 (1952).
221. K. Park, D. W. Hood, and H. T. Odum, *Publ. Inst. Marine Sci. Univ. Texas*, **5**, 47 (1958).
222. R. J. Beyers, *Ecol. Monographs*, **33**, 281 (1963).
223. R. J. Beyers, in *Primary Productivity in Aquatic Environments* (C. R. Goldman, ed.), Univ. California Press, Berkeley, 1966, pp. 61–74.
224. R. J. Beyers and H. T. Odum, *Limnol. Oceanog.*, **4**, 501 (1959).
225. R. J. Beyers and H. T. Odum, *Limnol. Oceanog.*, **5**, 228 (1960).
226. R. J. Beyers, J. L. Larimer, H. T. Odum, R. B. Parker, and N. E. Armstrong, *Publ. Inst. Marine Sci. Univ. Texas*, **9**, 454 (1963).
227. J. Verduin, *Limnol. Oceanog.*, **5**, 228 (1960).
228. J. Verduin, *Limnol. Oceanog.*, **6**, 82 (1961).
229. J. Lyman, *Limnol. Oceanog.*, **6**, 80 (1961).
230. E. P. Odum, *Fundamentals of Ecology*, W. B. Saunders, Philadelphia, 1953, p. 86.
231. G. V. Levin and J. Shapiro, *J. Water Pollution Control Federation*, **37**, 800 (1965).
232. N. T. de Saussure, *Recherches Chimiques sur la Végétation*, Nyon, Paris, 1804.
233. H. V. Marsh, Jr., J. M. Galmiche, and M. Gibbs, in *Biochemical Dimensions of Photosynthesis* (D. W. Krogmann and W. H. Powers, eds.), Wayne State Univ. Press, Detroit, 1965, pp. 95–107.
234. O. Holm-Hansen, in *Physiology and Biochemistry of Algae* (R. A. Lewin, ed.), Academic, New York, 1962, Chap. 2.
235. M. Calvin, *Proc. Federation Am. Soc. Exptl. Bio.*, **13**, 697 (1954).
236. R. M. Cassie, *U.S. Atomic Energy Comm. TID-7633*, 1963, pp. 163–171.
237. J. M. Teal, *Publ. Am. Assoc. Advan. Sci.*, **83**, 336 (1967).
238. J. H. Ryther, *Science*, **130**, 602 (1959).
239. G. A. Riley, *Bull. Bingham Oceanog. Collection*, **15**, 324 (1956).
240. T. R. Hellier, Jr., *Publ. Inst. Marine Sci. Univ. Texas*, **8**, 1 (1962).
241. T. R. Hellier, Jr., *Publ. Inst. Marine Sci. Univ. Texas*, **5**, 165 (1959).
242. C. R. Goldman, ed., *Primary Productivity in Aquatic Environments*, Univ. California Press, Berkeley, 1966.

243. *Yearbook of Fishery Statistics*, Vol. 18, Food Agr. Organ. United Nations, Rome, 1964.
244. C. H. Lyles, *U.S. Bur. Com. Fisheries Statistical Dig.*, **59**, 1 (1967).
245. J. L. McHugh, *Am. Fisheries Soc. Spec. Publ. 3*, 1966, p. 133.
246. J. E. Webb and M. B. Hill, *Phil. Trans. Roy. Soc. (London), Ser. B*, **241**, 355 (1958).
247. J. E. Webb, *Phil. Trans. Roy. Soc. (London), Ser. B*, **241**, 335 (1958).
248. C. E. Dawson, *Copeia*, **1965**, 505 (1965).
249. G. Gunter, *Texas A & M Res. Found. Publ.*, **2**, 1 (1949).
250. G. Gunter, *Publ. Inst. Marine Sci. Univ. Texas*, **2**(2), 119 (1952).
251. G. Gunter and H. Hildebrand, *Bull. Marine Sci. Gulf Caribbean*, **4**, 95 (1954).
252. J. H. Kutkuhn, *Am. Fisheries Soc. Spec. Publ. 3*, 1966, p. 16.
253. G. Gunter, *Geol. Soc. Am. Mem.*, **67**(1), 159 (1957).
254. E. G. Simmons, *Publ. Inst. Marine Sci. Univ. Texas*, **4**(2), 156 (1957).
255. E. H. Phillips, *Rice Inst. Pam.*, **41**(4), 1 (1955).
256. A. Calabrese and H. C. Davis, *Biol. Bull.*, **131**, 427 (1966).
257. D. E. Wohlschlag and J. E. Cameron, *Contrib. Marine Sci.*, **12**, 160 (1967).
258. *SFI Bull. No. 191*, The Sport Fishing Institute, Washington, D.C., January-February, 1968.
259. F. E. Clarke, *Am. Soc. Testing Mater. Spec. Tech. Publ.*, **416**, 100 (1967).
260. D. Uhlmann, *Verhandl. Intern. Ver. Limnol.*, **16**, 934 (1966).
261. R. T. Oglesby and W. T. Edmondson, *J. Water Pollution Control Federation*, **38**, 1452 (1966).
262. C. C. Davis, *Limnol. Oceanog.*, **9**, 275 (1964).
263. C. N. Sawyer, *J. Water Pollution Control Federation*, **37**, 1122 (1965).
264. R. A. Ragotzkie, *Publ. Inst. Marine Sci. Univ. Texas*, **6**, 146 (1959).
265. R. Patrick, Proc. Annual Industrial Waste Conference, 21st, Purdue University, May, 1966.
266. W. N. Torpey, *J. Water Pollution Control Federation*, **39**, 1797 (1967).
267. L. E. Cronin, *Publ. Am. Assoc. Advan. Sci.*, **83**, 667 (1967).
268. P. Doudoroff and M. Katz, *Sewage Ind. Wastes*, **22**, 1432 (1950).
269. P. Doudoroff and M. Katz, *Sewage Ind. Wastes*, **25**, 802 (1953).
270. C. H. Hoffman, in *Public Health Serv. Publ. 999-WP-25*, U.S. Dept. Health, Education, and Welfare, Cincinnati, 1965, pp. 253–260.
271. H. P. Nicholson, in *Public Health Serv. Publ. 999-WP-25*, U.S. Dept. Health, Education, and Welfare, Cincinnati, 1965, pp. 260–262.
272. R. W. Risebrough, D. B. Menzel, D. J. Martin, Jr., and H. S. Olcott, *Nature*, **216**, 589 (1967).
273. D. E. Ferguson, D. D. Culley, W. D. Cotton, and R. P. Dodds, *BioScience*, **14**(11), 43 (1964).
274. C. E. Boyd and D. E. Ferguson, *Mosquito News*, **24**(1), 19 (1964).
275. S. B. Vinson, C. E. Boyd, and D. E. Ferguson, *Science*, **139**, 217 (1963).
276. P. A. Butler and P. F. Springer, *Trans. N. Am. Wildlife Nat. Resources Conf., 28th, Detroit, 1963*, p. 378.
277. P. A. Butler, *J. Appl. Ecol.*, **3** (Suppl.), 253 (1966).
278. P. A. Butler, *Am. Fisheries Soc. Spec. Publ. 3*, 1966, p. 110.
279. E. Naylor, *Advan. Marine Biol.*, **3**, 63 (1965).
280. F. F. Ross, *Inst. Sewage Purif. J. Proc.*, **1**, 16 (1959).
281. B. Nelson, *Science*, **158**, 755 (1967).
282. D. L. Hetrick and R. L. Seale, *Science*, **159**, 376 (1968).
283. *Oceanology International*, **3**(1), 11 (1968).

284. D. L. Hetrick, Univ. Arizona, private communication, 1968.
285. J. Cairns, Jr., *Ind. Wastes*, **1**, 150 (1956).
286. M. W. Busby, U.S. Geol. Surv., Albany, N.Y., private communication, 1968.
287. J. Mauchline and W. L. Templeton, *Oceanog. Marine Biol. Ann. Rev.*, **2**, 229 (1964).
288. W. A. Chipman, in *Public Health Serv. Tech. Rept. W60-3* (C. M. Tarzwell, ed.), U.S. Dept. Health, Education, and Welfare, 1960, p. 8.
289. L. R. Donaldson, in *Public Health Serv. Tech. Rept. W60-3* (C. M. Tarzwell, ed.), U.S. Dept. Health, Education, and Welfare, 1960, p. 1.
290. H. T. Odum and E. P. Odum, *Ecol. Monographs*, **25**, 291 (1955).
291. W. L. Templeton, in *Disposal of Radioactive Wastes into Seas, Oceans and Surface Waters*, International Atomic Energy Agency, Vienna, 1966, pp. 847–859.
292. *Disposal of Radioactive Wastes into Seas, Oceans and Surface Waters*, International Atomic Energy Agency, Vienna, 1966.
293. B. Åberg and F. P. Hungate, eds., *Radioecological Concentration Processes*, Pergamon, Oxford, 1967.
294. H. Caspers, *Geol. Soc. Am. Mem.*, **67**(1), 801 (1957).
295. L. Zenkevitch, *Biology of the Seas of the U.S.S.R.*, George Allen & Unwin, London, 1963.
296. S. A. Zernov, *General Hydrobiology*, Akad. Sci., Moscow, 1949 (in Russian).
297. H. T. Odum, W. McConnell, and W. Abbott, *Publ. Inst. Marine Sci. Univ. Texas*, **5**, 65 (1958).
298. B. J. Copeland and R. S. Jones, *Texas J. Sci.*, **17**, 188 (1965).
299. B. J. Copeland (ed.), *Contrib. Marine Sci.*, **12**, 202 (1967).

Chapter 3 Chemical, Physical, and Biological Characteristics of Irrigation and Soil Water

J. D. Rhoades and Leon Bernstein
U. S. SALINITY LABORATORY
U. S. DEPARTMENT OF AGRICULTURE
RIVERSIDE, CALIFORNIA

I. Introduction

Any discussion of the properties of irrigation and soil water must reflect to some degree the interests and specialization of the authors. To soil and plant scientists, the water properties of paramount interest are those that modify plant growth. To public health, industrial, and recreational water experts, other water properties are paramount. An inclusive discussion of water properties is attempted herein, however, even though some aspects, of necessity, receive less than thorough treatment. The broad approach is justified because agricultural processes affect water properties of concern to all classes of water users, and these effects must at least be recognized.

Soil waters may derive primarily from irrigation waters in arid zones, but in humid zones rainfall is the main, and usually the sole, source. If we discussed only the soil waters of arid zones, we would fail to consider some of the effects that are pronounced in areas of high rainfall. Again, we choose the broader approach and discuss humid zone soil waters as well as those of the arid zones.

We deal with a portion of the hydrologic cycle that involves no phase change insofar as the residual water under investigation is concerned. The penetration of rainwater or irrigation water into the soil results in soil–water interactions that involve not only the chemistry, physics, and biology of the soil but also the effects of the aerial environment, both soil air and atmospheric air. Although the water we study is liquid water, its properties will be appreciably modified by evaporation and transpiration that convert much of the liquid soil water into vapor and release it into the atmosphere. This evapotranspirational loss of water is one of the main processes that affect the properties of soil water and that cause the product groundwater, or eventual surface waters originating from groundwaters, to differ from the water originally falling on or applied to the soil.

Total salinity and the specific ion content of irrigation waters are the primary characteristics that concern the agriculturist, since these properties directly affect crop growth and yields. When irrigation waters are reduced in volume by evapotranspiration, sparingly soluble salts present in the waters tend to precipitate. At the same time, soil minerals are being weathered and are releasing soluble salts. Soil mineral surfaces are charged, so that ions are adsorbed. As penetrating waters equilibrate with the soil, an exchange of ions between the water and the soil can occur, so that specific ion composition of the soil water may change as the water moves through the soil. Salts are also added to soils as fertilizers and soil amendments, and these are solubilized to varying extents and may then enter into the exchange and precipitation reactions. Finally, other chemicals added to control plant pests and diseases may dissolve in the soil water and modify the water properties. Pesticide and fertilizer solubilization is usually of little concern to the agriculturist so long as premature loss of desired activity does not occur, but it may be of great concern to other water users.

Thus, the water penetrating the soil is modified in its solute content by evapotranspirational loss of part of the water, by various precipitation and soil solubilization reactions, and by the introduction of fertilizers and soil amendments and pest-regulating chemicals. But not all the water that impacts the soil necessarily penetrates. Under conditions of high intensity rainfall and some poorly managed irrigation enterprises, some water runs off the surface, carrying suspended soil particles into rivers and streams. Such surface runoff with its suspended solids may contain chemicals that ordinarily would not be present in percolating water because of relative insolubility, adsorption on the soil, or precipitation. These may include some fertilizer as well as pesticide materials. Surface runoff occurs primarily in high intensity rainfall areas, limiting certain water quality problems to these areas.

Soils contain diverse organisms that affect the characteristics of soil water by altering inorganic or organic materials in the soil. In a well-aerated soil, the reactions are primarily oxidative. When the soil is water-saturated or flooded, reduction reactions may occur and result in the production of volatile constituents such as nitrogen or hydrogen sulfide. Organic materials may be metabolically degraded by soil organisms; the resistance of some pesticides to soil microbial action is currently receiving much attention. Soil, because of its finely divided state and large surface area, is an excellent adsorptive and filtering medium. Percolating waters containing suspended solids, pathogenic organisms, and some solutes are purified by passage through the soil. Thus, improvement as well as degradation of water quality may result from passage through soil. This chapter discusses irrigation water and soil water properties as well as the processes that ultimately determine the properties of product groundwaters and surface waters.

II. Properties of Waters Involved in or Affected by Agriculture

A. Sampling and Analytical Techniques for Irrigation, Soil, and Drainage Waters

Since the techniques commonly employed in the sampling(*1,2*) and analysis(*3,4*) of irrigation and drainage waters are the same as those described for surface waters and groundwaters (see Chaps. 1, 10, and 19) no further discussion is required. The analysis of soil water also employs the same methods, but sampling of soil waters presents special problems and requires special techniques.

Soil solutions are extremely heterogeneous. On a microscale, the water close to the negatively charged adsorbing soil surfaces contains higher concentrations of cations and lower concentrations of anions than the water at a greater distance from the surfaces. The degree to which soil water sampling removes the closely sorbed water layers will influence the composition of the soil solution obtained. Soil water is also discontinuous, and the water in small pores may not have the same composition as the water in adjacent larger pores. Soils are often heterogeneous even within relatively small areas. This may result from localized differences in soil materials or soil texture. Percolation rates and, therefore, soil water concentrations may vary because of localized compaction, clay lenses, or similar structural features. Soil solutions will vary not only with depth and location in the field but also with time because rainfall, fertilizer applications, evaporation, absorption of salts by plants, salt precipitation, and solubilization will continually interplay to prevent the development of a stable soil solution. Spatial variations in soil solution composition require care in sampling if representative samples are to be obtained, and microvariations in soil solutions affect the choice of methods for extracting the soil water from the sample.

The number and location of sampling sites for analysis of soil water depend on the objectives of the study. Rarely is interest limited to the mean composition of the soil solution; the range in composition is of at least equal importance. Since soil solution composition will usually vary with depth, separate samples for each distinguishable soil horizon should be obtained. In soils lacking distinct profile development, sampling by depth is still desirable because changes in soil solution will occur as a result of cumulative water extraction by crop roots in a succession of soil depths. Samples are therefore frequently collected from a series of depths, such as 0–15, 15–45, 45–90, and 90–180 cm. Because of the lateral heterogeneity of soils, several separate soil cores from a given site are usually composited and mixed by depth increments to obtain representative samples for the site. The number of sampling sites within the general area of study will be determined by the study's areal extent, the major features of soil heterogeneity as determined by examination of soil profiles and crop responses, and the specific objectives of the study. One generally requires several composite samples to characterize a given soil situation in order to establish the variation as well as the average condition within the area, and to permit comparisons with other areas or conditions in the study (*5*).

Soil solutions are held by attractive forces of the soil which must be overcome to remove the soil water from the soil. Dilution, extraction by pressure or suction, displacement, absorption, and in situ sensing are some of the methods used to obtain soil water or to measure some property from which soil water concentration or composition may be inferred.

Most soil water data have been obtained by extracting a unit weight of dry soil with a fixed ratio of water; 1:1, 1:2, 1:5 soil:water ratios have commonly been employed. The soil and water are mixed, and since the mixture usually represents at least several times the saturation value of the soil, the solution may be readily filtered off. However, this greater-than-normal dilution of the soil water changes its composition and concentration. The higher the ratio of water to soil, the greater the total quantity of extractable solutes, especially in arid zone soils(*6*). Such soils often contain sparingly soluble salts (gypsum and calcium carbonate, especially), some of which are brought into solution as the ratio of water to soil is increased. Composition also changes as soluble : adsorbed cation ratios change with increasing dilution(*7–9*). The proportion of monovalent cations held on adsorption sites is decreased while that of divalent cations is increased with dilution. Anion adsorption is also affected by dilution(*10–12*). Although water extracts are made more readily the greater the dilution, errors in estimating soil solution solutes on the basis of the extracts increase.

The saturation extract technique adapted and popularized by the U.S. Salinity Laboratory(*2*) is one of the most frequently used extraction methods. In this procedure, distilled water is stirred into the soil sample until the soil is saturated, i.e., to the point at which further addition of water would result in the separation of water from the soil paste. After standing at least 4 hr, the paste is filtered by suction to obtain the soil saturation extract. Because the saturation percentage of soils is generally well under 100%, the soil:water ratios for the saturated pastes are considerably larger than for the fixed ratio extracts discussed above, and dilution errors are, therefore, smaller. The saturation extract represents more nearly a constant dilution of the soil water retained by soils in a field moisture range, since the saturation percentage for most soils approximates twice the water content at field capacity and four times that at the permanent wilting percentage(see Sec. II. D. 3). When extraction of the undiluted soil water itself is not feasible, the relatively constant and minimal dilution of the soil water achieved by saturating the soil represents the next best method.

Various techniques have been used to displace soil water from soils in the field moisture range. Although these techniques most nearly approach the goal of obtaining unaltered soil water, they are generally too time-consuming for routine soil water analysis. Displacement of the soil water by other liquids or gases (if the soil is saturated) was popular some 30 years ago(*13,14*) and is still occasionally used. By this method, the soil water is displaced from the larger pores only and may not have the same composition as that in the smaller pores. Soil water may be removed by compaction, but the pressures of thousands of pounds per square inch affect the solubility of certain carbonate minerals(*15*). Centrifugation, like compaction, reduces the volume of the soil mass, allowing separation of the soil water; but adequate centrifugation procedures for soils have not been developed(*15*).

Molecular absorption of the soil water in situ may be accomplished by placing dry filter paper(*16*) or ceramic points(*17*) in the soil. Solutes are then extracted from the absorbent and analyzed, but highly sensitive techniques are required because of the small amounts of soil solution extracted. Vapor transfer and chromatographic separation of solutes affect the accuracy of this method.

Samples of the soil water may be obtained by suction applied to a porous ceramic filter candle or any other membrane material that is permeable to water but not to air at reduced pressures(*18,19*). Such filter candles may be installed in the field, and repeated samples of the soil water may be extracted without removing soil samples. Although the soil water extracted by this method is that held at less than 1-bar suction, this is not too serious a limitation since most of the water movement in soils occurs in the 0 to 1-bar range, even though additional water may be extracted by plants up to suctions of 15 bars or greater. Some changes in the soil solution may occur because of adsorption of ions by the ceramic candle and also because of changes in partial pressure of carbon dioxide during extraction. Despite these effects, extraction by suction appears to be one of the most useful and valid methods for obtaining soil water samples.

Soil water samples may be obtained over a wider range of soil water contents by means of a pressure membrane apparatus(*20,21*). The cellulose membranes used in this apparatus, when wet and properly supported, withstand high gas pressures, allowing the soil water from the soil samples to pass through the membrane but preventing the passage of gas. The soil water may be expressed from soil samples over the whole range of available water contents up to 15 bars of pressure, and the effect of decreasing water content on the composition of the soil water can be determined. Limitations of the method include the special equipment required; the slowness of achieving equilibration, especially at the higher pressures; the effects of high pressure and carbon dioxide buildup from microbial respiration on the solubility of alkaline earth carbonates; and the retention of phosphate by the membranes. Negative anion adsorption and salt-sieving also affect the composition of the soil water extract. Because of anion exclusion by the repulsive forces of the negatively charged clay colloids, salt concentration in the water close to the colloidal surfaces is lower than in the water at greater distances. The salt concentration, therefore, decreases as water film thickness is reduced(*21,22*). Salt-sieving(*23*) reduces salt concentrations in extracts because salt does not move freely in the water films which are under the constraint of electrostatic forces. The magnitude of the negative adsorption and salt-sieving effects depends on the nature of the soil minerals, the proportion of mono- and divalent cations, the soil solution concentration, and the soil water content or film thickness. When soil solution concentration is greater than 10 meq/liter and extracting pressures are less than 5 bars, the error for most agricultural soils should be less than 10%. Eaton et al. *(24)* have proposed a modified pressure membrane method, which removes water held at a 0.1-bar tension, for routine analyses.

In situ salinity sensors employing ceramic electrical conductivity cells have recently been developed for monitoring the concentrations of soil solutions *(25–27)*. The composition of the soil solution is not determined, but variations in concentrations of soil waters with time can be readily followed. Such information is basic for many problems involving soil salinity and its control. Although still experimental, the salinity sensor has great promise.

B. *Rainwater Properties*

Rainwater properties are a factor when considering the characteristics of irrigation and soil waters (also see Chap. 1 for a section on rainwater).

1. Solutes in Rainwater

Rainfall effectively washes out and concentrates a great variety of inorganic and organic solutes found in trace amounts in the atmosphere. The average total salt content of rainwater is about 11 mg/liter and consists mainly of Na^+, Ca^{2+}, Mg^{2+}, K^+, NH_4^+, Cl^-, SO_4^{2-}, HCO_3^-, and NO_3^- *(28)*. Organic materials, although not studied in as much detail, appear to be present in rainwater in about the same total amounts as the inorganics *(29,30)*. Pesticides have also been found in rainwater *(31,32)*. Rainwater composition varies with locality, especially with distance from the sea. As oceanic air masses move inland, atmospheric salinity tends to decrease exponentially until, at some distance inland, a plateau of minimal concentration is reached. Depending on topography and rainfall pattern, the plateau concentrations for chloride in rainwater vary from about 0.2 mg/liter over the United States at more than 600 km from the coast to very nearly the coastal level of 8 mg/liter over England.

2. Radionuclides in Rainwater

Radionuclides in the atmosphere and rainwater arise from emanation of natural gaseous radionuclides from the earth's crust, from cosmic ray bombardment of the outer atmosphere, and from man's activity in producing artificial radionuclides. The amounts of emanation radionuclides found in rainfall vary, being higher over land than over sea, and higher in the northern hemisphere. The most abundant emanation radionuclides in rainwater are ^{210}Pb, ^{210}Bi, and ^{210}Po, contributing several picocuries per liter each(*33*). Cosmic radiation constantly produces radionuclides in the atmosphere by nuclear reactions with atmospheric gases. Among the most important cosmic radiation radionuclides are ^{7}Be, ^{10}Be, ^{14}C, ^{32}Si, ^{3}H, ^{22}Na, ^{35}S, ^{33}P, and ^{32}P. These are present in rainwater to the extent of 0.1–6 pCi/liter each(*34,35*). Nuclear bomb testing produces a variety of radionuclides. Most of these are unstable and pass through decay chains before becoming long-lived stable isotopes(*36*). The important long-lived radionuclides produced artificially are ^{131}Cs (28.8 years), ^{90}Sr (27.7 years), ^{144}Ce (285 days), ^{95}Zr (65 days), ^{89}Sr (51 days), and ^{140}Ba (12.8 days). Cesium-137 and strontium-90 are probably of greatest concern in rainwater because of their stability, their predominance among the artificial radionuclides(*37*), and their similarity to the biologically important ions, K^+ and Ca^{2+}(*38*), respectively. The rates of sedimentation, washout, and fallout of radioactivity depend on the size and the type (surface vs. atmosphere) of bomb tested. The radionuclides in rainwater will therefore vary with the test, sampling site, year, season, etc. Representative values before testing have ranged from a few to 20 pCi/liter per radionuclide, increasing after testing to 30–400 pCi/liter per radionuclide(*39–44*) (see Chap. 25 regarding analysis for radionuclides).

C. *Irrigation Water Properties*

1. General Considerations

Properties of surface waters and groundwaters are described in Chap. 1. Since irrigation waters are just those surface waters and groundwaters that are used to irrigate farmland, they would have no distinctive properties except for two features. Irrigation is commonly practiced in semiarid and arid zones, and waters in these climatic zones are often more saline than those of humid zones. Also, some river waters and, especially, groundwaters may not be usable for irrigation because of excessive salinity or specific element toxicity. Irrigation waters may therefore be considered as the groundwaters and surface waters primarily of semiarid and arid zones selected for their use potential for crops and soils. The present summary of irrigation water properties will therefore focus on those properties of surface waters and groundwaters of special importance for irrigation agriculture, as well as on those which are especially influenced by irrigation agriculture.

To facilitate discussion of irrigation water properties, some concepts and indices need to be defined. The percentage of negatively charged sites in a soil that is occupied by Na^+ strongly influences the physical properties and plant nutritional capacities of the soil. When a soil is in exchange equilibrium with an irrigation water, its percentage of exchangeable sodium will be closely related to the calculated sodium adsorption ratio (SAR) of the water (*2*):

$$\text{SAR} = \frac{\text{Na}^+}{\sqrt{(\text{Ca}^{2+} + \text{Mg}^{2+})/2}} \tag{1}$$

The ionic symbols indicate concentrations of the ions in the irrigation water in milliequivalents per liter. Calculation of the SAR for a given water provides a useful index of the sodium hazard of that water for soils and crops. The greater the SAR of a water, the greater is the sodium hazard.

The sodium hazard is also increased by the tendency for irrigation waters to lose calcium by precipitation of calcium carbonate in the soil. The pH_c of the water has been used as an index for this tendency (see Sec. II.D.2.b for discussion):

$$pH_c = (pK_2' - pK_s') + p\text{Ca}^{2+} + p\text{Alk} \tag{2}$$

pH_c is defined as the pH that an irrigation water with a given Ca^{2+} and HCO_3^- concentration would have when in equilibrium with solid calcium carbonate and atmospheric carbon dioxide. It is calculated by Eq. (2) from the negative logarithms of the second dissociation constant of H_2CO_3 (pK_2'), the solubility product constant of calcium carbonate (pK_s'), the molar concentration of calcium (pCa^{2+}), and the equivalent concentration of CO_3^{2-} plus HCO_3^- (pAlk). Values of pH_c below 8.3 indicate an increasing tendency for precipitation of calcium carbonate from the water, and values above 8.3 indicate increasing tendencies for the water to dissolve any solid calcium carbonate present.

2. Major Inorganic Solutes in Irrigation Waters

The major chemical constituents of surface waters have been studied extensively (*45–50*). Compositions of representative river waters used for irrigation and of rivers in general (*51*) are given in Table 1. The total solute content for the rivers of the world (weighted mean average) is 120 mg/liter, but the range in concentrations is wide. Most rivers contain between 50 and 1000 mg/liter, but extreme values of 13 and 9200 mg/liter have been observed (*52*). One would expect rivers used for irrigation to contain higher solute contents than rivers in general because the former commonly drain arid regions that have greater sources of inorganic solutes than do humid zone regions. This is generally true, but some rivers in irrigated areas of the Northwest and California are quite low in salinity.

TABLE 1
Composition of Representative Irrigation Waters and Average River Waters of the World

Water	EC[a], μmho/cm at 25°C	Total concentration mg/liter	Total concentration meq/liter	B, mg/liter	Ca	Mg	Na	K	CO_3 + HCO_3	SO_4	Cl	NO_3	SAR	pH_c	References
					meq/liter										
Individual rivers															
San Joaquin, Biola, Calif.	60	46	0.52	0.10	0.23	0.08	0.19	0.02	0.36	0.06	0.09	0.02	0.5	9.5	
Rogue, Grants Pass, Ore.	82	74	0.82	0.03	0.39	0.23	0.20	—	0.79	0.04	0.06	0.01	0.4	8.9	
Sacramento, Knights Landing, Calif.	148	99	1.42	0.10	0.65	0.41	0.34	0.02	1.10	0.14	0.17	0.01	0.5	8.6	
Snake, Heise, Idaho	371	218	3.85	0.05	2.34	0.98	0.48	0.05	2.69	0.90	0.31	0.02	0.4	7.7	
Missouri, Nebraska City, Neb.	694	456	7.26	0.13	2.89	1.49	2.74	0.14	3.05	3.75	0.51	0.01	1.8	7.6	
Colorado, Lees Ferry, Ariz.	864	547	8.80	0.11	3.84	1.72	3.13	0.11	2.72	4.37	1.72	0.02	1.9	7.6	(46)
Arkansas, Ralston, Okla.	1260	724	11.75	—	2.74	0.94	8.07	—	1.77	1.98	7.95	0.05	5.9	7.9	
Salt, below Stewart Mt. Dam, Ariz.	1520	832	14.04	0.27	2.94	1.15	9.79	0.16	2.80	1.23	10.01	0.01	6.8	7.7	
Sevier, Lynndyl, Utah	1950	1190	20.37	0.31	3.76	5.80	10.66	0.15	5.44	6.23	8.46	0.14	4.9	7.4	
Pecos, Shumla, Tex.	3160	2000	31.48	0.22	7.18	5.51	18.79	—	2.54	8.59	20.57	0.03	7.4	7.4	
Gila, Gellespie Dam, Ariz.	8160	5620	87.02	2.60	14.87	10.94	60.90	0.31	3.18	27.90	56.96	0.34	17.0	7.2	
Average rivers															
North America	220	142	1.89	—	1.05	0.41	0.39	0.04	1.11	0.42	0.23	0.02	0.5	8.4	
Europe	270	182	2.28	—	1.55	0.46	0.23	0.04	1.56	0.50	0.19	0.06	0.2	8.1	
Australia	95	59	0.58	—	0.19	0.22	0.13	0.04	0.52	0.05	0.28	trace	0.3	9.4	(51)
World	190	120	1.42	—	0.75	0.34	0.27	0.06	0.96	0.23	0.22	0.02	0.4	8.9	

[a]Electrical conductivity.

Specific solutes in river waters also vary widely, but some general trends are evident. The proportions of sodium and chloride tend to increase as the total concentration increases, whereas the proportions of calcium and bicarbonate tend to decrease and those of magnesium and sulfate remain fairly constant. SAR tends to increase and pH_c tends to decrease with increasing total concentration. The suitability of rivers for irrigation purposes, therefore, deteriorates with increasing solute content, not only because of the greater salt burden but also because of the increased sodium hazard associated with changes of SAR and pH_c. These hazards are discussed in more detail in Sec. III.A.4.

Well waters in general are more saline than surface waters and also contain higher proportions of sodium, boron, and nitrate (*49,53*). According to Florea (*54*), groundwaters with soluble salt concentrations of less than 500 mg/liter are rare and those of 2000–8000 mg/liter are common, with reported concentrations of up to 84,000 mg/liter. Occasionally, waters containing several thousand milligrams per liter of soluble salts have been used for irrigation (*2*), but generally irrigation waters contain less than 1000 mg/liter, with nearly half of the irrigation waters (both surface and well waters) in the western United States containing between 175 and 500 mg/liter (*48*).

Rivers are complex, dynamic systems. Livingstone (*51*) has described the variations in composition of river waters with respect to time, location, and depth, among other factors. Since this subject is discussed in Chap. 1, only some aspects relating to irrigation agriculture are mentioned here. The effect of irrigation diversions and drainage returns on the quality of a river flowing through arid zone land is illustrated by the data for the Rio Grande between Otowi Bridge, near Santa Fe, New Mexico, and Fort Quitman, Texas (Table 2). Salinity, boron, and the proportions of Na^+ and Cl^- increase markedly, whereas the proportions of Ca^{2+} and HCO_3^- decrease with distance down this river. The proportions of Mg^{2+} and SO_4^{2-} remain nearly constant. This is the same general trend already noted for world rivers with increasing concentration. However, the relative magnitudes of these trends are often greater along rivers used for irrigation than for those that are not.

Minor elements in irrigation waters are seldom present in sufficient amounts to affect the chemistry of the soil; they are important, however, from the standpoint of plant nutrition because several are essential to plant growth or are toxic to plants or foraging animals if present in excessive amounts. Minor element concentrations in large North American rivers are given in Chaps. 1 and elsewhere (*50,55–60*). For a discussion of the chemistry of these minor elements in natural waters and the ranges expected for different kinds of waters, including groundwaters, see

TABLE 2

Changes in Irrigation Water Composition with Distance Downstream, Rio Grande River[a]

Distance and location downstream	EC^b, μmho/cm at 25°C	Total concentration mg/liter	Total concentration meq/liter	B, mg/liter	Ca	Mg	Na	K	CO_3 + HCO_3	SO_4	Cl	NO_3	SAR	pH_c
					meq/liter									
0 mile (Otowi Bridge)	280	191	2.85	0.03	1.67	0.49	0.69	—	1.72	0.97	0.14	trace	0.7	8.0
184 miles (San Marcial)	567	375	5.66	0.10	2.56	0.80	2.30	—	2.48	2.33	0.89	0.02	1.8	7.7
240 miles (Elephant Butte)	662	434	6.64	0.11	2.90	0.98	2.76	—	2.66	3.10	0.99	trace	2.0	7.7
268 miles (Caballo Dam)	740	478	7.41	0.11	3.13	1.08	3.20	—	2.89	3.24	1.40	0.01	2.2	7.6
318 miles (Leasburg Dam)	778	507	7.79	0.12	3.23	1.13	3.43	—	2.84	3.53	1.53	0.01	2.3	7.6
375 miles (El Paso)	1210	801	12.34	0.18	4.46	1.55	6.33	—	3.80	5.29	3.43	0.01	3.7	7.4
420 miles (County Line)	2880	1868	29.28	0.32	8.62	2.96	17.70	—	4.40	9.51	15.60	0.02	7.4	7.2
456 miles (Fort Quitman)	3200	2066	32.42	0.33	8.73	3.67	20.02	—	4.11	10.34	18.17	0.01	8.0	7.2

[a] Adapted from Ref. *45*.
[b] Electrical conductivity.

Hem(*47*) (also see Chap. 24). Other than boron, these minor elements are not included in most considerations of irrigation water quality. Specific data for southwest U.S. rivers indicate no greater content of these elements than found in river water in general, with the exception of molybdenum and lead. Selenium content of irrigation water has not been generally studied in surface waters, although 0–80 μg/liter in Colorado River water and 5–55 μg/liter in the Gunnison River, Colorado, have been reported(*56*). Groundwaters in certain areas may be some 10-fold higher in selenium content(*57*), causing selenium toxicity to plants and animals(*56*).

3. Sediments

Sediments may clog intake screens and pumps, fill channels, reduce reservoir capacity and useful life, deposit soil materials on crop lands or leaf surfaces, clog groundwater recharge facilities, and act as carriers for heavy metals, radioactive constituents, and pesticides. However, sediments in irrigation waters are seldom considered in evaluating water quality because they generally are very low or are removed before the water is used.

For the United States as a whole, low sediment concentrations tend to coincide with low dissolved solute concentrations(*61*). For half of the United States, the concentration of dissolved solutes is less than 230 mg/liter, and the discharge-weighted sediment content is less than 600 mg/liter. In 90% of the country, the prevalent concentrations are less than 900 and 8000 mg/liter, respectively.

Usually, the sediment concentration is greater during high flow than during low flow, whereas dissolved solute concentrations are usually lower during high flows. Concentrations of suspended sediments and dissolved solutes tend to be rather low in areas of high rainfall and high in areas of low rainfall(*61,62*). In humid regions, heavier vegetative cover protects soil from erosion; in arid regions, vegetative cover is sparse and occasional intense rainstorms cause severe erosion, resulting in high levels of suspended sediments. The amounts and kinds of sediments depend on the geology and topography of the drainage areas, as well as on rainfall and vegetative cover. Reservoirs tend to reduce variations in discharge and sediment contents. For a detailed treatise on factors influencing land erosion and its control, and on sediments in streams, estuaries, harbors, and reservoirs, see Ref. *63*; for an annotated bibliography on hydrology and sedimentation of the United States and Canada, see Ref. *64*. Kennedy(*62*) has published a detailed treatise on the mineralogy and cation exchange capacity of sediments in rivers.

4. Pesticides

Water pollution by pesticides probably began about 22 years ago when organic pesticides were introduced just after World War II. These materials replaced the previously used inorganic pesticide compounds because the new organics were effective at lower dosages and were supposedly not permanent in the soil. More than 650 different kinds of organic pesticides are presently on the market, and 200 are of major importance (*65*).

The first quantitative measurements of pesticide contents in surface waters were made in 1957 when DDT was detected in the Mississippi, Missouri, and Columbia Rivers (*66*). The resistant, long-lived, chlorinated hydrocarbons constituted 30% of all pesticides produced in the United States in 1965 (*67*) and have received the most study and attention. Weaver et al. (*68*) surveyed 56 rivers and three of the Great Lakes for chlorinated hydrocarbons; in 44 rivers and in Lake Michigan, the concentrations found ranged from zero to more than 0.118 μg/liter of pesticides. Even though only one-fifth as much dieldrin as DDT was used (*69*), dieldrin was dominant among the chlorinated hydrocarbon contaminants. DDT, or its metabolic product, DDE, and endrin had widespread occurrence, but DDD and aldrin were found in only one river each, and heptachloride, heptachloride epoxide, and benzene hydrochloride were not detectable in any of the rivers. (See Chaps. 23, and 30 regarding insecticide and herbicide analysis.)

According to Hansberry (*70*), one-fifth of the total pesticide use in the United States occurs in California on only one-fifteenth of the total cultivated farmland. Only DDT was found in relatively higher amounts in California rivers than in other rivers of the United States, indicating that factors other than intensity of use control the level of contamination of waters by pesticides.

5. Radionuclides

Discussion of radioactivity in surface waters and groundwaters is limited in this chapter to an indication of general levels of activity and the relationship to sediments in water. As noted in Sec. II.C.3, sediment loads depend on rainfall and land use patterns.

Gross beta activity of waters indicates contamination by man-made radionuclides. As may be expected, beta activity varied from a peak level of 105 pCi/liter following atomic weapon testing in the middle 1950's to less than 10 pCi/liter between testing periods. Strontium-90 exhibited similar fluctuations, with a range of 1.8–0.6 pCi/liter (*71*).

Gross alpha activity indicates natural radioactivity and varies from less than 1 pCi/liter in the eastern and northwestern United States to more than 10 or 20 pCi/liter in the streams of the Rocky Mountains where uranium-rich strata occur. Uranium is present in groundwaters usually in the range of 0.05–10 μg/liter but may reach values as high as 18 and 90 mg/liter in waters moving through uranium-rich rocks (*72,73*). Radium is generally less abundant than uranium because, like barium, it is more subject to precipitation and adsorption (*74*).

6. Nitrogen and Phosphorus

Reliable data on phosphate content of surface waters are scarce because this determination is not generally included in the chemical analysis of waters. Except in unusual instances, the amount present is small.

Feth (*49*) states that the concentration of nitrogen compounds in natural waters ranges from 0 to more than 100 mg/liter NO_3-N in surface water, and from 0 to more than 1000 mg/liter NO_3-N in groundwater. Hem (*47*) concludes that nitrate seldom exceeds 5 mg/liter in unpolluted surface waters and that often the concentration is less than 1 mg/liter. Feth (*49*) reported that two-thirds of the groundwaters sampled across the United States had concentrations in the range of less than 0.1 to 9.0 mg/liter NO_3-N for waters that had dissolved solids concentrations ranging from less than 100 to more than 10,000 mg/liter.

7. Miscellaneous Properties

There are numerous other water properties that are of little importance for irrigation water users but which may be important for domestic and industrial users. These include microbial content, temperature, dissolved oxygen, pH, biochemical oxygen demand, alkalinity, hardness, color, and turbidity. Rivers downstream from irrigation projects tend to be higher in alkalinity and hardness than rivers not used for irrigation. Few, if any, differences are evident between the two groups of rivers with regard to the other variables.

D. Soil Water Properties

1. General Properties

Although the literature contains many analyses of soil extracts, data on soil solutions are scarce, especially for humid zone soils. Limited data on major salt constituents of arid land soil solutions are available (*6,15,21,24,75*), but the irrigation waters used, if any, are not reported.

The relation between the concentrations of solutes in an irrigation water and in the resultant soil solution is complex, being determined primarily by the fraction of the applied water that passes through the root zone and, secondarily, by the water content of the soil (*76*). Precipitation of solutes, solution of soil minerals, and solute uptake by crops play lesser roles (*76*). The fraction of applied water that passes through the root zone over a period of time influences the amount of dissolved solutes retained in the root zone, whereas the water content of the soil determines the volume of water in which the solutes are dissolved. As the soil water content decreases, its solute concentration increases, although the relation is not necessarily linear (*6*). The fraction of applied water that passes through the root zone and the water content of the soil are, in turn, influenced by the frequency and amount of water application in relation to the amount of water lost by evapotranspiration, the position of the groundwater table, and the permeability of the soil. Thus, for a given irrigation water, the concentration of the soil solution may vary greatly, depending upon water management practices and soil, plant, and drainage conditions. In addition, the composition and concentration of the soil solution are functions of depth in the soil (*76*). In general, the soil solution in the upper zone of the soil in which the adsorbed constituents are in equilibrium with the irrigation water approaches the composition of the irrigation water, whereas at some lower depth, the solution phase may differ markedly from the irrigation water. In between is a zone where adsorption and cation exchange reactions are taking place, and the composition of the soil solution is intermediate between that of the upper and lower soil zones. With continuing irrigation, the upper zone migrates downward. Because evapotranspiration results in concentration of the soil solution and precipitation of some solutes, and because of solution of soil minerals, the composition of the soil solution at any depth is seldom the same as that of the applied irrigation water. The factors that cause these changes in composition are discussed below.

In spite of these complexities, the soil solution composition data in Table 3 illustrate the four major types of soil conditions frequently encountered in arid regions. This classification into normal, saline, sodic, and saline-sodic categories is commonly used to distinguish soils having different management and reclamation requirements (*2*). The compositions are given for two water contents within the field moisture range and for the saturation moisture percentage. These latter data are included for comparative purposes and are discussed below.

The composition of nonsaline-nonsodic soils (normal soil solutions) is characterized by solute contents of less than about 150 meq/liter and SAR values of less than 20 to 30 at the lowest water content that soils

TABLE 3
Soil Solution Compositions of Representative Arid Land Soils

Moisture %	Cations					Anions					SAR	References
	Ca	Mg	K	Na	Total	HCO_3	SO_4	Cl	NO_3	Total		
	meq/liter											
Normal soil – U.S.S.L.[a] soil No. 183; SP (28.4)[b], 1/3 bar (10.6)[c], 15 bar (4.7)[c]												
28.4% – Saturation extract	8.5	2.1	1.1	1.4	13.1	1.2	1.2	0.7	8.8	11.9	0.6	*(6)*
10.4% – Field range	18.3	4.4	1.7	3.3	27.7	3.8	2.8	3.4	18.6	28.6	1.0	*(21)*
8.1% – Field range	20.4	6.6	1.8	4.3	33.1	3.7	4.3	5.3	19.0	32.3	1.2	*(6)*
Saline soil – U.S.S.L. soil No. 57; SP (76.0), 1/3 bar (36.2), 15 bar (20.4)												
76.0% – Saturation extract	253.0	101.0	1.9	184.0	540.0	0.7	25.5	493.0	29.0	548.2	14.0	[d]
34.8% – Field range	584.0	247.0	2.7	439.0	1273.0	1.8	29.5	1173.0	66.0	1270.3	21.5	*(21)*
27.2% – Field range	787.0	317.0	2.9	561.0	1668.0	3.9	19.4	1553.0	89.0	1665.3	23.9	*(21)*
Sodic soil – U.S.S.L. soil No. 58; SP (41.4), 1/3 bar (21.6), 15 bar (6.1)												
41.4% – Saturation extract	1.0	0.5	0.7	29.0	31.2	6.0	14.5	5.1	3.6	29.2	33.5	*(6)*
19.7% – Field range	2.2	0.5	1.2	58.0	61.9	9.3	33.8	11.3	5.6	60.0	49.9	*(21)*
12.7% – Field range	6.4	1.7	2.1	100.0	110.2	11.2	66.2	28.1	9.0	114.5	49.7	*(6)*
Saline-sodic soil – U.S.S.L. soil No. 62; SP (33.0), 1/3 bar (25.1), 15 bar (8.8)												
33.0% – Saturation extract	78.0	58.0	39.0	1230.0	1405.0	2.9	109.0	1296.0	2.6	1410.5	149.0	[d]
21.9% – Field range	57.0	147.0	66.0	2002.0	2271.0	6.2	233.0	1999.0	14.1	2252.3	198.5	*(21)*
15.2% – Field range	50.0	190.0	97.0	2906.0	3243.0	4.7	284.0	2954.0	18.9	3261.6	265.3	*(21)*

[a] U.S. Salinity Laboratory, Riverside, California.
[b] SP = saturation percentage.
[c] Moisture percentages at designated tensions.
[d] Taken from U.S. Salinity Laboratory files.

might attain under field conditions. More often, however, field moisture is maintained at water contents in the upper half of the available water content; under these conditions, the total solute content in the soil solution of normal soils ranges from about 15 to 100 meq/liter with SAR values from less than 20 to 30. The calcium-plus-magnesium concentrations for a total solute concentration of 15 meq/liter are generally greater than 0.5 to 1.0 meq/liter, while sodium concentrations are less than 14.0 to 14.5 meq/liter. The calcium-plus-magnesium concentrations for a total solute concentration of 100 meq/liter are generally greater than 16.3 to 26.8 meq/liter, while sodium concentrations are less than 73.2 to 83.7 meq/liter. Potassium concentrations are generally less than 5.0 meq/liter, although occasionally higher values do occur. The pH of the soil solution for such soils is less than about 8.4. Carbonate concentrations are essentially nil, while bicarbonate concentrations are usually less than 5 meq/liter. Sulfate, chloride, and nitrate generally account for the remainder of the anions.

Saline soils have total soil solution salt contents in the range of 100–3000 meq/liter; such soils normally require reclamation before satisfactory yields of most crops can be obtained. The SAR is generally lower than about 30, and the pH is less than 8.4. For such soils, chloride is frequently the predominant anion, and bicarbonate concentrations are generally less than 10 meq/liter. Even if the soil is gypsiferous, sulfate seldom exceeds 50 meq/liter.

Sodic soils are characterized by total soil solution solute contents of less than 100 meq/liter, SAR values greater than 20 to 30, and pH values greater than 8.4. Carbonate may be as high as 5 meq/liter, and bicarbonate as high as 10–20 meq/liter. Sulfate, chloride, and nitrate in various proportions account for the rest of the anions.

Saline-sodic soil solutions are characterized by pH values of less than 8.4, total solute contents of greater than 100 meq/liter, SAR values greater than 20 to 30, and bicarbonate concentrations of less than 10 meq/liter. Sulfate, chloride, and nitrate may occur in various combinations.

The above generalizations should not be applied too rigorously. These ranges and descriptions are, however, reasonable and serve to "bracket" soil solution compositions for comparative purposes. On the basis of these data, soil solutions are obviously higher by a factor of 2–10 or more in total solute concentration (and sometimes greatly so) than most irrigation waters, for which the total solute content is generally less than 20 meq/liter. The same generalization holds for the individual major solute constituents, although upper limits of accumulation, as noted, do occur for carbonate, bicarbonate, and sometimes sulfate. Because of the paucity of data for minor elements, pesticides, and radionuclide contents of the soil solutions, further direct comparisons cannot be made, although some quantitative comparisons are possible by inference from an examination of soil drainage waters. These will be more obvious after the properties of soil drainage waters have been discussed.

Most determinations of the major solutes in soil solutions have been made on arid land soils to evaluate salinity by analysis of saturation extracts. In humid climates salts are quite low and of little concern, except in cases of saltwater intrusion, and similar solute content data have not been obtained. Representative saturation extract compositions for arid zone soils are given in Table 4, and the usual compositional range for non-salt-affected soils is given in Table 5. Soils having compositions in the range of values in Table 5 should not be deleterious except when extreme combinations occur (low calcium-plus-magnesium and high sodium). Soils with compositions similar to those of the salt-affected soils in Table 4 would reduce most crop yields and would therefore require reclamation.

The criteria for assessing salinity in the United States are based on the composition of saturation extracts. This composition is used for relating crop yields, management problems, need for reclamation, etc. The criteria and standards most frequently adopted are those proposed by the U.S. Salinity Laboratory (*2*). On the basis of years of experience with the properties of salt-affected soils, and from the accumulation of a large file of saturation extract analyses, the criteria and standards shown in Table 6 have been developed for classifying arid zone soils on the basis of their saturation extract compositions. The physical properties of these soil classes and the recommended management and reclamation procedures have been described in detail (*2*).

The effects of irrigating soils with waters of different composition are illustrated in Table 7 for four individual cases. The composition of the irrigation water that was used, the saturation extract of the soil before irrigation, and the saturation extract of the soil after irrigation are given. In case 1, irrigation with the water reclaimed a saline soil. In case 2, use of the irrigation water produced a saline soil. Note that this latter water contained less salt than the previous water which served to reclaim a saline soil. In case 3, use of the water on the soil produced a soil that was approaching a sodic condition. In case 4, use of the irrigation water on a normal soil produced a soil that was both saline and sodic. These data show that similar irrigation waters can either alleviate or produce salinity problems in soils, depending on such factors as soil properties, drainage properties, and water management.

The content of mineral nutrient elements in soil solutions has been reviewed by Reisenauer (*77*). Most of these data are old, but they are the

TABLE 4
Major Constituents in Soil Saturation Extracts of Typical Arid Land Soils

Soil type and U.S.S.L.[a] sample no.	EC[b], mmho/cm	Cations, meq/liter					Anions, meq/liter					SAR	pH	References
		Ca	Mg	Na	K	Total	CO_3	HCO_3	SO_4+NO_3	Cl	Total			
Normal														
2741	0.6	2.7	2.3	1.2	0.9	7.1	0	2.6	2.1	0.9	5.6	0.8	6.4	2
3567	0.6	2.9	0.8	0.2	1.4	5.3	0	3.0	1.0[c]	1.3	5.3	0.1	6.4	[d]
Saline														
3567	8.8	28.4	22.8	53.0	1.1	105.3	0	5.2	74.0	29.0	108.2	10.5	8.0	[d]
3581	7.3	48.3	21.1	16.1	1.8	87.3	0	4.7	72.1[c]	10.5	87.3	2.7	7.0	[d]
Sodic														
535	3.2	1.1	0.3	29.2	4.1	34.7	8.4	18.7	4.6	7.5	39.2	35.0	9.6	2
3581	2.2	2.4	0.6	19.6	0.3	22.9	tr	13.9	5.5[c]	3.5	22.9	16.0	8.4	[d]
Saline-sodic														
2739	9.2	6.7	9.9	79.5	0.5	96.6	0	2.4	20.1	72.0	94.5	27.6	7.3	2
3576	7.9	13.9	5.9	75.2	0.6	95.6	0.4	3.0	90.1[c]	2.1	95.6	24.0	8.1	[d]
3620	12.7	1.4	0.2	124.0	1.3	126.9	1.3	10.4	17.8[c]	97.4	126.9	139.0	8.9	[d]

[a]U.S. Salinity Laboratory, Riverside, California.
[b]Electrical conductivity at 25°C.
[c]By difference.
[d]Taken from U.S. Salinity Laboratory files.

TABLE 5
Concentration Ranges Found in Typical Nonsaline, Arid Land Soils[a]

Element	Usual range, meq/liter
CO_3	Trace–1.0
HCO_3	< 1.0–5.0
Cl	< 0.1–5.0
SO_4	< 1.0–20.0
Ca	< 0.1–10.0
Mg	< 0.2–5.0
Na	< 0.1.–5.0
K	< 0.1–1.0
Σ	< 0.5–30.0

[a]Adapted from Ref. *4*, by courtesy of University of California, Division of Agricultural Sciences.

TABLE 6
Criteria and Standards for Classifying Salt-Affected Soils[a]

	Soil condition (standards with reference to saturation extracts)			
Criteria	Normal	Saline	Sodic	Saline-sodic
EC[b], mmho/cm	< 4	> 4	< 4	> 4
SAR	< 13	< 13	> 13	> 13
pH	≤ 8.4	≤ 8.4	> 8.4	≤ 8.4

[a]Adapted from Ref. *2*.
[b]Electrical conductivity at 25°C.

only ones available. The range in concentrations and the most typical concentrations are indicated in Table 8 for the six major nutrients. Of chief interest with regard to water pollution is the fact that phosphorus is reduced to low concentrations in soil solutions, seldom exceeding a few tenths of a milligram per liter, while nitrate may attain several hundred milligrams per liter. Contribution of phosphates to water systems from return flow of internal irrigation drainage waters is unlikely. The same is not true of nitrates.

Data on the soil solution content of minor elements are rare, although extensive studies are currently under way (G. R. Bradford, Department

TABLE 7
Saturation Extract Compositions of Soils Before and After Irrigation with Waters of Given Compositions[a]

Water analyzed	EC[b], mmho/cm	B, mg/liter	Ca (meq/liter)	Mg (meq/liter)	Na (meq/liter)	K (meq/liter)	CO_3 HCO_3 (meq/liter)	SO_4 (meq/liter)	Cl (meq/liter)	NO_3 (meq/liter)	SAR, units
Case 1											
Virgin soil SE[c]	6.11	2.23	28.70	15.71	28.59	2.06	4.56	41.28	17.38	12.62	6.07
Irrigated soil SE	2.33	1.67	9.72	6.62	9.37	0.83	2.81	19.86	1.88	2.79	3.28
Irrigation water	1.85	1.10	6.60	8.18	6.86	0.11	2.61	17.33	1.63	0.36	2.52
Case 2											
Virgin soil SE	1.39	—	3.12	0.51	9.68	0.33	4.37	1.44	7.80	0.03	7.19
Irrigated soil SE	4.24	2.89	24.92	3.72	24.47	0.51	1.73	51.53	1.80	0.43	6.47
Irrigation water	1.67	2.14	3.79	3.18	10.42	0.12	3.44	12.73	1.50	0.08	5.57
Case 3											
Virgin soil SE	0.55	0.94	2.60	0.71	2.15	0.41	4.31	1.14	0.77	0.04	1.67
Irrigated soil SE	2.88	5.43	7.23	2.20	19.87	0.72	3.59	19.62	7.82	0.03	9.17
Irrigation water	2.72	3.97	6.14	5.58	16.62	0.09	3.39	14.41	10.90	trace	6.87
Case 4											
Virgin soil SE	1.88	—	5.12	1.29	13.21	0.46	4.17	11.83	3.40	0.04	7.38
Irrigated soil SE	4.49	—	6.89	2.05	36.77	0.53	3.01	25.32	18.95	0.02	17.43
Irrigation water	2.85	1.62	2.17	0.70	24.35	0.08	2.79	11.46	13.40	trace	20.22

[a]Adapted from U.S. Salinity Laboratory files.
[b]Electrical conductivity.
[c]Saturation extract.

TABLE 8
Mineral Nutrient Contents of Soil Solutions[a]

Element [no. of samples]	Concentration range, mg/liter	Percentage of observations within given concentration limits	
		%	mg/liter
Ca [979]	0–> 1000	77.7	0–100
Mg [337]	0–> 1000	85.2	26–200
N as (NO_3) [879]	0–> 1000	85.8	26–200
P as (PO_4) [149]	0–> 0.50	73.2	0–0.15
K [155]	0–> 200	54.8	0–50
S as (SO_4) [693]	0–> 2000	94.7	0–100

[a]Adapted from Ref. *77*, p. 507, by courtesy of Federation of American Societies for Experimental Biology.

of Soils and Plant Nutrition, University of California at Riverside, personal communication).

2. Reactions Between Infiltrating Waters and Soils

a. Concentration

One of the most obvious changes that takes place after an irrigation water is applied to a soil is an increase in the total solute concentration. This results from two processes, evaporation and transpiration, which remove water from the soil while salt is retained. This reduces the volume of soil water and produces an increased concentration. Plant transpiration has the greatest effect, with growing plants transpiring some 300 to 900 lb of water for every pound of dry matter produced(*78*). To meet this requirement, some 2–5 acre-ft of water are commonly required per acre of irrigated crop land per year(*48*). Since the fraction of applied water that is consumed in evapotranspiration commonly ranges between 0.4 and 1.0, the soil solution may be concentrated some 2.5-fold or more by this process.

Since irrigation waters generally contain 0.1–5 tons of salt per acre-foot of water (*48*), a considerable quantity of added salts is introduced into the soil, and the resulting concentration buildup in the soil solution will eventually reduce crop yields unless an excess of water is applied to leach out the salt. This additional water requirement is termed the "leaching requirement" (*2*). The fraction of water applied that appears after each irrigation as drainage water also influences the ultimate concentration of the soil solution.

b. Precipitation

Another process that changes the composition of an irrigation water after it is introduced into the soil is precipitation of slightly soluble salts such as calcium carbonate and calcium sulfate. Precipitation is accentuated by the reduction in volume resulting from evapotranspiration. If the degree of concentration of the soil solution were not limited by plant tolerance factors, many evaporite salts could potentially be precipitated from soil solutions. However, the soil solution, in practice, is seldom allowed to concentrate to this extent because plant growth is seriously reduced at total soil water potentials of −10 to −20 bars (*79*). A major part of this potential in saline soils is caused by soluble salts (osmotic potential). To prevent yield reductions, the soil solution is seldom allowed to remain at this potential for any appreciable period of time. Therefore, concentrations exceeding 300–500 meq/liter will seldom occur in soil solutions of agricultural soils. This limits the number of evaporites that commonly precipitate in soils to calcium carbonate and gypsum (*79*). However, at the soil surface where evaporation of the soil solution occurs, other evaporites may form as a result of capillary movement of water and deposition of salts. A shallow water table and highly saline groundwaters (*80, 81*) must be present before this occurs to any appreciable extent. Evaporation from raised soil surfaces, such as ridges between furrows, also favors surface salt deposition (*2*).

Predicting the degree of salt precipitation from irrigation waters upon reaction with soils is difficult because soil solutions are subject to influences not present in bulk water. Adsorptive forces associated with mineral surfaces, electrical force fields emanating from such surfaces, chelating effects of soil organic matter, matric suction, buffering of soil pH, cation exchange reactions, variable CO_2 pressures, etc., all influence the soil solution system and its ultimate composition. Because of these complications, only a few attempts have been made to predict quantitatively the ultimate composition of the soil solution that will result from the use of an irrigation water of known composition, and these predictions are either untested proposals (*79*) or are limited to very special conditions (*82–85*).

The precipitation of Ca^{2+} and HCO_3^- as calcium carbonate is of special importance in determining the ultimate composition of soil water. The importance of this reaction in affecting the suitability of waters for irrigation purposes has been well established (*76,86–88*). A means for quantitatively predicting the composition of a soil solution after irrigation with a water having potentially precipitable calcium carbonate has not been developed. However, several proposals have been advanced in an attempt to assess qualitatively the tendency of waters to undergo this reaction (*86,88–92*). Only the most satisfactory of these proposals, in the authors' opinion, are discussed herein.

Langelier (*93*) devised an index, termed the saturation index, for indicating the extent to which waters flowing through pipe will precipitate or dissolve calcium carbonate. This index is defined as the actual pH of a water (pH_a) minus the pH (calculated) which the water would have in equilibrium with calcium carbonate [pH_c; see Eq. (2)], viz.:

$$\text{saturation index} = pH_a - pH_c \tag{3}$$

Wilcox and Bower (*89*) have modified the Langelier index for use under soil conditions. The modification involves the use of the pH of the soil in place of the actual pH of the water, since poorly buffered irrigation waters are brought to the pH of the well-buffered soils after application. Their data show that calcium carbonate does not precipitate in the soil if the modified saturation index is negative. If the index is positive, the percentage of bicarbonate in the water that precipitates is correlated with the value of the index. Further work (*76,90–92*) has demonstrated that the tendency of irrigation waters to precipitate or dissolve calcium carbonate when applied to soils can be related to a calculated pH value of the water (pH_c) and to the leaching fraction.

c. *Weathering of Soil-Silicate Minerals*

It has been believed for some time that weathering of soil minerals contributes to the solute content of soil solutions (*94*), and the relative weathering rates of the more common minerals have been studied (*95–100*). However, it is still virtually impossible to predict the quantities of solutes that will be contributed to a percolating solution by this process. Factors such as rock structure, mineral composition, crystal size, degree of crystallinity, type of percolating water, time of contact, moisture content, carbon dioxide pressures, etc., all affect the weathering rate and the kinds and amounts of solutes released. As a result, and also because of lack of appreciation of the significance of this phenomenon, little consideration has been given to weathering as a factor contributing to soil solution composition, especially in water quality considerations. Recently, the effect of soil mineral weathering on the sodium hazard of irrigation waters has been studied under conditions somewhat comparable to those in the field(*101*). These studies have shown that soil mineral weathering may have appreciable effects on the ultimate soil solution composition. The minerals involved and the weathering mechanism are such that calcium plus magnesium concentrations generally increase by a few milliequivalents per liter in the soil solution, thereby lowering the sodium hazard potential (SAR) of applied irrigation waters. These findings demonstrate the importance of research in this little-studied area.

d. Adsorption and Exchange Reactions

An important property of soils is their ability to sorb or exchange large quantities of materials; this property greatly influences the ultimate composition of soil solutions. The two factors that seem to largely influence the extent of adsorption reactions are surface area and cation exchange capacity (CEC). These properties arise chiefly from the presence in soils of inorganic and organic colloidal fractions. Inorganic colloids in soils consist mainly of clay minerals (kaolinite, halloysite, chlorite, illite, vermiculite, montmorillonite, and interstratified and intergrade modifications of these clay minerals) and hydrous oxides. A feature common to most of the clay minerals is their high specific surface and the excess of net negative charge of the particles, which largely accounts for the CEC of these minerals. Positive charges, generally of lesser magnitude, may also occur, associated with partially hydrolyzed iron and aluminum complexes which commonly interact with clay minerals in weathered soils. These are much more abundant in humid zone soils than in arid zone soils. Positive charges associated with iron and aluminum sesquioxides and hydrous oxides also occur in soil. Soil organic matter may contribute either negative or positive charge sites. The positive charges in soils account for some of the anion exchange capacity, with the rest involving substitution of anions for certain anionic constituents found on or associated with the surfaces of solid soil material. The positive charge contribution from the iron and aluminum sesquioxides or interlayer hydroxy compounds and organic matter is highly pH dependent, increasing with decreasing pH.

Representative data on the CEC and surface properties of soil minerals, soil organic matter, and soils as a whole are given in Table 9. These two properties give rise to a tremendous sorptive capacity in soils. To gain an appreciation of their magnitude, consider an acre-plow layer of a soil, approximately 2 million lb, having intermediate CEC and surface area

TABLE 9
Representative Cation Exchange Capacity (CEC) and Surface Area of Soils and Soil Constituents[a]

Material	Physical property	
	CEC[b] meq/100 g	Surface area, m^2/g
Organic matter	200–400	500–800
Vermiculite	100–150	600–800
Montmorillonite	80–150	600–800
2:2 Integrades	10–150	50–800
Illite	10–40	65–100
Chlorite	10–40	25–40
Kaolinite	3–15	7–30
Oxides and hydroxides	2–6	100–800
Arid zone soils	< 10–50	50–200
Humid zone soils	< 10–30	50–200

[a]Adapted from Ref. *102* by courtesy of The American Chemical Society.
[b]Cation exchange capacity.

values. Such a soil would have 20 million acres of internal surface area and a cation exchange capacity equivalent to 10,000 tons of $CaCO_3$ (*103*). This tremendous sorptive reservoir interacts with pesticides, radionuclides, nutrient elements, and other inorganic solutes and greatly influences the fate of such materials in the soil system. The extent to which these materials are or are not retained in soils and their potential contribution to environmental pollution depend primarily on how they interact with the sorptive forces acting upon them—coulombic, van der Waals, or hydrogen bonding.

Several authors (*102,104–106*) have reviewed the sorptive interactions between soils and pesticides. Greenland (*105*) has classified the properties that influence the absorption of organic compounds by clays as: properties of the adsorbing surface, properties of the organic adsorbate, and properties of the solvent medium. The important properties of the adsorbing surface include surface area, accessibility of surface (size and tortuosity factors), chemical nature of atoms on the surface, pattern of surface charge (spatial density and distribution, origin of charge), kinds of exchangeable ions occupying the surface, and surface configuration. The important properties of the organic compounds include their charge, size, polarity, polarizability, shape, and flexibility. The important properties of the solvent are its solvent capacity for the adsorbate and its competition for adsorption on the clay surfaces. Additional factors such as soil water content, temperature, soil reaction, and formulation of the pesticide may also influence the adsorption of pesticides by soil systems(*102,104*). Texture, organic matter content, and rainfall are also regarded as important factors(*106*). The following discussion describes how some of these factors influence the adsorption of pesticides in soils, according to Bailey and White(*102*).

The influence of the adsorbing colloid has been shown to have a controlling effect on pesticide adsorption in soils. Soils with high organic matter or clay contents decrease the effectiveness of pesticides through adsorption(*102*) and also decrease their leachability(*102*). Correlation studies on pesticide absorption versus soil properties show high positive relationships to organic matter content and cation exchange capacity, surface area, and mineralogy(*102*). Because soils vary in their colloid content, and because soil colloids vary in their physical properties, a wide range of adsorptive properties may occur in soils. The magnitude and specificity of pesticide adsorption therefore vary among soils. The nature of the adsorbate influences the adsorption on soil materials(*102*), and pesticides are highly individualistic in this regard. Solubility and adsorption tend to be inversely related within a family of compounds but not between families. The exact chemical nature of the compound is important, especially the kinds of functional groups it contains, its electron distribution, and its acidity.

Pesticide adsorption depends strongly on soil pH(*102*); in general, adsorption increases as pH decreases, although the pH for maximum adsorption depends on each compound and adsorbent. The degree of dissociation or charge of both the adsorbate and adsorbent is influenced by pH.

Cation species occupying the exchange sites of soils affect the adsorption of pesticides, with increasing adsorption in the order sodium $<$ calcium $<$ magnesium among the common bases(*102,107,108*). In general, adsorption of pesticides decreases with increasing base saturation.

The water content of soils has an appreciable effect on the adsorption of pesticides(*102*), although the content can either increase or decrease adsorption of various pesticides.

The nature of the formulation (i.e., whether the pesticide is applied as a solution or a suspension in water or oil, an emulsion, wettable powder, granule, or dust) affects the adsorption of pesticides(*102*). In addition, some herbicides can be in acid, salt, or ester form, influencing their behavior in the soil.

Pore size and pore size distribution of soils influence the extent of adsorption(*102*) since these parameters affect the rate at which water enters and moves through the soil.

It seems apparent that the forces involved in the adsorption of pesticides by soils result from specific interactions and cannot be quantitatively predicted or precisely generalized. Even so, some order is being established despite the complexity of the subject.

For a review of the adsorption interactions of insecticides, fungicides, and herbicides with soils and clay minerals, see Kunze(*106*). For a detailed discussion of the chemical and structural properties of these pesticides, see the reviews by Fleck(*109*), Freed(*110*), and Rich(*111*).

Radionuclides also interact with and are adsorbed by soil constituents. Robinson(*112*) has reviewed the literature on the interactions between radionuclides and ion exchange minerals. Clay minerals in soils possess so large a capacity for sorption of radionuclides that minerals and soils are frequently used for radioactive waste disposal systems(*112*). Studies with clays(*113*) have shown that ^{141}Ce, ^{144}Ce, ^{144}Pr, ^{95}Zr, ^{95}Nb, ^{140}Ba, ^{140}La, ^{90}Sr, ^{90}Y, and ^{137}Cs radionuclides are effectively removed from solution by adsorption processes, while ^{106}Ru, ^{106}Rb, and ^{131}I are less effectively removed. The removal from solution of cesium, strontium, yttrium, and cerium was nearly complete when the initial concentration was less than 0.01 of the soil's saturation capacity(*114*). These adsorption reactions are pH dependent, however(*113,114*). Higgins(*115*) reported on the adsorption capacities and distribution coefficients (K_d) for cesium, strontium, yttrium, plutonium, cerium, and ruthenium for 15 soil types. K_d values found were generally in the order of 100 to several thousand, so at least 100 times more of these radionuclides are absorbed than are in solution. This is good evidence that these radioactive nuclides should be quite immobile in soils. The K_d values for a given isotope correlate with the exchange capacities of different minerals in the order: quartz $<$ kaolin $<$ illite $<$ montmorillonite. K_d values for a given mineral or soil and for different radionuclides were generally in the order: ruthenium $<$ strontium $<$ cesium $<$ yttrium $<$ plutonium $\approx$ cerium. The most mobile element in soil systems appears to be ruthenium. These generalizations and studies were made in pure-water systems; other investigators have demonstrated the modifying influences of pH and salt concentrations (*112*). The effect of added salt is not the same for all ionic species, but in general, K_d decreases logarithmically with increasing salt concentration. Varying pH has a very small effect between pH 2 and about 9, and most soil waters fall in this range. Higgins(*115*) concludes that ion exchange removes radionuclides so effectively that within a meter or so all the radioactivity is adsorbed from solution. The great retentivity of soils for cesium, cerium, strontium, and other radionuclides has been noted repeatedly(*116–118*). Most radionuclides are retained in the top several centimeters of soil to which they are applied, and only ruthenium moves appreciably through the soil(*119,120*).

Adsorption mechanisms in soil systems also exert an important control over the amounts of alkali and alkaline earth cations and boron, fluoride, and, to a lesser extent, sulfate present in soil solutions(*76,121–126*).

Fluoride is adsorbed by soils at both high concentrations(*127–129*) and low concentrations(*124*). Soils high in hydrous oxide minerals are the most active in fluoride adsorption in the concentration range of most interest for agriculture (1–15 mg/liter)(*124*). The soils studied were estimated to be capable of removing fluoride from solution at 8 mg/liter from 9 to 140 acre-ft of water per acre-foot depth of soil. Minerals range in their capacity to absorb F from trace amounts to as high as 32,000 mg/kg of mineral.

Eaton(*130*) showed that an equilibrium exists between the soluble and adsorbed phases of boron in soils, and observed that boron injury occurred sooner in coarse-textured soils than in fine-textured ones, also reclamation occurred in the same order. The amount of leaching required for reclamation was found to exceed that required to remove excessive amounts of chloride and sulfate salts(*131*). Soils vary in their capacity to adsorb boron, but in most cases, the amount of adsorbed boron at equilibrium is greater than the amount in solution(*123*). This is a beneficial property of soils, since plants respond to the boron concentration of the soil solution rather than to the amount of adsorbed boron present(*132*).

Sulfate has been shown to undergo adsorption reactions in soils(*125, 126,133–135*). These studies suggest that the concentration of sulfate in soil solutions and movement through soil profiles are controlled to some extent by adsorption processes. The soil constituents thought to adsorb sulfate are the positively charged edges of clays, interlayer aluminum and iron hydroxy compounds, and iron and aluminum sesquioxides(*135*). Exchange equilibrium studies of sulfate in soils have shown that several different sites for sorption are involved and that the sorbed sulfate may be either highly resistant to leaching or weakly so, although the latter is more common by far(*125,126,133,134*).

Other anions may be adsorbed and exchanged in soils but to a lesser degree than sulfate(*134–138*). This phenomenon is of little practical concern for most anions except phosphorus, which is discussed later.

Since Way(*139*) demonstrated that an exchange of cations took place when soils were treated with various electrolytes, it has been known that soils have cation exchange capacities and retain large amounts of cations; in fact, more cations are held in the adsorbed phase in soils than are in the solution phase except for highly saline soils. The cation exchange chemistry of soils is a primary factor in soil-plant-water relations. The change in the composition of adsorbed cations as a result of reaction with irrigation waters is one of the chief concerns in evaluating the suitability of waters for irrigation. Bower(*76*) has suggested that in predicting the effects of waters on the composition of the dissolved and adsorbed constituents of soils (the two are interdependent), one might usefully consider the irrigation of a soil to be a column process involving the adsorption and exchange of constituents, especially cations, during the downward flow of water. Many equations have been proposed to describe the phenomenon of cation exchange in soils. These have been reviewed for arid soils(*140–143*) and acid soils(*144*). This topic is too complex and extensive for detailed treatment here. Suffice it to say that all cations are adsorbed by soil materials and that an equilibrium exists between the adsorbed and solution phase constituents, with the following empirical rules being generally followed, according to Kunin(*140*):

(1) At low concentrations (aqueous) and ordinary temperatures, the affinity of adsorption increases with increasing valency of the exchanging cation ($Na^+ < Ca^{2+} < Al^{3+} < Th^{4+}$).

(2) At low concentrations (aqueous), ordinary temperatures, and constant valence, the affinity of adsorption increases with increasing atomic number of the exchanging cation ($Li^+ < Na^+ < K^+ < Rb^+$; $Mg^{2+} < Ca^{2+} < Sr^{2+} < Ba^{2+}$).

Organic matter and minerals differ greatly in exchange properties, and within each group there are many exceptions to these general rules.

Nitrogen, potassium, and phosphorus—all essential nutrient elements—undergo adsorption reactions in soils that greatly reduce their concentrations in the soil solutions and their availability to plants. Reitemeier (*145*) has reviewed the literature up to 1950 on the fixation of potassium and ammonium. (The term "fixation" is often used for the adsorption of ammonium, potassium, and phosphorus because the process is difficultly reversible and because the adsorbed nutrients are relatively unavailable to plants.) The adsorption of NH_4 and potassium by clay minerals is caused by the same clay minerals and is essentially the same(*146,147*), so that the two ions can be discussed together. Hoagland and Martin(*148*) were probably the first to demonstrate K fixation by soils in California. The soils they examined fixed 47–99% of the K they applied. Since then, many others have reported on the fixation of ammonium and potassium ions by soils(*149–153*).

Probably two mechanisms are responsible for the irreversible adsorption of K^+ and NH_4^+ in soils. One is the specific adsorption of K^+ and

NH_4^+ ions in the wedge-shaped regions around the edges of mica-like particles(*154–156*). This mechanism accounts for only about 5% or less of the total fixing capacity(*154*). The other, and apparently more significant, mechanism for K fixation is the trapping of K and NH_4 ions in the interlayer spaces of expanding 2:1 layer-silicate minerals, especially vermiculite(*157–163*). The size of the ion and its energy of hydration (*164*) are very important, and K, NH_4, Cs, and Rb ions all meet the requirements and are fixed by these types of clay minerals. The net result of this strong adsorption reaction of K and NH_4 by certain soil clay minerals is that K and NH_4 are generally quite low in soil solutions(*2,165,166*).

Phosphorus, a major anion nutrient element, is also adsorbed tenaciously by soils so that its concentration in soil solutions and its availability for plant uptake are greatly reduced(*167*). Adsorption plus other fixation reactions of phosphorus in soil tend to keep most added phosphorus near the point of addition(*168*). Of 3005 lb of phosphorus added per acre in an 11-year period, only 0.1 lb of phosphorus per acre annually was lost by leaching through the 18-in. depth of soil(*169*). The loss of only a trace of phosphorus in the drainage water from a sandy loam soil in Scotland has also been reported(*170*). Probably the only soils from which leaching loss of phosphorus is significant are sands and organic soils that have little tendency to react with phosphorus and are fertilized heavily(*171,172*).

3. Physical Status of Soil Water

Several reviews of the physical status of soil water have been published (*103,173–176*). Soil water often exceeds 30% of the soil weight; it functions in many soil processes—as a solvent, as a leaching agent, as a reactant, as a medium for chemical reactions, as a plasticizing agent, and as a reservoir for plant-required nutrients and waters. Soil water always contains dissolved substances. In addition to the inorganic components discussed in the preceding section, dissolved organic substances are present, and gases (nitrogen, oxygen, and carbon dioxide) in various combinations commonly total up to 25 ppm.

The soil is a porous medium with considerable colloidal material (clays and organic matter, chiefly). This results in considerable surface area accessible to water, ranging from less than 1000 cm^2/g for coarse soil to more than 1,000,000 cm^2/g for clay soils. The latter figure includes large amounts of "internal" surface, i.e., interlayer area between clay mineral platelets. This area generally accommodates only thin layers of water. When only limited amounts of water are present (air-dry soil), the interlayer spaces are filled and the external surfaces are covered by thin films. The interstitial space not occupied by liquid water contains atmospheric gases, including water vapor. When greater amounts of water are introduced, the water layers' external surfaces become thicker and, at the same time, wedges of water form at the points of contact of adjacent soil particles. In fully saturated soils, all of the interstices are occupied by water.

In the past, the water present in the soil in its various stages of wetting has been described functionally by such terms as hygroscopic, capillary, and gravitational water. More recently, the relationships between water content and physical variables such as vapor pressure, matric suction, and capillary or hydraulic conductivity have been found to be continuous, with no sharp distinctions in the water properties at different stages of wetting; thus, the previous classifications have little significance(*173*).

The bulk soil water differs from pure water due to the presence of dissolved electrolytes, while the state of the water in the solid-liquid interfacial regions of the soil differs from that in the bulk solution because of the interactions between the two phases. Four kinds of interactions extend some distance outward from the interfaces: (1) hydration of adsorbed cations; (2) formation of an electrical double layer at the solid surface, producing a high osmotic pressure near the interface; (3) formation of hydrogen bonds linking water molecules to the solid surfaces of the particles; and (4) existence of van der Waal's forces. These forces contribute to the adhesion forces between liquids and solids. The most outstanding influence of these forces on the behavior of water in soil is the rapid decrease of the partial molar-free energy (water potential) which occurs with decreasing water content(*173*).

Clay particles have electrostatic fields extending outward from the particle surfaces. The extent of this force field depends strongly on the nature of the adsorbed ions (especially on their valence) and on the concentration of salts dissolved in the water. The distribution of ions is governed, in turn, by the electrostatic field. The concentration of adsorbed ions varies roughly from 2 to 5 M at the solid surface, to 1 M at 5 Å distance, and to less than a few tenths molar at 15 Å from the surface (*177–179*). The property of the water in the vicinity of the solids is greatly influenced by these high concentrations of cations.

Hydrogen bonding and van der Waal's forces have also been cited as special features of the clay water surface. Low(*180*) presented data showing that water adjacent to the soil solid phase has a lower density than does the water in the bulk, and that the zone of decreased density extends outward as far as 60 Å. Kemper(*23*) and Low(*181*) showed that ions are less mobile in such layers because the water is more viscous near clay surfaces than in the bulk(*182,183*). Low proposed that hydrogen bonding occurs between the water and oxygen atoms of the clay surface, causing the adjacent water molecules to assume a tetrahedral coordination and inducing additional layers of water to do likewise by extension of the hydrogen-bonding mechanism; this apparently results in a quasi-crystalline structure of the water, which accounts for the decreased density of the water, the greater viscosity, and the reduced rates of ionic diffusion. The osmotic action of the exchangeable cations may also be responsible for these observations.

In addition to matric forces (those which result from the presence of the solid phase) and osmotic forces (those caused by the presence of dissolved solutes), the soil water experiences body forces (inertial and gravitational forces). These react in such a way that the affinity with which water is held is a reciprocal of water content. Thus, at higher water contents, water will move out of the soil by gravity, whereas at very low water contents, water is retained even against extracting forces several thousand times that of gravity(*174*). At such water contents, soil water behaves as if it were under a negative pressure; this is commonly referred to as the soil suction.

When water enters the surface of a dry soil, it moves downward under the influence of gravity and attractive forces. If the amount of water added is insufficient to cause the wetting front to reach the lower end of the soil, or to contact a zone of water saturation, the major downward movement will stop within a few hours after water entry at the surface is discontinued. The water content in this portion of the soil is known as field capacity (FC). This parameter represents, for practical purposes, the upper limit of soil water available for plant use. For many soils of medium texture, this is approximately equal to the water retained against a suction of $\frac{1}{3}$ bar. The lower limit is known as the permanent wilting percentage (PWP). This is a biologically determined parameter representing the water content at which test plants fail to recover from wilting under standard conditions. For many soils, the PWP is approximately equal to the water held against a suction of 15 bars.

The water content of soils on which plants are grown is generally kept between these two soil water limits (FC and PWP), preferably closer to FC than PWP so that yields are not reduced because of water stress within the transpiring plants. The water potential within these limits, which the plant must overcome to meet its transpiration requirements, is composed of matric and osmotic potentials. The interplay of these two potentials has been reviewed by Hayward and Bernstein(*184,185*). The increased osmotic pressure of soil water due to the presence of solutes restricts growth and reduces the uptake of water by plants(*186*), and different salts produce equivalent depressions at equal osmotic pressures except in instances of specific ion effects. In addition, the matric potential affects water availability to plants. Studies on the interactions of soil salinity and water tension(*187,188*) suggest that the two effects make separate contributions and that they are additive, so that the total potential controls soil water availability to plants regardless of the relative contributions of osmotic potential or matric potential. Osmotic potential is the predominant factor under saline soil and low tension conditions, while matric potential predominates under nonsaline and high tension conditions.

4. BIOLOGICAL PROPERTIES OF SOILS

The carbon, nitrogen, phosphorus, and sulfur cycles in soils have been reviewed by Stevenson(*189*), Alexander(*190*), and Waksman(*191*).

The mineral nutrient cycles (nitrogen, phosphorus, sulfur) in soils are very similar in regard to the nature of the microbial transformations involved. The cycle for each of these nutrient elements consists of mineralization (decomposition of large organic molecules into simple inorganic compounds: NH_4^+, NO_3^-, NO_2^-, N_2, $H_2PO_4^-$, HPO_4^{2-}, PO_4^{3-}, SO_4^{2-}, S^{2-}, etc.) and immobilization processes (assimilation of the simple inorganic forms into microbial tissue). The organisms involved differ, as do some of the pathways within the individual cycles, but they are similar in principle.

In the mineralization of nitrogen in soils, the initial reduction of organic nitrogen compounds to NH_4^+ constitutes ammonification, while oxidation of NH_4^+ to NO_3^- is termed nitrification. The nitrate produced may be lost from the soil by leaching or by biological denitrification whereby nitrates are reduced to free nitrogen. During or subsequent to mineralization, a large amount of nitrogen is incorporated and temporarily immobilized in microbial tissue. This soil organic fraction functions as a reservoir of nitrogen which can be gradually mineralized and recycled through the continuing activities of microbes. Since 90% of soil nitrogen occurs in the organic form(*192,193*), the importance of the nitrogen cycle is apparent.

Mineralization of nitrogen from organic residues is dependent on a number of environmental factors including soil water, pH, aeration, temperature, the total nitrogen status of the soil, and the inorganic nutrient supply(*189*). Also, both mineralization and immobilization of nitrogen in soil are markedly influenced by the carbon:nitrogen ratios of added organic residues. The active decomposition of organic matter requires that the amount of nitrogen in the decomposing tissue exceed that immobilized as microbial protoplasm; for this, the critical carbon:nitrogen ratio falls between 20:1 and 25:1, with wider ratios favoring immobilization and narrower ratios, mineralization(*194,195*). Thus, if high nitrogen-containing organic residues, such as obtained from legume cover crops, are incorporated into the soil, a large part of the nitrogen will be recovered in mineral form. On the other hand, cereal grain straws having wide carbon: nitrogen ratios will result in the almost complete immobilization of available nitrogen as microbial protoplasm.

Nitrogen may be gained in the soil from microbial activities through the process of nitrogen fixation. Both free-living microbes and symbiotic bacteria associated with the roots of Leguminosae are capable of nitrogen fixation. Estimates of the amount of nitrogen fixed by *Rhizobium* average 80–100 lb/acre per year(*196*). Apparently, some of this nitrogen is excreted into the soil from the nodules(*197*) and contributes to the soil water nitrogen content.

The fate of nitrogen applied to soils has been the subject of considerable research(*198–204*). Nitrate or NH_3-N added to soil as fertilizer or in water does not long remain unchanged in the soil. If not assimilated by microbes or higher plants, it will usually be lost either by volatilization or leaching. The extent of leaching depends chiefly on soil texture and on the quantity of water that penetrates the soil before the crop can assimilate the nitrogen. Volatilization losses include direct volatilization of ammonia in alkaline soils and of nitrogen produced by denitrifying microbes. However, it is at present impossible to draw up an accurate soil nitrogen balance sheet because the operation of all these processes depends greatly on specific climatic and soil conditions.

Soil phosphorus undergoes microbial transformations similar to those described for nitrogen. As in the case of nitrogen, a considerable fraction of total soil phosphorus resides in organic forms; these organic forms have been estimated to range between 25 and 80% of the total(*192*). In the mineralization of organic phosphorus, dephosphorylation by the phosphatase enzymes of many soil bacteria and fungi releases inorganic phosphate to solution, although some of the phosphorus-containing nucleic acids and protein moieties are adsorbed by clays, reducing dephosphorylation markedly(*192,205,206*). Also, in an acid environment, phytin reacts to form insoluble iron and aluminum phytates which are resistant to dephosphorylation(*205–207*), and the addition of lime causes increased mineralization of the more soluble calcium phytate(*205,208*).

Phosphorus mineralization is sensitive to pH (enhanced at higher pH values), water content, cultivation practices, and the carbon: phosphorus ratio in the organic residues(*189*). If the carbon: phosphorus ratio is wide, the greater part of the phosphorus will be immobilized; if the ratio is narrow, excess phosphorus will be liberated as inorganic phosphate.

The sulfur cycle in nature is very similar to the cycles of nitrogen and phosphorus, undergoing many transformations as a result of activities of microbes(*209,210*). As with nitrogen and phosphorus, a high fraction of soil sulfur (50–70%) resides in organic compounds(*211*). Thus, the microbial transformations play an important role in the soil solution concentration of sulfate. The soil sulfur cycle consists of mineralization, immobilization, and oxidation reduction transformations similar to those already described for nitrogen and phosphorus. The critical carbon : sulfur ratio, however, is more in the order of 50 : 1 (*211*).

The reduction transformation of sulfur in soils is probably more important in influencing the composition of the soil solution than are analogous nitrogen and phosphorus processes. Under anaerobic conditions, biological reduction of sulfur can lead to the formation of sodium carbonate in soils at the expense of calcium and sulfate(*212–214*). Sulfide produced by reduction of sulfate combines with iron to form insoluble ferrous sulfide, and the carbon dioxide released by biological oxidation of organic matter forms bicarbonate. As the solution concentrates or the carbon dioxide partial pressure is diminished, calcium carbonate precipitates with resultant predominance of a sodium bicarbonate solution of high pH.

The carbon cycle in soils has been reviewed(*189–191*) as well as the interaction of this cycle with pesticides(*215,216*). Soil organic matter is decomposed by many diverse microbes, including aerobic and anaerobic bacteria, actinomycetes, filamentous fungi, and higher fungi. The activity of these microbes is governed by a number of environmental conditions, such as temperature, aeration, moisture, pH, nutrient status, and the presence or absence of more readily available substrates. Organic matter decomposition serves two purposes for microbes – it supplies energy for growth and carbon for formation of new cell material.

Another important consequence of the carbon cycle in soils is the effect it has on pesticide persistence in soils. Because of the great diversity of pesticides, general statements with regard to their degradability are precluded. Each substance, or at best a small group, is best evaluated separately. To illustrate, however, two examples are given(*190*).

When 2,4-D (2,4-dichlorophenoxyacetic acid) is applied to soils, a segment of the microbial population develops that can cause the oxidation of this chlorinated hydrocarbon herbicide. The time for detoxification in normal, unsterilized soil, for field applications of several pounds per acre, is generally some 2–8 weeks, depending on such factors as temperature, soil texture, moisture, and pH. A very similar herbicide structurally, 2,4,5-T (2,4,5-trichlorophenoxyacetic acid), contains only one more chlorine atom on the benzene ring, yet this slight modification has a marked effect on its degradability. Soils treated with 2,4,5-T still contain the pesticide after all vestiges of toxicity due to equivalent quantities of 2,4-D have disappeared. Even under optimum conditions, 2,4,5-T remains in nonsterile soil for periods of 6 to more than 12 months(*217*).

Other herbicides show a range in persistence of from 2 weeks (2,2-dichloropropionic acid) to more than 2 years (2,3,6-trichlorobenzoic acid)(*190*).

Alexander(*190*) has summarized the effect of structure on the degradability of pesticides as follows: (1) High molecular weight aliphatic hydrocarbons are more quickly metabolized than the small molecules, and aromatic compounds with aliphatic side chains have their chains more readily transformed if they are short; (2) unsaturation favors more ready degradation among the aliphatics, but branching has the reverse effect; (3) type, number, and positions of substituents affect the rate of turnover of organic pesticides. Kaufman(*215*) discusses the effect of structure in detail.

Pesticides may affect both the composition and size of the soil microbial population(*190,216*). As inhibitors of biological systems, pesticides may decrease the saprophytic soil population, affecting mineral nutrient transformation and hence the soil solution and drainage water compositions. Soil chemical properties may be temporarily altered by accumulation of residual chemicals or their decomposition products, by increases in the solubility or concentration of manganese, tin, copper, phosphate, calcium, ammonium, chloride, bromide, and sulfate, and by the possible formation of toxic or stimulating organic substances. However, the results of many investigations have shown that herbicides applied at recommended field rates (in the order of a few parts per million) generally have no harmful effects upon the microbial population or its biochemical activities. Similar results have been observed with most insecticides. Fumigants and fungicides do markedly affect the saprophytic population, although with time the population is reestablished. The exact effect of these influences on soil solution composition and drainage effluents is not known.

E. Drainage Water Properties

1. Surface Drainage Waters

a. Erosional Losses of Sediments and Adsorbed Constituents

Sediment is by far the largest single pollutant present in rivers(*218*), although actual concentrations tend to be highly variable(*61,62*). Not only is sediment pollution important in itself, but sediments also may act as reservoirs or carriers of pesticides, radionuclides, nutrient elements, and inorganic cations(*62,219–221*).

The slow flow of irrigation waters on nearly level irrigated lands tends to keep the sediment level of surface drainage waters relatively low. Major contributions of sediments result rather from runoff of high intensity rain on fallow sloping lands, such as the wheat lands of the Northwest(*222*). In Montana, for example, sediment loads of somewhat more than 200 mg/liter were measured in streams below the point of addition of surface waste waters, compared to 25 mg/liter above the point of such additions. In contrast, as much as 672 metric tons of soil materials have been lost from 1 hectare in one season, and average losses were estimated at 22.4 metric tons/ha per year, of which one-fourth contributed to sediment burden of the streams, causing sediment concentrations in streams of 20,000–82,000 mg/liter. The extent of water erosion depends upon climate, topography, vegetation, soil, and human activities(*223*). Wadleigh(*224*) stated that 50% of the sediments in streams come from agricultural lands, 30% from geologic erosion in tributaries, and 5–10% from range lands and forests. The Mississippi River carries about half of the entire sediment burden of the United States. In contrast to the high losses from agricultural lands, sediment losses from land covered with native vegetation are a few metric tons per hectare per year and are usually in the range of 0.11–0.67 metric ton/ha per year in the humid regions of the United States(*225*). Permissible losses from agricultural lands, according to soil conservationists, may range from 1.1 to 13 metric tons/ha per year.

The adsorptive capacity of sediments not only for cations but also for various radionuclides, pesticides, and nutrient elements may be inferred from Kennedy's data for 21 streams of the United States. Kennedy(*62*) found that the ratio of cations adsorbed on sediments to cations in solution may reach a maximum of about 0.8 in the eastern streams but may be 3.0 or more in some western streams. A range in concentration of adsorbed cations of 0.1–18 meq/liter was reported. According to the data of Rainwater(*61*), and assuming an average sediment cation exchange capacity of 20 meq/100 g, 90% of U.S. rivers contain less than 1.6 meq/liter of cations adsorbed on sediments (one-tenth as much as is present in solution), while for 50% of the rivers, the adsorbed cations are less than 0.1 meq/liter, or only 3% of the level in solution. The high (more than 1.6 meq/liter) sediment-adsorbed cation contents cited by Kennedy, therefore, occur in 10% or less of U.S. rivers. The semiarid Southwest probably accounts for a high proportion of the sediment-rich rivers, which are also primary sources of irrigation water. Sediments in such rivers would influence the cation status of the water by exchange reactions, increasing the proportion of monovalent cations (sodium) and decreasing the proportion of divalent cations (calcium and magnesium) in solution relative to the sediment-free condition.

Sediments are important in carrying radionuclides into surface waters (*226–231*). Data for alpha and beta activity as suspended or dissolved constituents in representative U.S. rivers are given in Ref. *59*. The suspended fraction commonly accounts for 13.8–68.5% of the total radioactivity. Such values do not necessarily indicate the relative initial contributions of solution and adsorbed phases, however, since exchange between these fractions probably occurs.

Sediments may carry variable but significant amounts of nutrient elements in the surface waters (*219,232*). Organic matter losses of 378–1290 kg/ha and estimated nitrogen losses from crop soils of 2–75 kg/ha have been cited, based on measurements from plots generally 40 m^2 in size (*219*). Allison (*198*) estimates erosional losses of nitrogen from crop lands in the United States at about 27 kg/ha (as nitrogen) per year compared to leaching losses of 26 kg/ha (as nitrogen) per year. However, Feth (*49*) questions these values and presents data showing erosional losses of less than 0.3 to 6.4 kg/ha (as nitrogen) per year. These latter data are from river samples, while Allison's were from lysimeter plots of only a few square meters in size. Data for European rivers are similar to Feth's, with losses of about 1.6 kg/ha (as nitrogen) per year (*51*).

Plot tests on land with a 5% surface slope, fertilized with 224 kg of nitrogen/ha and subjected 1 hr later to 13 cm of simulated rain in 2 hr (an intensity occurring naturally only once in 100 years), caused the losses shown in Table 10 (*233*).

Although the fallow plot eroded badly, only 2.3% of the applied fertilizer was washed away. The first few inches of rain apparently moved the fertilizer downward where it was protected from erosion. Little erosion and fertilizer loss occurred on the sod plot. Moe et al. (*234*) observed losses of 2.5–14.5% of the applied nitrogen from 39-m^2 test plots on a 13% slope with a simulated rainfall of 13 cm in two 1-hr storms. Considerable quantities of organic nitrogen were lost from these plots. Erosional losses of other nutrient elements are shown in Table 11 (*219,232*).

Wadleigh (*224*) states that each 1000 metric tons of suspended sediment carries about 500 kg of adsorbed phosphorus, with only traces of phosphorus in solution (less than 1 ppm).

TABLE 10

Plot	Water lost, cm	Soil lost, metric tons/ha	N fertilizer lost, %
Fallow	10.5	39.2	2.30
Sod	1.6	0.33	0.15

TABLE 11

Element	Erosional losses, kg/ha	
	Soluble	Total
Ca	6.7–11.2	86–514
Mg	1.5–4.0	105–201
K	< 1–10.1	29–1398
S	12.9–29.7	47–113

The erosional losses of pesticides in the adsorbed state have been studied in a 1036-km^2 basin in northern Alabama(*221*). The economy in this basin of gently rolling hills (15–60% slopes) is predominantly agricultural, with cotton receiving most of the insecticides (toxaphene, DDT, and BHC). Sediments were more important in transporting DDT and DDE than toxaphene or BHC. The latter were transported in suspension but were not adsorbed on sediment particles. Maximum concentrations of 20 and 90 mμg/liter of DDT and DDE, respectively, were found in solution 1 year after large amounts of DDT were employed. Surface drainage also brings pesticides into streams. Nicholson(*65*), reviewing the findings on pesticides in runoff from agricultural lands, concluded that toxaphene, dieldrin, BHC, endrin, DDT, and parathion can enter water courses along with runoff water and that the addition can be more or less continuous throughout the year. Rarely were pesticide concentrations in waters greater than 1 μg/liter.

b. Inorganic Solutes, Salts, and Nutrient Elements

A brief description of the extent of irrigation agriculture and its operation will aid the general reader in understanding how this type of agriculture affects water quality. The 17 western states of the United States have about 33 million acres of irrigated land which receive about 128 million acre-ft of water per year(*235*). The return flow from this acreage is about 42.6 million acre-ft, or one-third the volume applied. This return flow was about equal to the combined municipal and industrial waste water flow from the same area(*235*).

Surface water irrigation systems operate as follows: Water is diverted to canals from which it flows to laterals serving farm units and thence to the land surface. Return flow of the nonconsumed portion occurs by overflow, runoff, and seepage. Overflow is the excess water needed to maintain a head on the laterals; this returns directly to the canal with little change in quality. Runoff is the excess water applied to land that returns to the surface drain system. This may be high in turbidity and suspended organic matter and may contain adsorbed nutrient elements and pesticides. Seepage, or infiltrated drainage, is the water that passes through the soil and may reenter the surface stream or accumulate in underground strata. This water is generally considerably higher in total salt content than the applied water.

A very complete study in the Yakima River basin provides valuable data on the changes in water properties in a major irrigation project (*236*). Of 6.6 acre-ft/acre of low salinity Yakima River water diverted to 375,000 acres, 4.25 acre-ft were actually applied per acre of land, of which 2.6 acre-ft were lost by evapotranspiration, the remainder becoming drainage water. Table 12 compares the properties of the applied, surface, and

TABLE 12
Average Compositions of Applied Water and Surface and Subsurface Drainage Waters During the Irrigation and Nonirrigation Seasons, 1959–1960, Yakima Valley, Washington[a]

Constituent or characteristic	Applied water	Subsurface drain water	Surface drain water
Temp, °C	16.0	13.3 (10.9)[b]	17.9 (8.1)[b]
Dissolved oxygen, mg/liter	10.2	6.8 (7.5)	9.0 (11.2)
pH	8.1	7.7 (7.7)	8.2 (8.0)
HCO_3 alk. mg/liter as $CaCO_3$	46.0	218.0 (248)	138.0 (223)
CO_3 alk. mg/liter as $CaCO_3$	1.0	0 (0)	2.0 (3)
Hardness as $CaCO_3$	46.0	186.0 (226)	121.0 (205)
Turbidity units	37,0	12.0 (8)	130.0 (35)
Color units	22.0	12.0 (7)	38.0 (16)
EC[c], μmho/cm	83.0	420.0 (441)	283.0 (406)
Cl, mg/liter	1.0	12.0 (19)	8.0 (16)
NO_3 as N mg/liter	0.25	2.5 (2.5)	0.8 (1.9)
Total Kjeldahl N mg/liter	0.27	0.32 (0.3)	0.3 (0.3)
Chem. oxygen demand, mg/liter	7.0	9.0 (4)	10.0 (5)
Soluble PO_4, mg/liter	0.21	0.7 (0.7)	0.6 (0.6)
Total PO_4, mg/liter	0.32	0.9 (0.8)	0.8 (0.8)
SO_4, mg/liter	5.4	39.0 (50)	37.0 (67)
Ca, mg/liter	10.0	44.0 (51)	31.0 (45)
Mg, mg/liter	5.0	20.0 (24)	12.0 (22)
Na, mg/liter	4.1	38.0 (46)	26.0 (46)
K, mg/liter	1.4	4.7 (5.3)	5.3 (6.0)
Coliforms/100 ml	1070.0	103.0 (90)	10600.0 (1170)

[a]Adapted from Ref. *236*.
[b]All figures in parentheses refer to the nonirrigation season.
[c]Electrical conductivity.

subsurface drainage waters. Most of the land in the Yakima Valley is drained by open drains into which most of the subsurface drains also empty. A precise distinction, therefore, between surface and subsurface drainage waters is not possible. This is true of most irrigation projects. Nevertheless, the properties of the waters in open drains are notably different from those of the subsurface drains. The number of coliform bacteria, the color, and the turbidity increase markedly in the surface drain water but decrease in the subsurface drain water as compared with the applied water.

Tile drain waters were found to be 10 times more saline than surface drain waters in the San Joaquin Valley of California(*237*) (Table 13).

TABLE 13
Chemical Composition of Agricultural Waste Waters from Irrigated Fields of the San Joaquin Valley, California[a]

Constituent	Concentration, mg/liter	
	Tile drains	Open drains
Ca	440	57
Mg	260	32
Na	2000	120
K	—	4
HCO_3	200	220
SO_4	3600	170
Cl	1200	130
B	28	0.8
Total dissolved solids	6400	650
Total N	29	6.5
Total P	0.3	0.5

[a]Adapted from Ref. *237*.

Drain waters from the rice-growing area of the Colusa basin, California, represent surface drainage waters that have moved through a series of contoured fields (Table 14). Although such drainage waters probably represent a minimum admixture of subsurface drainage, the prolonged ponding causes higher evaporative losses than would usually occur with briefer irrigation runs. The data in Table 14, therefore, approximate upper limits for increased salt content of surface drainage waters. Deliberate attempts to effect reclamation by removing surface salt accumulations from saline lands resulted in only a 750-ppm increase in the salt content of the water(*131*). From the foregoing and from data in Tables 12–14,

TABLE 14
Chemical Composition of Colusa Basin Drain Water and Sacramento River Water at Knights Landing, California

Constituent	Range in concentration, mg/liter	
	Colusa Basin Drain[a]	Sacramento River[b]
Total dissolved solids	246–792	95–131
SO_4	46–243	1.9–12.0
Cl	21–113	3.9–9.2
Total N	0.7–2.1	0–0.4
Total P	0,4–0.8	—

[a]Adapted from Ref. *238*.
[b]Adapted from Ref. *46*.

the increase in total salt content in surface drainage waters may be expected to range from about 50 to 750 mg/liter, but it should more often approximate the lower value. From the data in Tables 12–14, and according to Refs. *239,240*, total nitrogen and phosphorus concentrations in agricultural runoff water may be approximately 0.5–7 and 0.2–0.8 mg/liter, respectively.

c. Pesticides in Agricultural Surface Runoff Waters

The pesticide content of agricultural surface runoff waters has recently been reviewed(*241–243*). Nicholson(*244*) found no DDT in a river draining a 400-sq-mile cotton-growing area of Alabama, although DDT accounted for 25% of the applied insecticides. The other major pesticides used, BHC and toxaphene, were found in concentrations of 0.007–1.004 and 0.010–0.270 μg/liter, respectively. The insecticide found in the river, although in solution, was attributed to surface runoff water. In the Columbia basin, Washington, irrigation return flow was found to contain (in μg/liter) aldrin (0.0001–0.0020); DDT (0.00002–0.016); 2,4-D isopropyl ester (0.0004–0.018); and endrin (0.004–0.057). In addition, DDT was found in the river bottom sediment (56–144 μg/liter)(*245*).

Higher concentrations of pesticides in surface waste waters have been reported(*242*): aldrin and dieldrin, less than 0.025 to 0.6 μg/liter (Indiana) and 1.3 μg/liter (Arkansas); DDT, 0.05–0.6 μg/liter (Indiana); endrin, 0.66 μg/liter (Arkansas).

California would appear to be a useful source of information on pesticide contamination of surface waters because its use of pesticides amounts to one-fifth of the total national usage(*70*). However, many analyses have not included sediment-adsorbed pesticides; therefore, very low levels reported (parts per trillion) do not represent the total pesticide content of the drainage waters (*243*). Very low rainfall in many irrigated areas of California would also minimize runoff and pesticide losses.

Pesticide contamination in the Mississippi River delta was found only downstream from pesticide packaging or manufacturing plants (*246*), and similar negative evidence has been reported for the Mississippi and Missouri River systems (*247*). It has been pointed out that the activated carbon filter method used did not include pesticides adsorbed on suspended material (*248*). Reports of fish and wildlife kills have been frequent (*249,250*). Most of the pesticide may be associated with suspended material, primarily organic matter (*250*).

Conflicting views on the importance of agricultural runoff water and pesticide pollution stem largely from the sometimes inadequate analytical procedures used and partly from differences in standards. A factor causing further uncertainty in this regard is the phenomenon of biological cycling of persistent pesticides. What may appear to be a safe level of pollutant under most instances may under other circumstances become so concentrated, sometimes hundreds of thousands of times above levels in the environment, by biological cycling in a food chain that toxic levels are attained (*251*).

2. INFILTRATED DRAINAGE WATERS

Solute Composition and Content

Data on the composition of soil-infiltrated waters have already been presented (Tables 12 and 13). Tile effluents in the San Joaquin Valley have been studied especially to determine sources of nitrate-N (*252*). Representative data in Table 15 indicate variability in total and specific solute content between fields, even though the same water was applied to both. Total solute concentrations were 9–19 times greater than the solute concentration of the applied irrigation water. Even after several years of tile operation, more salts were being removed from the fields than were applied in the water. Gypsum in the soil apparently accounted for the high sulfate concentration of the effluents. The specific source of the nitrate was not determined, although fertilizers appeared to be a contributing factor.

In a review of tile drainage waters in the San Joaquin Valley, mean total solutes of 5000 mg/liter and nitrogen and phosphorus levels of 25 and 0.08 mg/liter, respectively, were found (*253*), with a range for nitrogen of 1.8–62.4 mg/liter and for phosphorus of 0.053–0.23 mg/liter (*254*). Large amounts of applied nitrogen seem to have been lost to the drainage water, but phosphorus losses were insignificant (*255*).

Inorganic salt losses from irrigated fields to streams and rivers have been estimated in kilograms per hectare per year. The values for nitrogen

TABLE 15
Compositions of Irrigation and Tile Drainage Waters in the San Joaquin Valley, California[a]

Water	EC[b], mmho/cm	B, mg/liter	HCO_3	Cl	SO_4	NO_3	Na	Ca	Mg	K
			meq/liter							
Irrigation water	0.7	0.1	1.6	3.7	1.4	0.12	3.7	1.1	1.6	0.2
Tile effluent, field A	6.0	6.5	5.1	18.0	52.0	1.64	32.7	27.4	14.7	0.4
Tile effluent, field B	9.7	13.5	5.2	19.8	95.7	3.18	79.5	21.9	23.0	0.5

[a] Adapted from Ref. *252*.
[b] Electrical conductivity.

are as follows: 8(*256*), 25(*257*), 43–186(*239*), and 76(*236,258*). Corresponding estimates for phosphorus are: 0.02(*259*), 0.4(*256*), 2.5(*236, 258*), and 2.8–10(*239*). Sylvester and Seabloom(*236,258*) estimated, in addition, losses of other constituents from the Yakima River basin, in kilograms per hectare per year: HCO_3, 1405; Cl, 112; SO_4, 280; Ca, 230; Mg, 95; Na, 280; and K, 18.

Feed lots may also contribute to agricultural drainage waters, both as runoff(*260*) and as seepage(*257*). In Colorado, average total nitrogen contents per 6.7 m of soil depth under various land uses were, in kilograms per hectare: alfalfa, 70; native grassland, 81; nonirrigated, cultivated dry land, 233; irrigated fields other than alfalfa, 452; and feed lots, 1282.

Pesticides have also been detected in subsurface drainage waters (*253,261*). The pesticide contents in tile drain effluents in the San Joaquin Valley of California and in the Panoche drain (a large open drain in the San Joaquin Valley) are given in Tables 16 and 17. Relatively small amounts of chlorinated hydrocarbons were found in the tile drainage effluent. Higher concentrations were found in the open drain, where both surface and subsurface drainage waters were collected.

The total quantity of insecticides in the tile drainage effluent was generally less than the quantity applied in the irrigation water, despite the fact that the depth of irrigation water was 5.2 times the equivalent depth of drainage water. Thus, if there were no removal or addition of pesticide residues with application of irrigation water and percolation through the soil, the concentration of the effluent should have been five to six times

TABLE 16

Chlorinated Hydrocarbon Pesticides in 66 Analyses of Tile Drainage Effluents in the San Joaquin Valley, California[a]

Pesticide	No. of times detected	Concentration, ng/liter		
		Max	Min	Average
BHC	2	1500	300	900
DDE	16	130	10	25
DDD and/or DDT	62	600	10	78
Dieldrin	13	130	10	35
Heptachlor epoxide	25	40	10	14
Lindane	22	340	10	35
Toxaphene	13	950	130	528
TCBC	1	50	50	50

[a]Adapted from Ref. *261* by courtesy of The American Geophysical Society.

TABLE 17
Pesticides Identified in 61 Analyses of Panoche Drain, San Joaquin Valley, California[a]

Pesticide	No. of times detected	Concentration, ng/liter		
		Max	Min	Average
Chlorinated hydrocarbons				
DDE	12	150	10	64
DDD and/or DDT	54	5,700	60	606
Dieldrin	1	120	120	120
Heptachlor epoxide	6	100	10	24
Lindane	7	220	10	67
Toxaphene	60	7,900	100	2,009
Thiodan	1	210	210	210
Methoxychlor	1	450	450	450
Summation of identified chlorinated hydrocarbons	61	11,400	100	2,548
Thiophosphates				
Baytex	4	160	30	87
Ethion	11	1,200	15	265
Malathion	7	320	55	119
Methyl parathion	3	6,400	300	2,530
Parathion	19	3,600	20	520
Thimet	1	30	30	30
Summation of identified thiophosphates	33	6,400	15	654

[a]Adapted from Ref. *261*, by courtesy of The American Geophysical Society.

that of the applied water. Surface runoff water, however, had pesticide concentrations 7–12 times as high as the applied water when DDT was applied to the land, and as much as 85 times more pesticide than the irrigation water when lindane was applied. Relatively large concentrations of pesticides were found in the surface soil of the area studied, even when no pesticide had been applied directly to the soil. This indicates that spray drift or adsorptive removal of pesticides from the irrigation water was involved. The former could be a major source of pesticide contamination of surface waters as well. DDD and/or DDT were the pesticides most frequently detected in tile drainage waters (Table 16), whereas toxaphene was the most frequently found pesticide in open drains (Table 17). Pillsbury(*253*), in a study of pesticide movement in a soil not previously treated with pesticides, found that traces of pesticides were passed through the soil profile and into tile drainage waters. Lindane, in particular, passed through the soil quite readily. Tile drainage effluents had about 2.5 times as high a pesticide concentration as the irrigation waters. Surface tail water or drainage water from the field contained a maximum of 62,600 ng/liter and required 4 months to decrease to 200 ng/liter, a level comparable to that of tile drain water. Although such large losses would not normally occur under field conditions, Pillsbury's study does demonstrate that most of the applied pesticides will not pass through the soil.

3. Salt Balance in Irrigation Projects and Salinity of River Systems

Half to nearly all of the applied irrigation water is lost by evapotranspiration. Most of the salts present in the water are left behind in the soil. As a result, the remaining volume of water increases in salt concentration. With repeated irrigations, the salt concentration would soon prohibit further plant growth. To prevent this, irrigation water must be applied in excess of evapotranspiration in order to leach out the accumulated salts. This leads to the concept of "salt balance," i.e., in an irrigated area, the quantity of salt carried out by the drainage water must equal or exceed that brought in by the irrigation water, to avoid a progressive accumulation of salt in the soil(*262*). As water is used by successive downstream diversions from a river, its salt concentration generally increases because of the return flow of more saline drainage waters to the river. This process has led to the "equivalent service concept" (*263*):

> Whenever the quality of water applied to land for irrigation is modified by an increase or decrease in the concentration of one or more ions detrimental to plant growth, maintenance of the same effective concentration of the soil solution, by increasing or decreasing the amount of water applied, will constitute equivalent service for the growing of crops (*263*).

Pillsbury et al. (*264*), on the basis of similar reasoning, proposed the "degradation ratio" (DR) to express the fractional equivalence of degraded water to good water. This concept assumes that a water becomes essentially valueless for irrigation when it has a salt content of 4800 ppm, equivalent to an electrical conductivity of 7.5 mmho/cm. Thus,

$$DR = \frac{(7.5 - EC_{iw})}{7.5} = \frac{(4800 - IW_{ppm})}{4800} \tag{4}$$

in which IW_{ppm} represents parts per million of salt in the irrigation water and EC_{iw} the electrical conductivity of the irrigation water. Use of a water of degraded quality requires more frequent and liberal irrigation, with increased costs for labor and capital investments such as drainage facilities.

Several salt balance studies of irrigation projects and river systems indicate the usefulness of the salt balance concept in assessing the salinity status of projects and the effects of irrigation on river systems. Bower et al.(*265*) utilized data on inputs of water and salt by irrigation and outputs of water and salt by drainage for the period 1957–1965 to examine the salt balance conditions in the Coachella Valley, California. The authors concluded that the salt balance index (output of salt/input of salt) was highly related to both the area of irrigated land having tile drainage and the leaching percentage (the percentage of water entering the soil which passes through the root zone to become drain water). In this example, the index became close to 1 when about half the irrigated land was tiled and the leaching percentage increased to about 30% (see Table 18). Comparison of the compositions of irrigation and drainage waters at salt balance indicates that about 10% of the applied salt precipitated in the soil, largely as calcium carbonate, and that calcium and magnesium in the water replaced exchangeable sodium in the soil during the percolation of the water through soil (see Table 19). The salt content of the drainage water was increased about threefold by consumptive use of water.

TABLE 18
Relationship of Salt Balance in Coachella Valley to Tiled Land Acreage and Leaching Percentage[a]

Year	Acres irrigated	Acres tiled land	Input water		Output water		Leaching percentage	Salt balance index[b]
			acre-ft	tons salt/acre-ft	acre-ft	tons salt/acre-ft		
1957	52,329	10,835	299,590	1.23	32,578	3.07	11.0	0.27
1958	53,592	14,585	348,590	0.97	46,467	3.37	13.2	0.46
1959	55,527	19,115	358,641	0.97	47,188	3.53	13.2	0.48
1960	54,333	22,285	368,926	0.99	61,327	3.72	16.6	0.63
1961	53,990	24,857	366,315	1.07	75,597	3.67	20.6	0.71
1962	53,443	27,071	389,877	1.12	101,169	3.43	26.0	0.79
1963	57,773	28,984	370,014	1.06	110,627	3.64	29.8	1.03
1964	60,053	30,686	359,989	1.13	113,104	3.66	31.4	1.02
1965	59,890	32,080	341,165	1.19	124,128	3.42	36.5	1.04

[a]Adapted from Ref. *265*, by courtesy of The American Society of Civil Engineers.
[b]Salt balance index = output of salt/input of salt.

Haney and Bendixen(*266*) reviewed salt balance studies for several irrigated valleys. Salt balance was not achieved in the El Paso and Yuma Valleys during the period studied (prior to 1940), even though in the former case (Table 20), the leaching percentage was as high as that which achieved salt balance in the Coachella Valley (Table 18). Salt balance was achieved in the Mesilla Valley with a leaching percentage of 60%.

Wilcox(*267*) has presented data on the salt balance for the lower Rio

TABLE 19
Average Composition of Colorado River Water and Drainage Waters in Coachella Valley, California (Average During 1963–1966)[a]

Water	Equivalent percentage of cations and anions							
	Ca	Mg	Na	K	CO_3^+ HCO_3	SO_4	Cl	NO_3
Irrigation	36.7	18.7	43.4	1.3	19.8	52.4	27.5	0.2
Drainage	25.4	8.5	65.0	1.1	15.0	48.5	36.2	0.8

[a]Adapted from Ref. *265*, by courtesy of The American Society of Civil Engineers.

TABLE 20
Salt Balance Data for Several Irrigated Valleys[a]

	Mesilla Valley, New Mexico-Texas	El Paso Valley, Texas	Yuma Valley, Arizona
Irrigated area, acres	80,000	60,000	40,000
Irrigation water, acre-ft	744,380	496,113	259,917
Drainage water, acre-ft	496,113	173,219	57,095
Leaching percentage	66.6	34.9	22.0
Salts in irrigation water,			
tons/acre-ft	0.805	1.226	0.956
mg/liter	592	900	704
Salts in drainage water,			
tons/acre-ft	1.226	2.666	1.993
mg/liter	900	1970	1470
Output of salts,			
% of input	101.5	75.9	45.8
Output salt conc.			
% of input	152	218	208

[a]Adapted from Ref. *266*, by courtesy of The American Water Works Association, Inc.

Grande Valley (see Table 21). In this basin, during the 20-year period 1934–1953, the flow at Fort Quitman was about one-fifth of the initial flow at Otowi Bridge, a distance of about 450 miles, while the salt concentration increased by 2 tons/acre-ft. Between these two sampling sites, there are several diversions and drainage returns, with a total irrigated acreage of about 250,000 acres. Under conditions of salt balance, the salt burden of the stream was approximately the same at all gauging stations below San Marcial.

TABLE 21

Salt Burdens and Compositions of Water at Seven Stations on the Rio Grande for the 20-Year Period 1934–1953[a]

Successive downstream stations	Mean discharge, thousands of acre-ft/year	Dissolved solids (weighted mean), tons/acre-ft	Dissolved solids, thousands of tons/year	Cations		Anions		Total salts, meq/liter
				% Ca	% Na	% HCO_3	% Cl	
Otowi Bridge	1079	0.30	324	57	25	59	6	3.28
San Marcial	853	0.61	520	44	41	—	—	6.82
Elephant Butte	790	0.65	514	42	43	35	17	7.29
Caballo Dam	781	0.70	547	42	44	35	19	7.81
Leasburg Dam	743	0.75	557	41	44	33	21	8.36
El Paso	525	1.07	562	35	52	28	29	12.10
Fort Quitman	203	2.30	467	28	60	13	54	26.37

[a]Adapted from Ref. *267*.

Other rivers may not show this pattern in relation to irrigation diversions (*268*). Table 22 shows salt burden data for successive downstream stations in the Yakima, Snake, Columbia, and Colorado Rivers. Dilution of return flow, by additional tributaries and salt precipitation, and mineral solubilization effects distort the inverse volume concentration relationship observed in the lower Rio Grande Valley. Subsurface drainage returns in the Yakima project, for example, show about a fivefold increase in salts when only 60% of the applied water was consumptively used (Table 12), suggesting mineral solubilization contributions. Even when total salt burden is relatively constant, as on the lower Rio Grande, compensating factors operate, the proportion of calcium and bicarbonate decreasing while sodium and chloride are increasing (Table 2). The significance of water quality degradation associated with consumptive use of water by irrigation projects depends on the initial salt status of the river. Although the Yakima project shows a greater than expected salt increase, the drainage water from the Yakima is lower in salinity (Table 12) than

TABLE 22
Discharge and Dissolved Solids at Successive Downstream Stations on the Yakima, Snake, Columbia, and Colorado Rivers[a]

River station	Discharge, annual thousands of acre-ft	Dissolved solids	
		Weighted mean tons/acre-ft	Annual thousands of tons
Yakima River:			
Cle Elum	1,238	0.05	62
Kiona	2,222	0.20	444
Snake River:			
Heise	4,091	0.31	1,268
King Hill	6,845	0.45	3,080
Clarkston	30,890	0.23	7,105
Columbia River:			
International Boundary	75,350	0.12	9,042
Grand Coulee Dam	81,790	0.12	9,815
Maryhill Ferry	129,600	0.14	18,144
Colorado River:			
Glenwood Springs	1,510	0.35	528
Cameo (irrig. area)	2,715	0.55	1,493
Cisco (irrig. area)	5,903	0.74	4,368
Lee's Ferry	13,200	0.61	8,052
Grand Canyon	13,530	0.75	10,148
Below Hoover Dam	14,450	0.92	13,294

[a]Adapted from Ref. *268*.

that from the Colorado River at Otowi Bridge. Low initial salinity permits appreciable relative increase without markedly degrading the water quality. Comparison of river water quality before and after large-scale irrigation diversions (see Table 23) further illustrates the quite different effects that may occur in different river systems (*269*). Increase in irrigated acreage in the past 50 years or so has had little effect on the water quality of the Colorado and Missouri Rivers but has markedly affected the Rio Grande.

TABLE 23
Chemical Composition (meq/liter) of River Water Before and After Intensive Irrigation Developments[a]

	Colorado River		Missouri River		Rio Grande River	
Constituent	1893	1962	1906	1953	1893	1945
Ca	3.3	4.5	3.2	3.1	2.95	4.5
Mg	1.1	2.4	1.6	1.5	0.67	1.5
Na	8.5	5.8	2.1	2.6	2.85	6.3
$CO_3 + HCO_3$	2.8	2.7	3.3	3.1	2.26	3.6
Cl	5.2	3.5	0.3	0.4	1.52	3.4
SO_4	4.8	6.8	3.5	3.7	2.61	5.3
Total solute contents, mg/liter	774	814	514	523	380	850
Millions of irrigated acres and year:	0.9 (1902)	3.5 (1950)	2.9 (1902)	5.1 (1950)	0.5 (1902)	2.5 (1950)

[a]Adapted from Ref. *269*, by courtesy of The American Society of Civil Engineers.

III. Quality Criteria and Standards for Irrigation Water

A. Criteria of Irrigation Water Quality

1. General

The usual criteria for irrigation water quality are salinity, sodicity, and toxicity. Under the last heading, only a few solutes that occur naturally in waters are usually considered. Many others that may be introduced in industrial or municipal wastes have heretofore been ignored. In many of these cases, reliable tolerance data are scarce, but available information is summarized herein.

Missing from the list of the usual irrigation water quality criteria are many criteria important in assessing water quality for other uses: odor, taste, color, turbidity, temperature, BOD, pathogenic organisms, pH, hardness, nitrates, phosphates, etc. Some of these are irrelevant to irrigation water quality, but some are significant and are therefore discussed in this general section.

Odor, taste, and color are esthetic properties of a water that do not usually affect its suitability for irrigation. Partially reclaimed sewage waters that may be esthetically unacceptable for other uses can be used for irrigation of at least some crops. In the process, filtration and adsorption by soil remove most of the undesirable contaminants, and most organic materials are oxidized and decomposed by soil microorganisms.

Turbidity, an esthetic problem for domestic water, may be significant for irrigation. Solid particles can settle out and clog water conveyance and distribution systems and may affect soil permeability to air and water. Sprinkler irrigation systems may be clogged by sediments which may also cover sprinkled leaf surfaces. Special provision for removal of sediments may be required. Permissable sediment loads or particle size limits depend on the characteristics of the sprinkler units employed, and no general recommendations are available. Filtration through soil effectively removes turbidity and sediments.

Although temperature is not usually a specified criterion for irrigation waters, under special circumstances it can be very important. Low water temperatures inhibit the development of rice, and water-holding basins are recommended for warming the irrigation water before release onto the paddy (*270*). Most crops are much less intensively irrigated than rice, and water temperature is modified by soil and ambient temperatures. A wide range of water temperatures of 10–35°C may therefore be tolerated. Irrigation, by fostering energy absorption, may increase the temperature of return flow waters from irrigation districts (*236,258*). Such warmer waters may raise the temperature of streams and affect other water uses.

Biological oxygen demand (BOD), as may be inferred from the use of partially reclaimed sewage water, is not a critical factor for irrigation water. However, in poorly drained and, therefore, poorly aerated soil, the additional depletion of oxygen by oxidizable organic matter may be deleterious. Certainly, the use of water of high BOD for irrigation should be restricted to well-drained, well-aerated soils.

Hardness is not a relevant criterion for irrigation water since calcium and magnesium are beneficial ions in the soil. The precipitation of calcium carbonate is rarely directly deleterious to agriculture. Such problems as may result therefrom are discussed in connection with bicarbonate. Despite precipitation of some calcium carbonate, hardness generally increases in soil water as the result of the increase in total salt concentration and, therefore, of Ca + Mg, by evapotranspiration.

The pH of irrigation water is not a usual criterion for water quality because a wide range (pH 4.5–8.5) is tolerated by most crops, and the soil also tends to buffer the pH of the soil water. Injury associated with excessively high pH values of irrigated soil is usually related to exchangeable sodium (treated below) rather than pH of the irrigation water per se. The use of very acid (pH 2–4) waters, such as sometimes result from percolation through mine tailings or which may sometimes be associated with recently reclaimed marine soils, would not be recommended. In addition to low pH, such acid waters may also lead to toxicity by increased solubility of heavy metals.

Pathogenic organisms are rarely a problem in irrigation water from rivers and wells. When sewage-derived waters are used for irrigation, pathogenic organisms may contaminate the surface of growing plants. If these plants are to be consumed raw, a minimum time must elapse between the last irrigation and the harvest of the crop(*271*).

Nitrates and phosphates provide nutrients for plant growth and reduce the fertilizer requirement. Waters that may be unsuitable for release into streams because of eutrophication and that may cause health problems by contaminating drinking water supplies with nitrate present no problem to irrigation agriculture by reason of such nutrient content. The absorption of some nutrients by crops (nitrate) and the precipitation or adsorption of others in the soil (phosphate) can effectively reduce the eutrophication problem in nutrient-rich waste waters.

2. Salinity

Salinity is one of the major criteria for irrigation water quality. The salts present in irrigation water provide some of the nutrients essential for plants, but excessive salt concentrations inhibit plant growth. Inhibition by salinity is, in most cases, primarily a function of the osmotic pressure of the soil water. Inhibition by specific ions is usually of secondary importance. Since there is an excellent correlation between the electrical conductivity (EC) of soil waters and their osmotic pressure, the readily measured electrical conductivity has been adopted as a measure of the salinity hazard of irrigation water(*2*). Salinity is sometimes reported as total dissolved solids in milligrams per liter. In most irrigation waters, salts comprise virtually all of these soluble solids.

3. Specific Element Hazards

Some solutes may specifically inhibit plant growth. Since the proportions of the various ions in irrigation waters vary greatly, total salinity does not indicate the specific element hazard. Woody plants, including most tree, vine, and berry plants, are sensitive to accumulations of chloride or sodium in the leaves. Characteristic necroses in the form of marginal, tip, or other burn patterns develop when the leaves accumulate more than about 0.5% Cl or 0.25% Na on a dry weight basis(*185*). Such leaf injury, if uncorrected, often leads to death of the plants. The chloride and sodium contents of irrigation water, therefore, constitute distinct hazards to woody plants, and permissible limits for these elements must be specified.

Boron is essential for plant growth, but when the concentration in the root medium exceeds about 0.5–1.0 mg/liter, boron toxicity often results (*272*). In some crop plants, leaf injury symptoms may not be evident, but plant damage still occurs.

The aforementioned elements—chloride, sodium, and boron—and also carbonate plus bicarbonate are the only ones regularly considered in judging the quality of irrigation water. Bicarbonate and carbonate are considered primarily because they affect the concentration and proportions of cations in soil water. In some areas, other elements have caused serious local problems. Toxic levels of selenium occur in well waters of Wyoming and adjacent states, and in these areas selenium content is an important criterion of water quality(*57*). In some limited areas of California, lithium in well water has been reported to be toxic to citrus, and standards have been proposed(*273*). For the many other elements that occur naturally in waters or that may be introduced as contaminants, there has been little concern in the past, despite a rich literature on toxic effects of various elements and mineral compounds in solution culture studies and in the field. The main reason for this apparent neglect is that water has not been the carrier for these toxic materials in the past. Under field conditions, toxic accumulations of such elements as copper and molybdenum have resulted from repeated or excessive applications of these elements to correct nutritional deficiencies. Prolonged use of some agricultural chemicals, such as copper fungicides and arsenical insecticides for pest control, have caused toxic accumulations of copper and arsenic(*274*). Many other elements may occur in soils as normal products of mineralization and soil formation but may achieve toxic levels only under certain conditions. Increased solubility at low pH is responsible for widespread aluminum toxicity in acid soils and nickel toxicity in serpentine soils(*274*). In some instances, industrial pollution of natural waters may contribute toxic concentrations of these same elements and others; however, the mere occurrence of toxic levels of aluminum, copper, nickel, or some other heavy metal in a water resource need not make such a water unfit for agriculture. Upon permeating the soil, the element may be precipitated, adsorbed, or otherwise fixed in the soil. This will often occur if a soil contains sufficient lime or bases to raise the pH of the water above 5. Liming of acid soils generally corrects heavy metal toxicity by decreasing the solubility of aluminum, copper, manganese, and zinc. Whether a potentially toxic concentration of an element in water will actually cause toxic effects on plants depends on soil water interactions, and these can be modified by soil amendments and fertilizers. Prediction of toxicity is also complicated by the interactive effects among trace nutrients. Copper and zinc toxicities can cause iron deficiency, and improving iron nutrition can remove these toxic effects. In addition to direct toxic effects on plants, potential toxicity of plants to consuming animals must be considered. Molybdenum, selenium, and nitrate toxicities are of primary concern because of their effects on animals or humans even when plants may not be directly affected(*274*).

4. Sodium Hazard

It is generally agreed that an excessive proportion of sodium relative to calcium and magnesium in the irrigation water is hazardous to soils and crops, and that the sodium hazard may be estimated from the proportion of sodium equivalents as a percentage of the total exchangeable cation equivalents in soils. Sodium toxicity in sensitive fruit crops may occur when an exchangeable sodium percentage (ESP) in soil is as slow as 5% (*275*). Higher exchangeable sodium percentages (10–20%) associated with decreased percentages of exchangeable calcium and magnesium affect the physical structure of soils and also the nutrition of crops growing on them. It is pertinent to note that with regard to sodium, domestic and industrial uses have requirements diametrically opposite to those of agriculture. Soft waters, relatively high in sodium and low in calcium plus magnesium, are preferred for domestic and most industrial uses because of reduced lime deposition in pipes and boilers. Municipal water supplies, therefore, are often softened by removal of some of the divalent cations. When the same water supply must also be used for gardens, landscape plantings, and agriculture, the conflict in requirements necessitates a compromise, and the degree of water softening must be limited to a level that can be tolerated in irrigation waters(*276*).

An obvious requirement for assessing the sodium hazard of an irrigation water is a precise, predictive relation between the cationic concentrations of the water and the proportion of exchangeable sodium in the soil. Since this relationship is concentration dependent as well as ratio dependent, various modifications of adsorption equations have been used. The sodium adsorption ratio [SAR; see Eq. (1)], based on the Gapon equation, is now widely used. SAR and ESP are related by the empirical equation:

$$ESP = \frac{100(-0.0126+0.01475\ SAR)}{1+(-0.0126+0.01475\ SAR)} \quad (5)$$

This equation indicates an unequivocal relationship between the chemical composition of an irrigation water and the resultant exchangeable sodium status of the soil if the soil water has the same composition as the applied irrigation water. However, as already noted, irrigation waters undergo various salt concentration changes. Evapotranspiration, alone, will result in increased concentration of the cations, and since the SAR equation is concentration dependent, an increase in concentration will, ipso facto, increase the SAR and hence the ESP of the soil. Furthermore, the reduction in water volume by evapotranspiration will result in precipitation of the less soluble salts, primarily calcium carbonate, also increasing the SAR. In evaluating the sodium hazard of an irrigation water, possible changes in the proportions of cations in the soil water relative to the irrigation water must be taken into account. This is discussed further in Sec. III. B. 3.

5. Other Criteria

a. Pesticides and Agricultural Chemicals

Pesticide chemicals should not adversely affect irrigation agriculture if the materials are properly used. In any case, such adverse effects as have occurred have not been associated with excessive concentrations of pesticides or agricultural chemicals in irrigation waters, but rather with improper use of chemicals on sensitive crops or with airborne transport of dusts and volatile compounds to areas adjacent to the treated fields (*243*). Recent intensified analyses of river waters for pesticides have, indeed, revealed widespread contamination, but at such low concentrations as to affect the more sensitive aquatic fauna only or higher organisms of the food chain (*249*). Since wildlife, domestic animal, and human tolerances toward pesticides will obviously be more restrictive than crop tolerances per se, it is not likely that pesticides in irrigation waters will require specific criteria. Since most waters used for irrigation are also used for other purposes, the pesticide criteria for other water users will adequately protect irrigation agriculture.

Although irrigation agriculture may not be a plaintiff in questions of pesticide contamination, it is obviously a principal defendant. Loss of pesticides from treated fields is held to be a major contributor to pesticide contamination of river waters and estuaries. The other main source of contamination is release of pesticides from chemical plants producing or formulating and packaging the materials. Irrigation agriculture as opposed to humid zone agriculture should be a minimum contributor to pesticide problems except for some of the more soluble pesticides or in areas subject to seasonal high intensity rainfall that causes high runoff and erosion.

b. Sediments

Sediments in water may damage or clog conveyance systems, especially sprinkler irrigation systems. Permissible sediment loads, therefore, depend on the nature of the water transport and application systems. Sediments may also adversely affect surface soil structure by filling soil pores and decreasing permeability to air and water. Such problems in the past have been handled by individual irrigators or irrigation districts, and sediment burden has not been specified as a general criterion for irrigation water quality.

c. Radionuclides

Direct damage to plants by radionuclides has been observed only under extreme conditions in the laboratory or in areas subject to extreme radiation(*277*). As with pesticides, the lower tolerance of plant-consuming animals and humans to radionuclides will determine permissible levels of radioactive materials in plants and, therefore, in soils and irrigation waters. Plants can accumulate some minerals, and concentrations in the plants can be higher than the concentration in the root medium. Moreover, the radionuclides of primary concern, because of long half-life and relative abundance, are cations such as ^{90}Sr and ^{137}Cs. These elements, because of adsorption on the exchange complex, may be highly concentrated in the surface layers of soil and may initiate a sequence of increasing concentrations in the plant-animal food chain. Because they are adsorbed and not readily leached, decontamination of surface soils has been investigated, but no effective, practical remedies have been found. Since soils and plants can affect successive concentrations of radionuclides, radionuclide content in irrigation waters should be carefully considered as an important criterion for irrigation water quality. Unfortunately, relatively little is known regarding the magnitude of the accumulation effects. Radionuclides have heretofore been introduced primarily in fallout, rather than in irrigation water. Studies have been concerned with relative absorption of radionuclides by plants. Plants do not differentiate between radioisotopes of Sr and Ca, and accumulation in plants is in proportion to relative abundance of the two elements in the soil. Liming acid soils decreases uptake of Sr by increasing the supply of calcium. With neutral or alkaline soils, however, liming is ineffective in reducing Sr uptake(*278*). Although radionuclide content may become an important criterion for irrigation water quality, the available information appears to be insufficient for setting specific standards (*271*).

d. Microorganisms

River and well waters used for irrigation have generally posed no serious problems associated with microbial contamination. Such contamination may be a factor in spreading plant disease from field to field if runoff water is reused; these problems can be handled by improving farm management. Of more general concern is the possibility of contaminating edible farm products with disease organisms carried in irrigation water. Increasing use of municipal waste water and partially reclaimed sewage waters for irrigation may well require the introduction of criteria for microbial content. Studies on the persistence of disease organisms in irrigated fields of edible crop plants have recently been reviewed in a federal report on irrigation water quality standards (*271*). As would be expected, contact between edible plant parts and contaminated water supplies, whether by sprinkler or flood irrigation, is a primary source of contamination in harvested produce. However, since contamination is limited to the surface of the plant, persistence of the contaminants in the air and light on aerial plant organs or even on edible root surfaces is not indefinite. Generally, if a few weeks elapse between the final irrigation and harvest, residual bacterial counts on exposed plant parts are usually reduced to a safe level. Since monitoring of irrigation waters for a wide range of specific disease organisms would be prohibitively difficult, standards based on coliform counts have been proposed (*271*).

B. Standards for Irrigation Water Quality

1. Salinity

Although total salinity is generally acknowledged to be one of the most important criteria for irrigation water quality, there is no general agreement on salinity standards. This is not surprising in view of the approximate 10-fold range in salt tolerance of crop plants. Districts producing sensitive fruit crops are restricted to much lower salinities than those producing more tolerant field and forage crops, and they therefore require waters of lower salinity, other factors being equal. Furthermore, there is no unique correspondence between salinity of irrigation waters and soil waters. The relationship is highly variable and is influenced by climatic, management, and soil factors. Thus, with a water of a given salinity one may be able, under one set of conditions, to maintain a low level of soil salinity, but under less favorable conditions soil salinity may be excessive. A system for predicting the impact of environment and management on soil salinity has recently been proposed(*279*).

Soil salinity depends not only on the initial salinity of the irrigation water but also on the fraction of the total applied water that moves below the root zone, carrying away the added salts. The leaching fraction (LF) depends on the amount of water introduced by irrigation in excess of the evapotranspirational consumption, that is, $D_d = D_i - D_e$, where D is the equivalent depth of water spread uniformly over the area and the subscripts refer to drainage, infiltration, and evapotranspiration, respectively. Since the depth of irrigation water infiltrating the soil can be expressed as It_i, or the average infiltration rate I (cm/day) times the irrigation time t_i (days), and the equivalent depth of water evapotranspired can be expressed by Et_c, or the average evapotranspiration rate E (cm/day) during the total irrigation cycle t_c (days), the leaching fraction for a given situation can be calculated by substitution in the basic formula $\mathrm{LF} = D_d/D_i$ to get

$$\mathrm{LF} = 1 - \frac{Et_c}{It_i} \tag{6}$$

If internal drainage is limiting, however, the leaching fraction becomes

$$\mathrm{LF} = \frac{O}{E+O} \tag{7}$$

where O is the average drainage rate at the bottom of the root zone (cm/day). Calculation of the leaching fraction provides an estimate of the maximum salinity developed in the root zone as a result of the concentrating effect of evapotranspiration of the soil solution. Since values for some of the parameters, such as infiltration rate (I), drainage rate (O), and irrigation time (t_i), may vary even within a single field, the values used will determine the applicability of the findings. If average values for the variable parameters are employed, the calculated leaching fraction will refer to the average conditions, but parts of a project or a field that have less favorable conditions (lower I, O, or t_i) will have lower leaching fractions and will develop higher salinities than those predicted for average conditions. The frequent occurrence of salt spots and variable salinity conditions in fields is the obvious result of variable conditions affecting salt accumulation.

From the computed leaching fraction, permissible salinities of the irrigation water (EC_i) are calculated using electrical conductivity as a measure of total salinity:

$$\mathrm{EC}_i = \mathrm{LF} \times \mathrm{EC}_d \tag{8}$$

The EC_d is determined from the maximum salinity in the root zone permitting acceptable crop production (Table 24). Permissible irrigation water salinities are tabulated for a range of leaching fractions and EC_d values (Table 25).

Current research on leaching requirements may result in some modification of prescribed EC_d values. Factors which need to be taken into account, but which are neglected in Eq. (*8*) which assumes conservation of salt in the soil water, are: no net loss or gain in salt by precipitation, solubilization of soil minerals, or salt uptake by the crop. Although these factors may largely cancel out in many instances, they cannot be ignored in others. Estimation of the leaching fraction also aids in predicting salt precipitation, since this precipitation is influenced by leaching fraction as well as by the composition of the water (*92*). It is evident from Table 25 that waters of a wide range of salinity may be satisfactory depending on soil, climate, and crop factors. It is far more realistic to assess the suitability of irrigation water in relation to specific conditions than to employ an arbitrary classification into "low," "medium," and "high" salinity hazards regardless of specific use conditions.

TABLE 24
Maximum Allowable Salinities, as Electrical Conductivity (EC) of the Soil Solutions, for Crops Representing the Full Range of Salt Tolerance Among Crop Plants. One mmho/cm is Equivalent, on the Average, to Approximately 640 mg of Salt per Liter in the Soil Solution[a]

Crops	Maximum permissible EC in soil solution	Maximum permissible EC at indicated sensitive stages[b]
Bermuda grass, tall wheat grass	18	—
Barley	18	10 (S)
Sugar beets	16	8 (G)
Cotton	16	—
Tomato, broccoli, spinach, alfalfa	8	—
Rice	8	4 (S, F)
Potato, corn, flax	6	—
Bean, clovers (*Trifolium* spp.)	4	—
Most pome and stone fruits, citrus	4	—
Strawberries, blackberries, boysenberries	2–4	—

[a]Taken from Ref. *279*, by courtesy of The American Society for Testing and Materials.

[b]G = germination, S = seedling, F = flowering.

TABLE 25
Maximum Permissible Electrical Conductivity (EC) in Irrigation Water for Given Leaching Fractions and Crop Tolerances to EC of Soil Water, mmho/cm at 25°C[a]

	Leaching fraction			
	0.10	0.20	0.30	0.40
Maximum permissible EC in soil water	Maximum permissible EC in irrigation water			
2	0.2	0.4	0.6	0.8
4	0.4	0.8	1.2	1.6
8	0.8	1.6	2.4	3.2
16	1.6	3.2	4.8	6.4

[a]Taken from Ref. *279*, by courtesy of The American Society for Testing and Materials.

2. Chloride Hazard

For chloride-sensitive woody plants (fruit crops, ornamental shrubs, and trees), the permissible chloride content in the irrigation water may be calculated by

$$Cl_i = LF \times Cl_d \tag{9}$$

where Cl_d is the maximum chloride concentration in the soil water permitting good plant growth (Table 26).

Permissible chloride concentrations in irrigation waters are tabulated in Table 27 for different leaching fractions and Cl_d values. A wide range of chloride concentrations may be permissible in irrigation waters, depending on use conditions.

3. Sodium Hazard

In evaluating the suitability of waters for irrigation, one important consideration is the extent to which soils will adsorb sodium from the water by cation exchange reactions. Because of the well-established relation between the exhangeable sodium percentage (ESP) of soil exchange complexes and the sodium adsorption ratio (SAR) of saturation extracts, the SAR value is frequently used as an index of irrigation water quality(*2*). This criterion, however, is valid only when a steady-state condition is attained in which the SAR value of the soil solution approximates the SAR value of the applied irrigation water (SAR_{iw}). Seldom is the SAR of the soil solution the same as that of the applied irrigation

TABLE 26
Maximum Permissible Chloride Contents in Soil Solution for Various Fruit Crop Varieties and Rootstocks[a]

Crop	Rootstock or variety	Limit of tolerance to chloride in soil solution, meq/liter
	Rootstocks	
Citrus	Rangpur lime, Cleopatra mandarin	50
	Rough lemon, tangelo, sour orange	30
	Sweet orange, citrange	20
Stone fruit	Marianna	50
	Lovell, Shalil	20
	Yunnan	14
Avocado	West Indian	16
	Mexican	10
	Varieties	
Grape	Thompson Seedless, Perlette	50
	Cardinal, Black Rose	20
Berries	Boysenberry	20
	Olallie blackberry	20
	Indian Summer raspberry	10
Strawberry	Lassen	16
	Shasta	10

[a]Taken from Ref. *279*, by courtesy of The American Society for Testing and Materials.

water. A major factor known to affect the final soil water SAR value (SAR_{sw}) is the loss of Ca from solution as the result of lime precipitation (*92*). An additional factor influencing the SAR_{sw} value is the introduction of Ca and Mg into the soil water from the weathering of soil minerals (*101*). Since it is the SAR of the soil water rather than that of the irrigation water, per se, that affects the exchangeable sodium content of soils, any attempt made to evaluate the Na hazard of waters for irrigation purposes should take these factors into account.

An equation has recently been proposed (*280*) to predict the SAR_{dw} (*dw* refers to water draining from the bottom of the root zone) value resulting from the use of an irrigation water having known SAR_{iw} and pH_c (see Sec. II. C. 1 for definitions) values under various leaching regimes (i.e., with given leaching fractions) and known soil mineral weathering

TABLE 27

Maximum Permissible Chloride Concentrations in Irrigation Waters for Given Leaching Fractions and Crop Tolerances to Chloride in the Soil Water, meq/liter[a]

	Leaching fraction			
	0.10	0.20	0.30	0.40
Maximum permissible Cl in soil water	Maximum permissible chloride concentrations in irrigation water			
10	1	2	3	4
20	2	4	6	8
30	3	6	9	12
40	4	8	12	16
50	5	10	15	20

[a]Taken from Ref. *279*, by courtesy of The American Society for Testing and Materials.

properties. This equation is

$$SAR_{dw} = \frac{y^{(1+2LF)}}{\sqrt{LF}} SAR_{iw} [1 + (8.4 - pH_c)] \quad (10)$$

The $(8.4 - pH_c)$ term is an index of the tendency of a water to either precipitate or dissolve lime when reacted with a calcareous soil. The term y is a soil mineral weathering coefficient. When a maximum permissible SAR_{dw} is assigned to Eq. (10), the value is designated by SAR'_{dw}. The choice of SAR'_{dw} would be the lower of two values: (a) that SAR_{dw} value at which the soil permeability becomes appreciably reduced at salt concentrations typical of drain waters, or (b) that SAR_{dw} value corresponding to the ESP value at which the crop in question would suffer from sodium toxicity(*279*). Under these conditions, the LF value required from Eq. (10) for any given SAR_{iw} and pH_c combination is called the leaching requirement for exchangeable sodium control (LR_{SAR}). This gives a method for estimating the management required to use waters under given soil and crop conditions. In addition, the equation can be used for water suitability evaluations with regard to Na hazards.

Table 28 shows SAR_{iw} values of irrigation waters calculated from Eq. (10) that are suitable for use for given pH_c values and SAR_{dw} values. The standards given in Table 28 are tentative and subject to revision as further research findings become available.

For noncalcareous soils and nonbicarbonate-containing waters, Eq. (11) should be used:

$$SAR_{dw} = \frac{y^{(1+2LF)}}{\sqrt{LF}} SAR_{iw} \quad (11)$$

TABLE 28
Permissible SAR_{iw} Levels for Irrigation Waters Calculated for Three Allowable SAR_{dw} Values, Four Leaching Fraction Values, and Four pH_c Values[a]

		SAR'_{dw}		
		7	17	27
LF	pH_c	Maximum permissible SAR_{iw}		
0.1				
	6.6	1.2	3.0	4.7
	7.6	1.9	4.6	7.3
	8.4	3.4	8.3	13.1
	8.6	4.2	10.3	16.4
0.2				
	6.6	1.8	4.5	7.1
	7.6	2.9	6.9	11.0
	8.4	5.1	12.5	19.9
	8.6	6.4	15.6	24.8
0.3				
	6.6	2.4	5.9	9.4
	7.6	3.8	9.2	14.6
	8.4	6.8	16.5	26.2
	8.6	8.5	20.6	32.8
0.4				
	6.6	3.0	7.3	11.6
	7.6	4.7	11.4	18.1
	8.4	8.4	20.5	32.5
	8.6	10.5	25.6	40.7

[a]Adapted from data given in Ref. *280*.

The data in Table 28 can be used for this equation as they were for Eq. (10) by assuming a pH_c of 8.4, since under these conditions Eq. (10) reduces to Eq. (11). For more details see Ref. *280*.

4. Other Standards

Since water-soil interactions affect the concentration of most of the minor salt constituents, any single set of standards for permissible concentration of these trace elements in irrigation water is unrealistic. It is preferable, instead, to consider root zone concentrations (soil water concentrations) at which plants are affected, as a guide for cautionary examination of the water and soil to determine whether soil water concentrations might reach damaging levels with any given set of soil and management conditions (Table 29).

TABLE 29
Root Zone Concentration Associated with Toxicity to Plants or Animals, and Corrective Soil Treatments for Reducing Toxic Concentrations or Effects

Element	mg/liter in root medium[a]	Toxic to:[b]	Corrective soil treatment[b]
Al	0.5	Plants	Liming to pH 5 or 5.5
As	0.5	Plants and animals	—
B	0.75	Plants	—
Cd	0.005	Humans (hypertensive disease)	—
Cr	5	Plants	—
Co	0.2	Plants	—
Cu	0.2	Plants	—
F	—	Plants and animals	Increase soluble Ca
Fe	—	—	—
Pb	5	Plants	—
Li	0.1	Citrus	Increase soluble Ca salts
Mn	2	Plants	Increase soluble Ca salts plus drainage
Mo	0.005	Animals	—
Ni	0.5	Plants	Liming
Se	0.05	Animals	—
Sn	—	Very low uptake by plants	—
Ti	—	Very low uptake by plants	—
V	10	Plants	—
Zn	5	Plants	Liming, P fertilizer

[a]Adapted from Ref. *271*.
[b]Adapted from Ref. *274*.

IV. Minimizing Pollutional and Degradational Effects of Agriculture on Water Quality

A. Decreasing Pollution by Surface Drainage

1. Reduction of Erosional Losses

It is well established that man's activities accelerate erosion. For many years the U.S. Department of Agriculture and state agricultural experiment stations have been providing technical advice on field management practices to reduce soil erosion and the movement of sediment from field to surface waters (*281*). Stabilization of the sediment source is the most direct approach to solving most sediment problems. Practices such as mulching, sod crops, strip cropping, contour cultivation, and modern terrace systems have been shown to be highly effective in reducing soil erosion on sloping farmlands in grain crops. They provide the elements essential for control – soil cover and surface runoff control.

On uplands, watershed management practices should be followed. Improved grazing practices, fire protection, seeding of perennial grasses, reforestation, replacing brush land with grasses, care in the construction of logging roads and skid trails, and fertilization to increase the native cover are all helpful.

Various types of engineering works can be used to reduce the amount of sediment that is delivered to downstream points when control on the land has been neglected, including stream-bank erosion controls and detention dams. Sediment-charged waters can be applied to groundwater recharge areas; the water is stored for use and the sediment is removed as well. Most of these practices are discussed by Wadleigh (*224*).

2. Placement of Soil Additives

Much of the data discussed herein have shown that surface runoff is the major source of pollution of rivers by radionuclides and pesticides. This is not true for nitrate nitrogen, however, because it is so soluble and unreactive in the soil that it penetrates quickly beyond the influence of surface runoff. If, instead of being broadcast, phosphorus fertilizers were placed below the soil surface, their removal in surface runoff could be appreciably diminished.

3. Improved Control of Pesticide Applications and Use of Alternative Control Measures

Some of the data presented herein have indicated that drift and volatilization may be important causes of environmental pesticide transport. Application systems that eliminate or curtail drift and volatilization loss should appreciably diminish pollution of surface waters with pesticide constituents. Ground spray rigs, for example, are more easily controlled in this regard than aerial application of herbicides and certain pesticides. Development of improved pesticides that are more readily degraded and that have minimum toxicity to humans, animals, fish, wildlife, and beneficial insects and microorganisms would help greatly.

Alternate control measures have been proposed to eliminate the need for pesticide chemicals. Most of the work is oriented to the development of biological or physical control measures, or attractants to which only the intended target will respond. The chief biological methods are based on the use of parasites, predators, or microbial pathogens which keep the pest under natural control.

Additional alternative control measures involve: (1) use of pest-resistant strains of crops, (2) sterilization to eliminate the pests by population decline, (3) bioenvironmental methods to make the environmental conditions less favorable to the pest, and (4) electromechanical devices for capturing and destroying pests. See Wadleigh(*224*) and Nicholson(*65*) for greater detail and discussion of specific cases in point.

B. Decreasing Pollution by Internal Drainage

1. Minimizing Pollutants in Drainage Water

a. Saline Springs, Aquifers, and Saline Strata

Dramatic increases in solute content of surface waters may result from the release of fossil salts to surface waters from geologic salt sources such as saline springs, aquifers, and stream bed strata(*79*). For example, the Arkansas River increases its salinity severalfold in southern Kansas as it crosses the truncated edge of a Permian salt deposit, just east of Hutchinson. McNeal and Bower(*79*) cite several other examples of increased salinity of waters by contact with saline strata as well as by drainage from saline springs and sedimentary strata containing saline connate waters. Good examples of the latter are the Green River formation of Utah and the Mancos shale of Wyoming, Colorado, and Utah. In certain cases, significant reductions in solute content might be achieved by lining the stream beds through those sections of the stream course that contain the fossil salts. Similarly, local sources of salines such as springs, etc., could be sealed or diverted from access to the stream.

Since plants absorb water preferentially and since the dissolved salts left behind in the soil must be removed if satisfactory crop yields are to be maintained, no significant decrease in total salt burden of subsurface drainage water is possible without adversely affecting agriculture. However, any reduction in the volume of water needed for a crop would reduce the amount of water consumed and hence the ultimate increase in salt concentration in the surface streams. Development of transpiration suppressants or crop varieties with lower water requirement, or the use of alternative crops with lower water needs would all reduce water consumption.

b. Nitrate and Other Agricultural Chemicals

Numerous studies have demonstrated the relative immobility of phosphorus in soils and its very slight transport in most subsurface drain waters. Nitrogen, on the other hand, is quite mobile, and methods to reduce its access to surface streams should be adopted. Since the ammonium ion is adsorbed in the soil, the use of ammonium–nitrogen fertilizers (where other considerations are favorable) in preference to nitrate–nitrogen fertilizers would help reduce nitrogen losses in drainage waters. The ammonium will of course be nitrified with time, but during this time less nitrate will be lost by leaching. The use of band-placed fertilizer and encapsulated pellets to slow the release of nitrate would also help; more of the nitrogen would be utilized by the crop with less of it lost. Recognition by farmers of nitrogen in many of our waters, especially well waters, which can supply a portion of their need and hence reduce the amounts they would otherwise supply, would help reduce excessive drainage burdens(*282*). Lining of feed lot or dairy areas and diversion from surface streams of high nitrogen-containing waters typical of such facilities are other ways to eliminate nitrate pollution.

2. Alternative Methods of Drainage Disposal

Many water pollutants from agriculture, such as pesticides, sediments, and nitrates, should be controlled primarily at the source through improved management of soils and pest control operations. These contaminants are truly pollutants in the sense that they are first introduced into waters as a result of agricultural operations and therefore should be controllable through improved or changed agricultural practices.

Increased salinity of drainage waters from irrigated areas represents a qualitatively different problem. Such salinity is not a pollutant in the usual sense because the salts may have been initially present in the original irrigation water and are concentrated only by the evapotranspirational processes involved in growing crops. Elimination of nonbeneficial evaporation will reduce the intensity of the effect and therefore mitigate the problem somewhat, but such beneficial results would lead only to a more extensive use of the conserved water supply, producing much the same end product—a reduced volume of water with an increase in salt concentration approximately proportional to the reduction in volume. So long as crop growth requires a high degree of evapotranspirational use of water, the problem will remain. Attempts to restrict transpiration without interfering with required gas exchange between plants and the atmosphere have yielded no promising results to date.

The problem thus reduces to the need for disposing of drainage waters of increased salinity and with perhaps some undesirable pollutants as well, without affecting the quality of the water supply in the area. The normal procedure in the past, of returning drainage waters to the rivers and streams from which downstream users must draw their water, has produced the aspect of water quality deterioration in irrigated regions of greatest concern to both agricultural and nonagricultural water users.

Alternative methods for disposing of drainage water without degrading river water quality are practical in some instances. It is always necessary to collect the drainage waters to be disposed. In tile-drained districts or areas drained by tube wells, such collection is already achieved. When drainage waters return to rivers by underground flow, special interceptor drains would be required to permit alternative methods of disposal. Instead of returning the collected drainage waters to a river from which others have to draw water, the drainage waters may be routed to the sea through a separate drainage channel or ditch. This solution has been adopted for disposal of some of the highly saline tube well waters from the Wellton-Mohawk district of Arizona. A master drain to the sea is also proposed for the San Joaquin project in California. Obviously such direct disposal is practical only for districts that lie close to the sea. Desalination techniques may be employed in the future to reclaim saline drainage waters, making it practical to transport the reduced volume of by-product brine to the sea from greater distances inland. Drainage waters of little or no reuse potential may contain only one-third the salt content of seawater or less. Desalination of such waters by reverse osmosis or electrodialysis rather than by distillation processes is attractive because desalination power costs by these former methods are approximately proportional to the initial salt content of the water (*283*).

Disposal of drainage waters in basins with no outlet to the sea is another method for waste water disposal. Drainage waters from the Imperial and Coachella Valleys of California flow into the Salton Sea where they are further concentrated by evaporation. If such inland seas are to be used for wildlife conservation and recreation, however, increased salinity much above that of oceanic strength begins, in time, to limit their usefulness. Salt must then be exported out of the basin, or at least separated from the main body of water, if salinity is to be controlled.

In some areas, it may be possible to infiltrate drainage waters into deep soils that have no usable underground water resources. Waste waters have also been pumped into depleted oil wells.

In general, it would seem that an increased scarcity of good quality water will make it more attractive to search for other methods of waste disposal than transport to the sea in rivers. Drainage wastes, as well as unusable industrial wastes, may be increasingly sent to the sea in outfall sewers or infiltrated into the ground when this will not damage underground water resources.

V. Agriculture as an Ameliorative Agent for Water Quality

Because filtration through soil can improve water quality through the action of soil organisms in degrading toxic or undesirable organic materials and by the adsorptive, precipitative, and filtering capacities of the soil itself, irrigation agriculture may effectively purify wastes produced by other water users.

Treated municipal waste waters may cause eutrophication of rivers and lakes because of their high content of such nutrients as nitrogen and especially phosphorus, which stimulate algal growth. These nutrients are beneficial for crops and their presence in irrigation water is usually not objectionable.

Sanitary engineers for some municipalities in southern California have requested calculations on the degree to which nutrient-rich waste waters could be disposed of by irrigation rather than by return to streams. About 4 acre-ft of water could be utilized in producing 6 tons of alfalfa per acre. This alfalfa would contain about 3.5% of total N and about 0.3% P. Assuming all these nutrients were supplied by the waste water used for irrigation, the crop would accumulate nitrogen and phosphorus equivalent to about 39 mg/liter N and 3.3 mg/liter P in the waste waters. Almost all of the waste waters applied could be evapotranspired with only a small percentage leaching through the soil. Since the waste waters contained about 10–20 mg/liter each of phosphorus and nitrogen, all of the nitrogen in the water would be absorbed by the crop, but only about one-third or less of the phosphorus would be so absorbed. The remainder of the phosphorus would, however, be fixed in the soil and would not be leached downward. Thus, by a combination of nutrient uptake by the crop and fixation in the soil, the nutrients would be effectively disposed of. Since municipal waste water production will not vary with the season as much as water requirements by crops do, the crop acreage required would need to be calculated on the basis of the minimal water use rate by the crop during the cool weather. Disposal of waste waters by irrigation of crops is practical the year round only when year-round growing conditions prevail.

Industrial plants may find irrigation an acceptable method of waste water disposal. Some pollutants that may be unacceptable in municipal or recreational waters may not affect the utility of the water for irrigation.

The feasibility of such disposal depends on the nature of the pollutant and whether it can be tolerated by plants or fixed by soils.

VI. Concluding Statement

The trinity of water quality requirements for irrigation agriculture are salinity, sodicity, and toxicity. With respect to some criteria, irrigation agriculture is more tolerant than are other major water users; for example, irrigation agriculture is not at all bothered by phosphorus levels that may damage recreational waters. Some of its requirements are diametrically opposite to those of other water users. Soft waters that may be desirable for municipal or industrial uses, for example, may present quite unacceptable sodicity (sodium) hazards for agriculture. The specific criteria and standards for irrigation water quality are sufficiently distinctive to require special consideration in the analysis of the quality of our environmental water.

Irrigation water quality must be considered not only with regard to its immediate effects on soils and crops but also with regard to the welfare of consumers. Pesticides, pathogens, radionuclides, and even some naturally occurring water constituents may not affect crops directly but may affect animals or humans and so are equally important criteria of water quality.

Irrigation agriculture affects water quality in diverse ways. A major effect, the high consumptive use of water by evapotranspiration, producing reduced volumes of more salt-concentrated waters, is a process that occurs throughout nature but is intensified by irrigation agriculture. In some irrigated areas with water of exceptionally low salinity, the evapotranspirational loss of a large fraction of the total available water may still leave water of acceptable salinity. In other areas, the residual water may be of such high salinity as to be of little further use unless reclaimed by a desalination process. Disposal of such irrigation waste waters by methods that would avoid mixing the waste water with usable water resources may be warranted in some areas, in order to preserve water quality.

Although little can be done to minimize the effect of consumptive use of water by agriculture, other aspects of water quality deterioration by agriculture are more amenable to control. Improved management can reduce erosional losses of soil and associated contamination of surface water with pesticides and phosphorus. Surface water and groundwater contamination by nitrate can be reduced by improved fertilizer practices.

Irrigation agriculture is likely to increase in importance as demands for food and fiber increase with population growth. A realistic appraisal of the effects of irrigation agriculture on our water resources is important in assessing the problems of the future, as well as for correcting the problems we face today.

REFERENCES

1. M. L. Jackson, *Soil Chemical Analysis*, Prentice-Hall, Englewood Cliffs, N.J., 1958.
2. U.S. Salinity Laboratory Staff, *U.S. Dept. Agr. Handbook 60*, 1954.
3. C. A. Bower and L. V. Wilcox, in *Methods of Soil Analysis* (C. A. Black, ed.), Part 2, American Society of Agronomy, Madison, Wis., 1965, pp. 933–951.
4. H. D. Chapman and P. F. Pratt, *Methods of Analysis for Soils, Plants and Waters*, Univ. of California, Riverside 1961.
5. A. H. Sayegh, L. A. Alban, and R. G. Petersen, *Soil Sci. Soc. Am. Proc.*, **22**, 252 (1958).
6. R. F. Reitmeier, *Soil Sci.*, **61**, 195 (1946).
7. F. M. Eaton and V. P. Sokeloff, *Soil Sci.*, **40**, 237 (1935).
8. R. K. Schofield, *Intern. Congr. Pure and Appl. Chem.* (*London*) *Proc.*, **11**, 257 (1947).
9. C. A. Bower and J. T. Hatcher, *Soil Sci.*, **93**, 275 (1962).
10. G. H. Bolt, *Kolloid Z.*, **175**, 33 (1961).
11. G. H. Bolt, *Kolloid Z.*, **175**, 144 (1961).
12. G. H. Bolt and B. P. Warkentin, *Kolloid. Z.* **156**, 41 (1958).
13. F. W. Parker, *Soil Sci.*, **12**, 209 (1921).
14. J. S. Burd and J. C. Martin, *J. Agr. Sci.*, **13**, 265 (1923).
15. O. C. Magistad, R. F. Reitemeier, and L. V. Wilcox, *Soil Sci.*, **59**, 65 (1945).
16. R. Gardner, R. S. Whitney, and A. Kezner, *Colo. Agr. Exp. Sta. Tech. Bull.*, **20**, 1 (1937)
17. D. Shimshi, *Soil Sci.*, **101**, 98 (1966).
18. L. J. Briggs and A. G. McCall, *Science*, **20**, 566 (1904).
19. R. C. Reeve and E. J. Doering, *Soil Sci.*, **99**, 339 (1965).
20. L. A. Richards, *Soil Sci.*, **51**, 377 (1941).
21. R. F. Reitemeier and L. A. Richards, *Soil Sci.*, **57**, 119 (1944).
22. C. A. Bower and J. O. Goertzen, *Soil Sci. Soc. Am. Proc.*, **19**, 147 (1955).
23. W. D. Kemper, *Soil Sci. Soc. Am. Proc.*, **24**, 10 (1960).
24. F. M. Eaton, R. B. Harding, and T. J. Ganje, *Soil Sci.*, **90**, 253 (1960).
25. W. D. Kemper, *Soil Sci.*, **87**, 345 (1959).
26. L. A. Richards, *Soil Sci. Soc. Am. Proc.*, **30**, 333 (1966).
27. J. D. Oster and R. D. Ingvalson, *Soil Sci. Soc. Am. Proc.*, **31**, 572 (1967).
28. D. Carroll, *U.S. Geol. Surv. Water-Supply Paper 1535G*, U.S. Dept. Interior, 1962.
29. F. A. Herman and E. Gorham, *Tellus*, **9**, 180 (1957).
30. G. H. Neumann, S. Forselius, and L. Wahlman, *Intern. J. Air Pollution*, **2**, 132 (1959).
31. G. A. Wheatley and J. A. Hardman, *Nature*, **207**, 486 (1965).
32. S. R. Weibel, R. B. Weidner, A. G. Christianson, and R. J. Anderson, *Advances in Water Pollution Research*, Water Pollution Control Federation, Washington, D.C., 1967, pp. 329–342.
33. P. King, L. B. Lockhart, R. A. Baus, R. L. Patterson, H. Friedman, and I. H. Blifford, *Nucleonics*, **14**, 78 (1956).
34. D. Lal, P. K. Malhorta, and B. Peters, *J. Atmospheric Terrest. Phys.*, **12**, 306 (1958).

35. N. Bhandari, S. G. Bhat, V. P. Kharkar, S. Krishna, D. Lal, and A. S. Tamhane, *Tellus*, **18**, 504 (1966).
36. S. Kateoff, *Nucleonics*, **16**, 78 (1958).
37. P. K. Kuroda, B. D. Palmer, M. Attrep, J. N. Beck, R. Ganopathy, D. D. Sabu, and M. N. Rao, *Science*, **147**, 1285 (1965).
38. A. T. Krebs and N. G. Stewart, in *Nuclear Radiation in Geophysics*, Academic, New York, 1962, pp. 241–294.
39. J. F. Bleichrodt, J. Blok, and R. H. Dekker, *J. Geophys. Res.*, **66**, 135 (1961).
40. J. F. Bleichrodt and J. Blok, *Tellus*, **16**, 135 (1964).
41. J. N. Beck and P. K. Kuroda, *J. Geophys. Res.*, **71**, 2451 (1966).
42. D. H. Peirson, R. N. Crooks, and E. M. R. Fisher, *Atomic Energy Res. Estab. Gt. Brit. Rept. R 3358*, 1960.
43. R. N. Crooks, R. G. D. Osmond, E. M. R. Fisher, M. J. Owens, and T. W. Evett, *Atomic Energy Res. Estab., 3349*, 1960.
44. M. P. Menon, K. K. Menon, and P. K. Kuroda, *J. Geophys. Res.*, **68**, 4495 (1963).
45. L. V. Wilcox, *U.S. Dept. Agr. Tech. Bull. 962*, 1948.
46. S. K. Love, *U.S. Geol. Surv. Water-Supply Papers U.S. Dept. Interior, 1264*, 1954; *1362*, 1955; *1430*, 1958; *1465*, 1959; *1485*, 1960; *1524*, 1960; *1575*, 1961; *1699*, 1963; *1946*, 1966.
47. J. D. Hem, *U.S. Geol. Surv. Water-Supply Papers U.S. Dept. Interior, 1473*, 1959.
48. L. E. Allison, *Advan. Agron.*, **16**, 139 (1964).
49. J. H. Feth, *Water Resources Res.*, **2**, 41 (1966).
50. L. V. Wilcox and W. H. Durum, in *Irrigation of Agricultural Lands* (R. M. Hagan, ed.), American Society of Agronomy, Madison, Wis., 1967, pp. 104–119.
51. D. A. Livingstone, *U.S. Geol. Surv. Prof. Paper 440-G*, U.S. Dept. Interior, 1963.
52. B. Mason, *Principles in Geochemistry*, 2nd ed., Wiley, New York, 1964.
53. D. E. White, V. D. Hem, and G. A. Waring, *U.S. Geol. Surv. Prof. Paper 440-F*, U.S. Dept. Interior, 1963.
54. N. Florea, *Pochvovedenie*, **7**, 11 (1956).
55. W. H. Durum and J. Haffty, *U.S. Geol. Surv. Circ. 445*, U.S. Dept. Interior, 1961.
56. H. W. Lakin, *U.S. Dept. Agr. Handbook 200*, 1961.
57. W. M. Miller, *Wyoming Agr. Exp. Sta. Mimeo. Circ. 64*, 1956.
58. M. W. Skougstad and C. A. Horr, *U.S. Geol. Surv. Circ. 420*, U.S. Dept. Interior, 1960.
59. U.S. Public Health Service, *National Water Quality Network, Annual Compilation of Data, U.S. Public Health Serv. Publ. 663*, 1957–1962 ed., U.S. Govt. Printing Office, Washington, D.C.
60. J. H. Wallace and H. B. Smith, *U.S. Geol. Surv. Bull. 1019-B*, U.S. Dept. Interior, 1955.
61. F. H. Rainwater, *U.S. Geol. Surv. Hydrol. Invest. Atlas HA-61*, U.S. Dept. Interior, 1962.
62. V. C. Kennedy, *U.S. Geol. Surv. Prof. Paper 433-O*, U.S. Dept. Interior, 1965.
63. U.S. Dept. Agriculture Research Service Staff, *U.S. Dept. Agr. Misc. Publ. 970*, 1965.
64. H. C. Riggs, *U.S. Geol. Surv. Water-Supply Paper 1546*, U.S. Dept. Interior, 1962.
65. H. P. Nicholson, *Science*, **158**, 871 (1967).
66. F. M. Middleton and J. J. Lichtenberg, *Ind. Eng. Chem.*, **52**, 99A (1960).
67. *The Pesticide Review*, U.S. Dept. Agriculture, Washington, D.C., 1966.
68. L. Weaver, C. G. Gunnerson, A. W. Breidenbach, and J. J. Lichtenberg, *Public Health Rept. (U.S.)*, **80**, 481 (1965).
69. O. Johnson, N. Krug, and J. L. Poland, *Chem. Week*, **92**, 118 (1963).

70. R. Hansberry, in *Pesticides and Their Effects on Soils and Water, Am. Soc. Agron. Spec. Publ. No. 8*, 1967, pp. 10–17.
71. L. Weaver, A. W. Hoadley, and S. Baker, *Radiol. Health Data*, **4**, 306 (1963).
72. D. A. Phoenix, *U.S. Geol. Surv. Prof. Paper 320*, U.S. Dept. Interior, 1959, pp. 55–64.
73. A. N. Tokarev and A. V. Shcherbakov, *U.S. Atomic Energy Comm. Rept. AEC-tr-4100*, 1956.
74. F. B. Barker and R. C. Scott, *Am. Geophys. J. Trans.*, **39**, 459 (1958).
75. O. C. Magistad and R. F. Reitemeier, *Soil Sci.*, **55**, 351 (1943).
76. C. A. Bower, in *Salinity Problems in the Arid Zones, Proc. UNESCO Arid Zone Symp., Tehran, Iran, 1961*, pp. 215–222.
77. H. M. Reisenauer, in *Environmental Biology* (P. L. Altman and D. S. Dittmer, eds.), Federation of American Societies for Experimental Biology, Bethesda, Md. 1966, pp. 507–508.
78. O. Meinzer, *Hydrology*, Dover, New York, 1942.
79. B. L. McNeal and C. A. Bower, in *Encyclopedia of Soil Science* (J. E. Gieseking, ed.), Vol. 1, Springer-Verlag, Germany, Chap. 8, in press.
80. W. R. Gardner, in *Arid Zone Research*, UNESCO, 1959, pp. 37–61.
81. W. O. Willis, *Soil Sci. Soc. Am. Proc.*, **24**, 239 (1960).
82. K. K. Tanji, G. R. Dutt, J. L. Paul, and L. D. Doneen, *Hilgardia*, **38**, 307 (1967).
83. G. R. Dutt, *Soil Sci. Soc. Am. Proc.*, **26**, 341 (1962).
84. G. R. Dutt and L. D. Doneen, *Soil Sci. Soc. Am. Proc.*, **27**, 627 (1963).
85. J. L. Paul, K. K. Tanji, and W. D. Anderson, *Soil Sci. Soc. Am. Proc.*, **30**, 15 (1966).
86. F. M. Eaton, *Soil Sci.*, **69**, 123 (1950).
87. L. V. Wilcox, G.Y. Blair, and C. A. Bower, *Soil Sci.*, **77**, 259 (1954).
88. C. A. Bower and M. Maasland, *West Pakistan Eng. Congr. Proc.*, 1963, pp. 49–61.
89. L. V. Wilcox and C. A. Bower, *Report to Collaborators, 1958*, U.S. Salinity Laboratory, Riverside, Calif., 1958, pp. 57–59.
90. C. A. Bower and L. V. Wilcox, *Soil Sci. Soc. Am. Proc.*, **29**, 93 (1965).
91. C. A. Bower, L. V. Wilcox, G. W. Akin, and M. G. Keyes, *Soil Sci. Soc. Am. Proc.*, **29**, 91 (1965).
92. C. A. Bower, G. Ogata, and J. M. Tucker, *Soil Sci.*, **106**, 29 (1968).
93. W. F. Langelier, *J. Am. Water Works Assoc.*, **28**, 1500 (1936).
94. W. P. Kelley, *Alkali Soils*, Reinhold, New York, 1951.
95. E. R. Graham, *Soil Sci. Soc. Am. Proc.*, **14**, 300 (1949).
96. J. E. McClelland, *Soil Sci. Soc. Am. Proc.*, **15**, 301 (1950).
97. V. E. Nash and C. E. Marshall, *Univ. Missouri Agr. Exptl. Sta. Bull. 613*, 1956.
98. V. E. Nash and C. E. Marshall, *Univ. Missouri Agr. Exptl. Sta. Bull., 614*, 1956.
99. W. D. Keller, W. D. Balgard, and A. L. Reesman, *J. Sediment. Petrol.*, **33**, 191 (1963).
100. W. D. Keller and A. L. Reesman, *J. Sediment. Petrol.*, **33**, 426 (1963).
101. J. D. Rhoades, D. B. Krueger, and M. J. Reed, *Soil Sci. Soc. Am. Proc.*, **32**, 643 (1968).
102. G. W. Bailey and J. L. White, *J. Agr. Food Chem.*, **12**, 324 (1964).
103. L. A. Richards and S. J. Richards, in *Soil, 1957 Yearbook of Agriculture*, U.S. Dept. Agriculture, Washington, D.C., pp. 49–60.
104. G. W. Bailey, in *Agricultural Waste Waters, Rept. No. 10*, Water Resources Center, Univ. of California, 1966, pp. 94–103.
105. D. J. Greenland, *Soils Fertilizers*, **28**, 415 (1965).
106. G. W. Kunze, in *Pesticides and Their Effects on Soils and Waters, Am. Soc. Agron. Spec. Publ. No. 8*, 1966, pp. 49–70.
107. D. C. Nearpass, *Soil Sci.*, **103**, 177 (1967).

108. R. W. Poehler and W. A. Young, *Clays Clay Minerals Proc. Natl. Conf. Clays Clay Minerals*, **9**, 468 (1962).
109. E. R. Fleck, in *Pesticides and Their Effects on Soils and Water*, Soil Science Society of America, Madison, Wisc., 1966, pp. 18–24.
110. V. H. Freed, in *Pesticides and Their Effects on Soils and Water*, Soil Science Society of America, Madison, Wisc., 1966, pp. 25–43.
111. S. Rich, in *Pesticides and Their Effects on Soils and Water*, Soil Science Society of America, Madison, Wisc., 1966, pp. 44–48.
112. B. P. Robinson, *U.S. Geol. Surv. Water-Supply Paper 1616*, U.S. Dept. of Interior, 1962.
113. W. J. Lacy, *Ind. Eng. Chem.*, **46**, 1061 (1954).
114. J. R. McHenry, D. W. Rhodes, and P. O. Rowe, *U.S. Atomic Energy Comm. Rept. TID-7517*, 1956, pp. 170–190.
115. G. H. Higgins, *J. Geophys. Res.*, **64**, 1509 (1959).
116. A. Walton, *J. Geophys. Res.*, **68**, 1485 (1963).
117. C. W. Thornthoxite, J. R. Mather, and J. K. Nakamura, *Science*, **131**, 1015 (1960).
118. J. R. Miller and R. F. Reitemeier, *ARS Res. Rept. 300*, U.S. Dept. Agriculture, 1957.
119. E. H. Essington and H. Nishita, *Plant Soil*, **24**, 1 (1966).
120. H. Nishita and E. H. Essington, *Soil Sci.*, **103**, 168 (1967).
121. W. P. Kelley, *Cation Exchange in Soils*, Reinhold, New York, 1948.
122. J. T. Hatcher and C. A. Bower, *Soil Sci.*, **85**, 319 (1958).
123. J. T. Hatcher, G. Y. Blair, and C. A. Bower, *Soil Sci.*, **94**, 55 (1962).
124. C. A. Bower and J. T. Hatcher, *Soil Sci.*, **103**, 151 (1967).
125. L. A. G. Aylmore, K. Mesbahul, and J. P. Quirk, *Soil Sci.*, **103**, 10 (1967).
126. N. J. Barrow, *Soil Sci.*, **104**, 342 (1967).
127. S. R. Dickman and R. H. Bray, *Soil Sci.*, **52**, 263 (1941).
128. H. R. Samson, *Clay Minerals Bull.*, **1**, 266 (1952).
129. P. M. Huang and M. L. Jackson, *Soil Sci. Soc. Am. Proc.*, **29**, 661 (1965).
130. F. M. Eaton, *U.S. Dept. Agr. Tech. Bull. 448*, 1935.
131. R. C. Reeve, A. F. Pillsbury, and L. V. Wilcox, *Hilgardia*, **24**, 69 (1955).
132. J. T. Hatcher, G. Y. Blair, and C. A. Bower, *Soil Sci.*, **88**, 98 (1959).
133. T. T. Chao, M. E. Harward, and S. C. Fang, *Soil. Sci. Soc. Am. Proc.*, **26**, 234 (1962).
134. E. J. Kamprath, W. L. Nelson, and J. W. Fitts, *Soil Sci. Soc. Am. Proc.*, **20**, 463 (1956).
135. M. L. Chang and G. W. Thomas, *Soil Sci. Soc. Am. Proc.*, **27**, 281 (1963).
136. S. Mattson, *Soil Sci.*, **32**, 343 (1931).
137. L. E. Ensminger, *Soil Sci. Soc. Am. Proc.*, **18**, 259 (1954).
138. W. A. Berg and G. W. Thomas, *Soil Sci. Soc. Am. Proc.*, **23**, 348 (1959).
139. J. T. Way, *J. Royal Agr. Soc.*, **11**, 313 (1850).
140. R. Kunin, *Ion Exchange Resins*, Wiley, New York, 1958.
141. K. L. Babcock, *Hilgardia*, **34**, 417 (1963).
142. C. E. Marshall, *The Physical Chemistry and Mineralogy of Soils, Vol. 1, Soil Minerals*, Wiley, New York, 1964.
143. R. C. Salmon, *J. Soil Sci.*, **15**, 273 (1964).
144. N. T. Coleman and G. W. Thomas, in *Soil Acidity and Liming* (R. W. Pearson and Fred Adams, eds.), American Society of Agronomy, Madison, Wis., 1967, pp. 1–42.
145. R. F. Reitemeier, *Advan. Agron.*, **3**, 113 (1951).
146. G. Stanford and W. H. Pierre, *Soil Sci. Soc. Am. Proc.*, **11**, 155 (1947).
147. C. A. Bower, *Soil Sci.*, **70**, 375 (1950).
148. D. R. Hoagland and J. C. Martin, *Soil Sci.*, **36**, 1 (1933).
149. F. E. Allison, M. Kefaurer, and E. M. Roller, *Soil Sci. Soc. Am. Proc.*, **17**, 107 (1953).
150. P. F. Pratt and B. Goulben, *Soil Sci.*, **84**, 225 (1957).

151. L. R. McKinnon and D. Lilleland, *Soil Sci.*, **31**, 407 (1931).
152. D. Lilleland, *Western Fruit Grower*, **13**, 9 (1959).
153. L. K. Stromberg, *Calif. Agr.*, **14**, 4 (1960).
154. G. H. Bolt, M. E. Sumner, and A. Kamphorst, *Soil Sci. Soc. Am. Proc.*, **27**, 294 (1963).
155. C. I. Rich, *Soil Sci.*, **98**, 100 (1964).
156. M. L. Jackson, *Clays Clay Minerals Proc. Natl. Conf. Clays Clay Minerals*, **11**, 29 (1963).
157. I. Barshad, *Am. Mineral.*, **33**, 655 (1948).
158. I. Barshad, *Am. Mineral.*, **35**, 225 (1950).
159. I. Barshad, *Soil Sci.*, **72**, 361 (1951).
160. I. Barshad, *Soil Sci.*, **78**, 57 (1954).
161. J. B. Page and L. D. Baver, *Soil Sci. Soc. Am. Proc.*, **4**, 150 (1940).
162. J. I. Wear and J. L. White, *Soil Sci.*, **71**, 1 (1951).
163. G. W. Kunze and C. J. Jefferies, *Soil Sci. Soc. Am. Proc.*, **17**, 242 (1953).
164. J. A. Kittrick, *Soil Sci. Soc. Am. Proc.*, **30**, 801 (1966).
165. P. F. Pratt, in *Methods of Soil Analysis* (C. A. Black, ed.), American Society of Agronomy, Madison, Wis., 1965, pp. 1022–1030.
166. J. M. Bremer, in *Methods of Soil Analysis* (C. A. Black, ed.), American Society of Agronomy, pp. 1179–1237.
167. L. T. Kardos, in *Chemistry of the Soil* (F. E. Bear, ed.), Reinhold, New York, 1964, pp. 369–394.
168. A. R. Midgley, *J. Am. Soc. Agron.*, **23**, 788 (1931).
169. M. F. Morgan and H. G. M. Jacobson, *Conn. Agr. Exp. Sta. Bull. 459*, 1942.
170. J. Hendrick and H. D. Welsh, *Proc. Intern. Congr. Soil Sci., 1st*, Washington D.C., *1928*, pp. 358–366.
171. J. R. Neller, *Soil Sci. Soc. Am. Proc.*, **11**, 227 (1947).
172. J. E. Larsen, R. Langston, and G. F. Warren, *Soil Sci. Soc. Am. Proc.*, **22**, 558 (1958).
173. P. R. Day, G. H. Bolt, and D. M. Anderson, in *Irrigation of Agricultural Lands* (R. M. Hagan, ed.), American Society of Agronomy, Madison, Wis., 1967, pp. 193–208.
174. M. B. Russell, *Water and Its Relation to Soils and Crops*, Academic, New York, 1960.
175. S. A. Taylor, D. D. Evans, and W. D. Kemper, *Utah State Univ. Agr. Exp. Sta. Bull. 426*, 1961.
176. L. A. Richards, in *Methods of Soil Analysis*, (C. A. Black, ed.), American Society of Agronomy, Madison, Wis., 1965, pp. 128–152.
177. R. K. Schofield, *Proc. Intern. Congr. Soil Sci., 3rd*, Oxford, England, 1935, p. 30.
178. G. H. Bolt and R. D. Miller, *Am. Geophys. Union Trans.*, **39**, 917 (1958).
179. B. P. Warkentin and R. K. Schofield, *J. Soil Sci.*, **13**, 98 (1962).
180. P. F. Low, *Advan. Agron.*, **13**, 269 (1961).
181. P. F. Low, *Soil Sci. Soc. Am. Proc.*, **22**, 375 (1958).
182. P. F. Low, *Clays Clay Minerals Proc. Natl. Conf. Clays Clay Minerals*, **8**, 170 (1960).
183. W. D. Kemper, D. E. L. Maasland, and L. K. Porter, *Soil Sci. Soc. Am. Proc.*, **28**, 164 (1964).
184. H. E. Hayward and L. Bernstein, *Botan. Rev.*, **24**, 584 (1958).
185. L. Bernstein and H. E. Hayward, *Ann. Rev. Plant Physiol.*, **9**, 25 (1958).
186. H. E. Hayward and W. B. Spurr, *Botan. Gaz.*, **105**, 152 (1943).
187. C. H. Wadleigh and A. D. Ayers, *Plant Physiol.*, **20**, 106 (1945).
188. C. H. Wadleigh, H. G. Gauch, and O. C. Magistad, *U.S. Dept. Agr. Tech. Bull. 925*, 1946.
189. I. L. Stevenson, in *Chemistry of the Soil* (F. E. Bear, ed.), Reinhold, New York, 1964, pp. 242–291.

190. M. Alexander, *Introduction to Soil Microbiology*, Wiley, New York, 1961.
191. S. A. Waksman, *Soil Microbiology*, Wiley, New York, 1952.
192. J. M. Bremmer, *J. Soil Sci.*, **2**, 67 (1951).
193. J. M. Bremmer, *Soils Fertilizers*, **19**, 115 (1956).
194. F. E. Broadbent, *Advan. Agron.*, **5**, 153 (1953).
195. G. W. Harmsen and O. A. Van Schreven, *Advan. Agron.*, **7**, 299 (1955).
196. E. K. Allen and O. N. Allen, in *Handbuch der Pflanzer Physiologie*, (W. Rugland, ed.), Vol. 8, Springer-Verlag, Berlin, p. 48.
197. T. W. Walker, H. D. Orchiston, and A. F. R. Adams, *J. Brit. Grassland Soc.*, **9**, 249 (1954).
198. F. E. Allison, *Advan. Agron.*, **7**, 213 (1955).
199. F. E. Allison, *Soil Crop Sci. Soc. Florida Proc.*, **21**, 248 (1961).
200. F. E. Allison, *Soil Sci.*, **96**, 404 (1963).
201. F. E. Allison, *Agr. Sci. Rev.*, **2**, 16 (1964).
202. F. E. Allison, *Advan. Agron.*, **18**, 219 (1967).
203. P. F. Pratt and H. D. Chapman, *Hilgardia*, **30**, 445 (1961).
204. H. H. Chang and L. T. Kurtz, *Soil Sci. Soc. Am. Proc.*, **27**, 312 (1963).
205. C. A. Bower, *Iowa Agr. Exp. Sta. Res. Bull. 362*, 1949.
206. J. B. Hemwall, *Advan. Agron.*, **9**, 95 (1957).
207. C. A. Black and C. A. I. Goring, in *Soil and Fertilizer Phosphorus* (W. H. Pierre and A. G. Norman, eds.), Academic, New York, 1953, p. 123.
208. W. H. Pierre, *J. Am. Soc. Agron.*, **40**, 1 (1948).
209. K. R. Butlin, *Research*, **6**, 184 (1953).
210. R. L. Starkey, *Soil Sci.*, **70**, 55 (1950).
211. T. W. Walker, *J. Brit. Grassland Soc.*, **12**, 10 (1957).
212. L. D. Whittig and P. Janitzky, *J. Soil Sci.*, **14**, 320 (1963).
213. P. Janitzky and L. D. Whittig, *J. Soil Sci.*, **15**, 145 (1964).
214. G. Ogata and C. A. Bower, *Soil Sci. Soc. Am. Proc.*, **29**, 23 (1965).
215. D. D. Kaufman, in *Pesticides and Their Effects on Soils and Water*, Soil Science Society of America, Madison, Wisc., 1966, pp. 85–94.
216. J. P. Martin, in *Pesticides and Their Effects on Soils and Water*, Soil Science Society of America, Madison, Wis., 1966, pp. 95–108.
217. H. R. DeRose and A. S. Newman, *Soil Sci. Soc. Am. Proc.*, **12**, 222 (1947).
218. L. M. Glymph and C. W. Carlson, *Cleaning Up Our Rivers and Lakes*, *Agr. Engr.*, **49**, 590 (1968).
219. H. L. Barrows and V. J. Kilmer, *Advan. Agron.*, **15**, 303 (1963).
220. J. L. Nelson, R. W. Perkins, and J. W. Nielsen, *U.S. Atomic Energy Comm. Rept. HW-83614*, 1964.
221. H. P. Nicholson, A. R. Grzenda, and J. I. Teasley, in *Agricultural Waste Waters* (L. D. Doneen, ed.), Water Resources Center, Univ. of California, Los Angeles, 1966, pp. 132–141.
222. W. E. Bullard, *J. Water Pollution Control Federation*, **38**, 645 (1966).
223. L. D. Baver, *Soil Physics*, Wiley, New York, 1956.
224. C. H. Wadleigh, *U.S. Dept. Agr. Misc. Publ. 1065*, 1968.
225. R. H. Smith and W. L. Stamey, *Soil Sci.*, **100**, 414 (1965).
226. D. E. Carrit and S. H. Goodgal, *U.S. Atomic Energy Comm. Rept. NYO-4591*, 1953.
227. J. M. Garner, O. W. Kochtitzky, and A. M. Asce, *J. Sanit. Eng. Div. Am. Soc. Civil. Engrs. Paper*, **82** (SA4, Paper 1051), 1 (1956).
228. J. M. Nielsen and R. W. Perkins, *J. Geophys. Res.*, **67**, 3584, 1962.
229. J. M. Nielsen, *U.S. Atomic Energy Comm. Rept. TlO-7664*, 1963, pp. 91–112.

230. J. L. Nelson, R. W. Perkins, and J. M. Nielsen, *U.S. Atomic Energy Comm. Rept. HW-83614*, 1964.
231. V. Johnson, N. Cutshall, and C. Osterberg, *J. Water Resources Res.*, **3**, 99 (1967).
232. J. H. Stallings, *Soil Conservation*, Prentice-Hall, Englewood Cliffs, N.J., 1957.
233. A. W. White, A. P. Barnett, W. A. Jackson, and V. J. Kilmer, *Crops Soils*, **19**, 28 (1967).
234. P. G. Moe, J. V. Mannering, and C. B. Johnson, *Soil Sci.*, **104**, 389 (1967).
235. E. F. Eldridge, *J. Water Pollution Control Federation*, **35**, 614 (1963).
236. R. O. Sylvester and R. W. Seabloom, *A Study on the Character and Significance of Irrigation Return Flows*, Univ. of Washington, Dept. Civil Engineering, 1962.
237. G. Walton, in *Agricultural Waste Waters* (L. D. Doneen, ed.), Water Resources Center, Univ. of California, Los Angeles, 1966, pp. 273–281.
238. W. Jopling, in *Agricultural Waste Waters* (L. D. Doneen, ed.), Water Resources Center, Univ. of California, Los Angeles, 1966, pp. 144–150.
239. R. O. Sylvester, *Public Health Serv. Rept. SEC TR W61-3*, U.S. Dept. of Health, Education, and Welfare, 1960, pp. 80–87.
240. E. J. Martin and L. W. Weinberger, *Eutrophication and Water Pollution*, Great Lakes Res. Div. Univ. Mich. Publ. 15, 1966.
241. W. E. Westlake, in *Agricultural Waste Waters* (L. D. Doneen, ed.), Water Resources Center, Univ. of California, Los Angeles, 1966, pp. 90–93.
242. B. I. Sparr, in *Agricultural Waste Waters* (L. D. Doneen, ed.), Water Resources Center, Univ. of California, Los Angeles, 1966, pp. 83–89.
243. J. E. Swift, *Agricultural Waste Waters* (L. D. Doneen, ed.), Water Resources Center, Univ. of California, Los Angeles, 1966, pp. 9–14.
244. H. P. Nicholson, *Limnol. Oceanog.*, **9**, 310 (1964).
245. E. Hinden, D. S. May, and G. H. Dunstam, *Residue Rev.*, **7**, 130 (1964).
246. W. F. Barthel, D. A. Parsons, L. L. McDowell, and E. H. Grissinger, in *Agricultural Waste Waters* (L. D. Doneen, ed.), Water Resources Center, Univ. of California, Los Angeles, 1966, pp. 128–144.
247. A. W. Breidenbach and J. J. Lichtenberg, *Science*, **141**, 899 (1963).
248. F. K. Kawahara and A. W. Breidenbach, in *Pesticides and Their Effects on Soils and Water*, American Society of Agronomy, Madison, Wisc., 1966, pp. 122–127.
249. E. H. Dustman and L. F. Stickel, in *Pesticides and Their Effects on Soils and Water*, American Society of Agronomy, Madison, Wisc., 1966, pp. 109–121.
250. J. O. Keith, M. H. Mohn, and G. H. Ise, *U.S. Fish and Wildlife Serv. Circ. 226*, U.S. Dept. Interior, 1965, pp. 37–40.
251. G. M. Woodwell, *Sci. Am.*, **216**, 24 (1967).
252. L. D. Doneen, *Univ. Calif., Davis, Water Sci. Eng. Paper 4002*, 1966.
253. A. F. Pillsbury, in *Agricultural Waste Waters* (L. D. Doneen, ed.), Water Resources Center, Univ. of California, Los Angeles, 1966, pp. 240–244.
254. W. R. Johnson, R. Ittihadieth, R. M. Dann, and A. F. Pillsbury, *Soil Sci. Soc. Am. Proc.*, **29**, 287 (1965).
255. G. H. Wagner, *Soil Sci.*, **100**, 397 (1965).
256. C. N. Sawyer, *J. New England Water Works Assoc.*, **61**, 109 (1947).
257. B. A. Stewart, F. G. Viets, G. L. Hutchinson, and W. D. Kemper, *Environ. Sci. Technol.*, **1**, 736 (1967).
258. R. O. Sylvester and R. W. Seabloom, *J. Irrigation and Drainage Div. Am. Soc. Civil Engrs.*, **89**, 1 (1963).
259. R. S. Englebrecht and J. J. Morgan, *U.S. Publ. Health Serv. Publ. SEC TR W61-3*, U.S. Dept. Health, Education, and Welfare, 1961, pp. 74–79.

260. J. R. Miner, R. I. Lipper, L. R. Fina, and J. W. Funk, *J. Water Pollution Control Federation*, **38**, 1582 (1966).

261. W. A. Johnston, F. T. Ittihadieth, K. R. Craig, and A. F. Pillsbury, *J. Water Resources Res.*, **3**, 525 (1967).

262. C. S. Scofield, *J. Agr. Res.*, **61**, 17 (1940).

263. R. E. Hill, *J. Irrigation and Drainage Div. Am. Soc. Civil Engrs.*, **87**, 1 (1961).

264. A. F. Pillsbury, M. Asce, and H. F. Blaney, *J. Irrigation and Drainage Div., Am. Soc. Civil Engrs.*, **92**, 77 (1966).

265. C. A. Bower, J. R. Spencer, and L. O. Weeks, *J. Irrigation and Drainage Div., Am. Soc. Civil Engrs.*, **95**, 55 (1969).

266. P. O. Haney and T. W. Bendixen, *J. Am. Water Works Assoc.*, **45**, 1159 (1953).

267. L. V. Wilcox, *Symposium on Problems of the Rio Grande*, New Mexico State Engineer Office, 1957, pp. 39–44.

268. L. V. Wilcox, *Proc. Conf. Water Quality in the Columbia Basin, Seattle, 1961*, pp. 137–141.

269. H. L. Parkinson and H. R. McDonald, *Irrigation and Drainage Speciality Conf. Billings, Mont., 1965*, American Society of Civil Engineers, pp. 159–175.

270. F. C. Raney, *Proc. Calif. Rice Res. Symp., 1959*, pp. 20–28.

271. *Report of Committee on Water Quality Criteria, Sec. IV, Agricultural Uses*, Federal Water Pollution Control Admin., U.S. Dept. Interior, 1968, pp. 111–184.

272. L. V. Wilcox, *U.S. Dept. Agr. Inform. Bull. 211*, 1960.

273. F. T. Bingham, R. R. Bradford, and A. L. Page, *Calif. Agr.*, **18**, 6 (1964).

274. H. D. Chapman, ed., *Diagnostic Criteria for Plants and Soils*, Div. Agr. Sci., Univ. of California, Berkeley, 1966.

275. L. Bernstein, in *Problems of the Arid Zone, Proc. Paris Symp. UNESCO, 1962*, p. 139.

276. H. E. Pearson and M. R. Huberty, *Proc. Am. Soc. Hort. Sci.*, **73**, 248 (1959).

277. G. M. Woodhill, *Science*, **156**, 461 (1967).

278. R. S. Russell, *Ann. Rev. Plant Physiol.*, **14**, 271 (1963).

279. L. Bernstein, *Am. Soc. Testing Mater. Spec. Tech. Publ. 416*, 1967, p. 51.

280. J. D. Rhoades, *Soil Sci. Soc. Am. Proc.*, **32**, 652 (1968).

281. W. H. Wischmeier and D. D. Smith, *U.S. Dept. Agr. Handbook 282*, 1965.

282. D. E. Longenecker, *Tex. Agr. Prog.*, **9**, 12 (1963).

283. Office Saline Water, U.S. Dept. Interior, in *Saline Water Conversion Rept.*, 1965, p. 183.

Chapter 4 Self-Purification in Natural Waters

Richard J. Benoit
ECOSCIENCE LABORATORY
DIVISION OF ENVIRONMENTAL RESEARCH & APPLICATIONS, INC.
NORWICH, CONNECTICUT

I. Introduction

Self-purification involves physical, chemical, and biological processes. These interact and occur simultaneously, needless to say, and all three are discussed in this chapter. Rivers, lakes, estuaries, and the ocean all have unique features—physiographically, geochemically, and biologically—but the important processes in all types of habitats are fundamentally similar. Our approach, therefore, does not separate the geographical realms but emphasizes the processes common to all habitats.

Waste materials introduced into a water body affect the biota both directly and indirectly. The effect may be simply lethal or it may be more subtle, affecting behavior or reproduction adversely. The toxicity of substances or mixed wastes is determined by means of bioassays using fish, insect larvae, and other invertebrate animals, or microbes (see Chap. 16). A variety of laboratory procedures are used, but the methods of Doudoroff et al.(*1*) have become standard(*2*).

In addition to laboratory procedures, it is a fairly common and suitable procedure to set out in the environment cages or other containers with numbers of animals for observation of their survival, growth, or other behavior in response to pollution. Benoit et al.(*3*) used the physiological condition of diatoms collected on special traps [diatometers, Patrick et al. (*4*)] to determine biological water quality in an impounded river receiving acid drainage from copper mines.

Toxicity and bioassay methods are discussed in other chapters of this work and certain general aspects are discussed later in this chapter (see Chaps. 13 and 14).

In a general sense we seek to understand what changes occur as a result of water pollution, how rapidly the changes proceed, and why they occur (the mechanisms involved). We are interested in which factors influence changes in the absolute sense and the rates of change. To cite a simple example, suppose we introduce a heavy metal salt into a stream in sufficient quantity to be toxic to some but not all the biota present. Some species will be killed to the point of local extinction, other species will be decimated but will still have numbers of survivors in the biota, and still other species will be hardly affected. If the surviving species are of the sort that concentrate heavy metals from solution in the environment, and this is a salient property of living things(*5*), the length of the affected stretch of stream will be shortened. How real is the situation just described? Are such biological processes in nature commonly more rapid than the chemical processes that remove heavy metals from solution? Answers to these and similar questions are not easily derived from known principles of the natural sciences because of the complexity of natural ecosystems; careful and exhaustive studies of actual stream sites are necessary, and safe, fruitful generalizations are seldom to be made.

It may seem that interest in self-purification processes will wane with passage of stricter laws and policies requiring a high degree of waste treatment in all situations, but for many regions and many streams the upper economic limit on degree of treatment will be reached in the very near future, and receiving streams will be required to carry the residual burden. The effects of the "irreducible" residual burden on water quality and on the assimilatory or self-purification capacity of streams will thus remain an important subject for scientific inquiry. From the standpoint of a waste producer, dependence upon self-purification capacity merely pushes the problem downstream and to some degree diminishes the water rights of downstream water users.

The sea represents the ultimate downstream locale for all streams, of course, and as the world population and urbanization burgeon, the self-purification capacity of the seas will become more important. The seas are vast, but not infinite, and already one sees their capacity obviously over-burdened locally.

A. Definitions

1. Self-Purification

Self-purification means the partial or complete restoration, by natural processes, of a stream's pristine condition following the introduction (usually through the agency of man) of foreign matter sufficient in quality and quantity to cause a measurable change in physical, chemical, and/or biological characteristics of the stream.

Lee(*6*) defined self-purification in terms of the reactions producing "transformations that result in the production of a chemical compound that has a less deleterious effect on water quality than the parent compound." He divided reactions into eight types: acid-base, precipitation, gas transfer, complex formation, oxidation reduction, photochemical reactions, sorption, and biochemical reactions.

2. Assimilatory Capacity

Assimilatory capacity in the broad sense means the extent to which a body of water can receive wastes without significant deterioration of beneficial uses. Suitability for a given use is defined in terms of quality criteria, and these are still to some extent arbitrarily designated. The criterion most widely used is dissolved oxygen, although that parameter is by no means always relevant. Assimilatory capacity with respect to dissolved oxygen is discussed in detail later.

3. Eutrophication

Kenton and Rohlich(*7*) gave a comprehensive review of the subject of eutrophication, and Rohlich(*8*) has published a brief, readable summary of that review. Eutrophication is the process whereby water bodies, especially lakes, undergo an increase in the quantity of available plant nutrients (notably phosphate and nitrogenous salts), resulting in an increase in biological productivity in the water body. In nature the process operates on a time scale measured in thousands of years, but the activities of man can accelerate the process enormously. The special significance of the process to the subject of this essay is that eutrophic conditions exist downstream from the stream reach in which the processes of self-purification have taken place. The production of algae and waterweeds in hypereutrophic bodies of water can be just as much of a nuisance as the primary pollutional degradation upstream. Sewage treatment that is complete in the context of present practice (primary, secondary, and disinfection) produces an effluent that is virtually undiminished in plant nutrients in comparison to the raw waste. The operational similarity between sewage treatment and natural purification has been pointed out

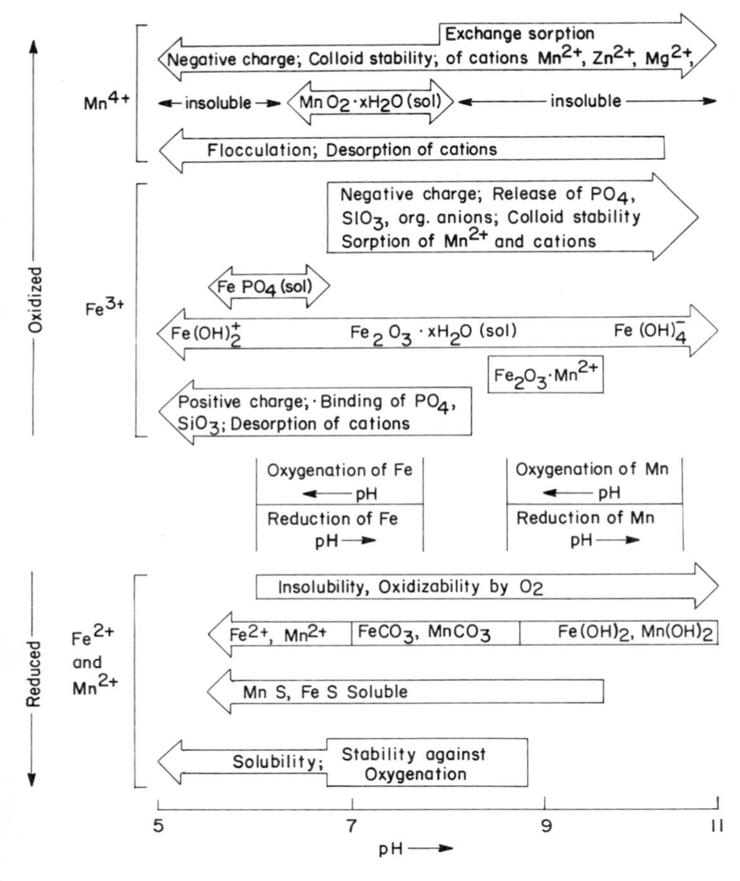

Fig. 1. Behavior of iron and manganese in natural waters. [After Ref. *34*, by courtesy of Pergamon Press.]

many times. Complete sewage treatment in its most modern context involves tertiary treatment for the removal of plant nutrients.

A depression in the earth filled with rainwater and open to the atmosphere will eventually develop a biota of microbes, plants, and animals. Which species come to be present is obviously dependent upon accidents of nature. In time the water approaches chemical equilibrium with the crustal materials, and the milieu becomes more congenial to life. All natural basins are subject to some inflow and outflow; they are dynamic systems in the simplest sense of that phrase. Basins act as traps of certain materials flowing in, and the accumulation of these materials changes the character of the water body as time passes. That process of secular change is termed eutrophication when the aspect of concern is biological productivity. The lowest levels of productivity are termed oligotrophic and the higher levels eutrophic; the term hypereutrophic has recently come into use to describe the extreme conditions observable when nutrient substances, primarily nitrogen species and phosphates, are, through man's activities, added in amounts exceeding those found under primitive conditions. It is by no means certain that a regular continuum exists in the evolution of a lake basin from oligotrophic to eutrophic; in fact, like many biological systems, this system has a strong autocatalytic character over the secular time scale [see especially Fig. 1 of Ref. (*7*)]. In addition, the sudden appearance in abundance of certain species in the phytoplankton and other sections of the biota, which has been reported in many lakes, can be considered to be a kind of biotic quantum jump. Furthermore, the plant and animal communities observed in lakes, taken to be representative of discrete stages or plateaus in the evolution of lakes, do seem to have a certain temporal stability, suggesting the existence of discrete stages of succession rather than a uniform continuum; but the whole bulk of knowledge and the time span for which appropriate knowledge is available do not permit pressing the argument too far.

II. Physical Processes

A. Dilution and Mixing

An important physical process in self-purification is dilution, the result of diffusion and transport phenomena. Molecular diffusion is a slow and unimportant process in water at the temperatures characteristic of natural surface waters.

Cowgill(*9*), however, showed that the vertical heat transfer coefficient in the middepth layer (5–7 meters) in Linsley Pond, North Branford, Connecticut, was consistent with molecular heat transport during summer stratification (June–August). The situation described by Cowgill no doubt represents an unusual case; in flowing waters advection and turbulence (eddy diffusion) dominate.

Mixing occurs mainly as a consequence of eddy diffusion, conceptually a process occurring because of density gradients or heterogeneous density fields, mainly due to temperature but also due to dissolved salts in some actual situations. In rivers, the slope of the bed can dominate the transport and mixing picture. In estuaries and along the seashore, the effects of tides are very significant. In the ocean and in lakes, the work of the wind can dominate mixing, especially in the surface layer. Ocean currents are the result of a complex of forces including wind, vertical density gradients, tides, and the Coriolis force. The major surface currents of the oceans are relatively well known(*10*), but subsurface currents (especially deep water currents), which are of great consequence in "ultimate" disposal of radioactive wastes, are much less well known(*11*,*12*). Of equal importance are the chemical and biological processes resulting in the concentration of dispersed radioactive elements. According to Revelle and Schaefer(*13*), our knowledge of physical, chemical, and biological processes governing the fate of radioactive wastes dumped into the sea was (in 1957) "woefully incomplete," and "it is urgently necessary to learn enough about these processes to provide a basis for engineering estimates." The whole subject of the 1957 NAS-NRC report(*13*) on radioactive wastes in the sea has been under intensive study over the last 10 years and a new report will be issued shortly under the title *Atomic Radiation in the Marine Environment*(*13a*).

In addition to artificial radionuclides, another class of man-made materials has become a potential pollution problem on a global scale. Chlorinated organic pesticides from the areas of the world where they have been extensively used, such as Europe and North America, have become distributed to the most remote parts of the world by mixing processes operating on a global scale.

Dilution and mixing, although commonplace phenomena, are complex processes; the general theories are conceptually straightforward but mathematically quite complex. Eckart(*14*) gave the conceptual framework in simple terms. Advection, the movement of one parcel of water with respect to another, increases the mean value of gradients of concentration or physical properties. This effect of advection Eckart called stirring, and he pointed out that stirring is independent of vorticity; that is, it occurs in laminar flow as well as turbulence. The effect of conduction or diffusion is to decrease gradients, which is the essence of mixing. In the early stages of a process in which both stirring and mixing occur, advection dominates, increasing the mean gradient so that mixing eventually dominates over stirring. Viscosity, to the extent that it is not counteracted by momentum, tends to stop the stirring process and thus inhibit mixing. Von Karman(*15*) reviewed the statistical theory of turbulence based on the classical papers of Dryden, Heisenberg, Kolmagoroff, Navier, Onsager, Prantl, Reynolds, Stokes, and G. I. Taylor. Other theoretical bases for mixing in rivers, estuaries, and the oceans are given in the proceedings of the Symposium on Diffusion in Oceans and Fresh Waters(*16*), which also contains a description of field studies of diffusion in major rivers and estuaries using fluorescent dye tracers. One conclusion of Kisiel(*17*) from dye tracer studies on the Ohio River is especially noteworthy: "More elaborate systems of waste discharge into the river are required than just placing an outfall at the shoreline. As far as the stream is concerned, per cent efficiency of the treatment process is not relevant. What is pertinent is the remaining strength of treated waste to be assimilated by the stream."

B. Sorption and Sedimentation

Sorption can be defined as the binding of ions or molecules in solution to solid-phase particles. In the bound form their behavior is altered, to be sure, but toxic ions bound to suspended particulate matter can affect filter-feeding animals ingesting the particles, and ooze feeders can be poisoned by sedimented particles with sorbed toxic ions. The sorption of one ion on a particle may displace another ion into solution, in which case ion exchange sorption is said to occur. Sorption phenomena are geochemically significant for some elements in the hydrosphere, as amply demonstrated by soil science. Quantitative differences in the relative importance of sorption versus chemical equilibrium processes in water and soil are no doubt due to the very different ratio of liquid phase to solid phase.

Sorption is usually considered and described as a physical process, but Stumm(*18*) pointed out that some so-called sorption phenomena can be just as readily explained on the basis of solubility equilibria. According to Hsu(*19*), the fixation of phosphate in acid soils is governed by sorption on amorphous aluminum hydroxides and iron oxides, rather than by the solubility of crystalline aluminum or iron phosphate species; however, the process can be treated as a chemical equilibrium system and Hsu discussed the effects of pH and concentrations of participating elements on phosphate solubility in acid soils. In neutral and basic soils, the solubility of crystalline calcium phosphate species seems to dominate over sorption phenomena.

In the case of polluted reaches of rivers and lakes, however, it is obvious that we are dealing, by definition, with departures from equilibrium. In addition, the activities of the biosphere as a whole couple solar energy with geochemical equilibria and impose steady-state departures from strictly chemical kinetic equilibria. That is to say, the fact that oxygen is found in the atmosphere at all is a consequence of sunlight-dependent photosynthesis, and the organic products of photosynthesis are used as fuel providing energy for so-called active transport of ions across cell and tissue membranes against concentration gradient. To carry the argument one step further, one might say that man, through pollution, disturbs these steady states, and the self-purification process restores them. The details of how pollution affects the biotic steady states are given in Sec. IV.

Sedimentation removes particulate wastes from suspension, and the sedimentation of both introduced and native suspended matter containing sorbed ions and molecules removes soluble material from the water column. Sedimented material affects the benthal biota in various ways, depending on the nature and quantity of the material. It can serve as food, promoting the growth of the benthal fauna, or it can serve as the grave for the fauna. It has been observed many times that photosynthesis or primary productivity in shallow flowing waters is preponderantly due to the benthic epifauna (the attached microalgae growing on bottom substrates) rather than the phytoplankton. Even a relatively small amount of sedimentation in streams can bury the epifaunal algae faster than it can grow.

The resuspension of sediment during rare periods of high stream velocity can have profound effects on the properties of the water column, a fact first emphasized by Imhoff in the 1930's. In fact, relatively stable organic matter with low short-term biochemical oxygen demand (BOD), after a period of anaerobic decomposition in the sediment, develops an increased short-term BOD. The resuspension of such sediments can suddenly impose a significant burden on dissolved oxygen(*20*).

Jansa and Akerlindh(*21*) examined the relationship between the laboratory test for settleable solids and the distance which suspended solids would be carried before settling in streams at various velocities. They pointed out that downstream from a sewage outfall there is a point where equilibrium exists with regard to sedimentation, i.e., a point where the BOD sedimented per day is equal to the amount oxidized per day. This assumption is straightforward and reasonable if only the oxidizable part of suspended solids is of concern, and surely it applies to the region in a polluted stream where resuspension of sediment by flow surges will be of no significance to dissolved oxygen status of the water. Jansa and Akerlindh showed that a term for oxygen depletion due to sediment BOD could be added to the Streeter–Phelps equation, which is discussed in Sec. III.E of this chapter:

$$\frac{dl_s}{dt} = k_s L_s e^{-k_s t} \tag{1}$$

where l_s is settled sludge BOD at time t, L_s is total quantity of settleable sludge BOD, k_s is rate constant of sedimentation, and t is time of sedimentation.

They found the relationship to be useful in describing the dissolved oxygen status of a sewage-polluted Swedish stream, where k_s was found to be on the order of 40 day^{-1}. They suggested further that settling of BOD-containing solids was negligible at stream velocities above 50 cm/sec, and as a first approximation, sludge-carrying capacity can be taken as proportional to velocity at low velocity and proportional to the square of velocity at high velocity (greater than 12 cm/sec). In any case, the BOD of suspended solids can be determined empirically, as can the settleability at zero velocity or even under simulated flow conditions; but one hopes that adequate settling of BOD solids will be carried out in treatment plants universally in the future, thus making the sediment term in the Streeter–Phelps equation of academic interest only.

Hoak and Bramer(*22–25*) gave a general model for the design of sedimentation basins showing the effect of basin geometry, flow velocity, and particle characteristics such as size, density, and quantity. That model should be applicable, with modification, to sedimentation in streams.

III. Chemical Processes

A. Solubility Equilibria

The general classification of chemical processes important in self-purification has been given above. Benoit(*26*) reviewed the recent work on the composition of fresh waters in relation to chemical equilibria between water and the predominant mineral species of the lithosphere.

Kramer(*27*) and Sutherland et al.(*28*) have provided a model for the composition of fresh waters based on solubility (or activity) equilibria for common lithosphere materials containing the various elements under consideration, namely, calcite, dolomite, kaolinite, gibbsite, Na- and

K-feldspars, and an atmospheric carbon dioxide concentration corresponding to $P_{CO_2} = 3.5 \times 10^{-4}$ atm. The model predicts the actual principal ionic contents of Great Lakes waters remarkably well, but Kramer(*27*) pointed out that there is no natural control on chloride and sulfate in lake waters, except that the amounts in rainwater set the minima. The addition of chloride and sulfate by man (principally as the sodium salts) affects the solubility equilibria in known ways according to the Kramer model, principally through the adsorption of sodium by clays and feldspar, causing a shift in pH and alkalinity, which affects the carbonate dissociation. The adsorption of sodium or its exchange with a less alkaline cation obviously decreases pH and alkalinity and drives the carbonic system to the left.

$$CO_2 \uparrow + H_2O \rightleftharpoons H_2CO_3 \rightleftharpoons H^+ + HCO_3^- \rightleftharpoons 2H^+ + CO_3^{2-} \qquad (2)$$

Another most important consequence of the addition of salts to natural waters was pointed out by Livingstone and Boykin(*29*) in connection with sorption phenomena involving phosphate. The partition of phosphate between solution and the solid phase, for example, lake mud, is a function of the total quantity of anions in the solution bathing the mud, or to put it another way, other anions like sulfate and chloride, even though they may not be as avidly bound to sorption sites as phosphate, nevertheless compete with it for the available sites and thus increase the amount of phosphate in solution at equilibrium. Thus, phosphate would tend to be released from muds formed under a regime of low total dissolved solids but bathed by water of higher ionic strength. The discharge of physiologically inert brines of common salt or sodium sulfate could thus promote eutrophication.

B. Acid-Base Reactions

The simplest way of looking at the assimilation of acid wastes in streams is through the reaction of mineral acids with feldspars, the commonest mineral species in crystalline rocks. Similar reactions can be written for clays, the hydrous weathering products from feldspars, and other silicates. In limestone regions, the acid assimilation reactions involve limestone and dolomite:

$$\underset{\text{Feldspar}}{2(Na,K)AlSi_3O_8} + \underset{\text{acid}}{4H_2SO_4} \rightarrow (Na,K)_2SO_4 + Al_2(SO_4)_3 + 6SiO_2 + 4H_2O \qquad (3)$$

or

$$\underset{\text{limestone}}{(Ca,Mg)CO_3} + \underset{\text{acid}}{H_2SO_4} \rightarrow (Ca,Mg)SO_4 + H_2O + CO_2 \qquad (4)$$

Caustic wastes are assimilated primarily through the reaction with silica, bicarbonate, and free carbonic acid:

$$\underset{\text{caustic}}{2(Na,K)OH} + \underset{\text{silica}}{SiO_2} \rightarrow (Na,K)_2SiO_3 + H_2O \tag{5}$$

and

$$\underset{\text{caustic}}{2(Na,K)OH} + CO_2 \rightarrow (Na,K)_2CO_3 + H_2O \tag{6}$$

These reactions in nature take place in a complex aqueous ionic milieu in contact with the atmosphere and a variety of mineral species. From the standpoint of the capacity of a river to assimilate wastes, the rate at which the reactions occur is of more practical interest than the reactions themselves, for it governs the dimensions of the river reach degraded by unassimilated acid or caustic.

In the absence of significant quantities of mineral acids or bases, the pH of waters is governed by the carbonic system, as discussed in great detail by Hutchinson(*30*).

The solution-dissociation reactions have already been given in Eq. (2), but one commonplace situation not described by Eq. (2) occurs when carbon dioxide is passed into alkaline solutions:

$$CO_2 + OH^- \rightleftharpoons HCO_3^- \tag{7}$$

Hutchinson(*30*) pointed out that carbon dioxide is about 200 times more soluble than oxygen in water. The amount found in natural waters, however, is largely a consequence of the amount found in air, and air has only about 0.03% CO_2 while it contains about 20% oxygen. Thus, while natural waters contain on the order of 5–10 mg of O_2/liter, they contain only about 0.5–1.0 mg of CO_2/liter. Hutchinson stated that the passage of carbon dioxide into solution from the air following its removal from the water by photosynthesis is a slow process; it takes "a long time, of the order of days or weeks . . . to achieve equilibrium with the atmosphere across a normally disturbed film." Attempts to measure the so-called invasion coefficient have resulted in values varying over a 200-fold range, according to Hutchinson. The time scale for the other equilibrium reactions in Eq. (2), however, is on the order of minutes. This fact apparently escaped the attention of Keuntzel(*31*), whose paper on the significance of bacterial CO_2 production in algae blooms is discussed in Sec. IV.D.2. The proportions of the various carbonic chemical species over the pH range of biological interest are given in Table 1.

The solubility of calcium carbonate in water in equilibrium with ordinary air is also of interest. Hutchinson(*30*) gave 55–60 mg of $CaCO_3$/liter as the solubility at 20°C, corresponding to 70–77 mg of HCO_3^-/liter.

TABLE 1
Molecular Proportions of Carbon Dioxide, Bicarbonate, and Carbonate at Various pH Values[a]

pH	CO_2	HCO_3^-	CO_3^{2-}
4	0.996	0.004	$\sim 10^{-9}$
7	0.21	0.79	$\sim 10^{-4}$
9	0.003	0.966	0.03
10	10^{-4}	0.76	0.24

[a]Reprinted from Ref. *30*, p. 657, by courtesy of John Wiley & Sons, Inc.

Even though there are a number of different qualities determined analytically and a great deal of confusion in terminology, it is clear that solubility of $CaCO_3$ and the partial pressure of CO_2 in air poise the carbon dioxide status and the pH of most natural waters. A recent paper (*31a*) illustrates these effects on the CO_2 partial pressure in the Columbia River. For a full discussion of the terminology used in the older literature and especially the foreign literature, the reader is referred to Hutchinson's admirable treatise (*30*).

Photosynthetic and respiratory activities of the biota directly affect the carbon dioxide content of waters, and diurnal variation in pH is commonly ± 0.5 units, as pointed out by Lee and Hoadley (*32*). Where there are intense growths of algae, the pH at the end of the day can approach the upper physiological limit. The simplest way of depicting the change that takes place is by the following simple reaction, which takes into account only the calcium bicarbonate out of all the salts in fresh waters:

$$Ca(HCO_3)_2 - 2CO_2(\text{photosynthesis}) \rightarrow Ca(OH)_2 \tag{8}$$

Such a reaction obviously will result in high pH. On the other hand, in the deep waters of lakes and in soil, especially waterlogged soils, the total content of carbonic species can become much higher than the level indicated by solubility equilibria due to anaerobic metabolism of bacteria [see Eq. (17)].

C. *Organic Matter in Natural Waters*

Organic matter exists in natural waters in the form of the bodies and cells of living plants and animals, the nonliving particulate remains of plants and animals, known as detritus, and soluble organic residues and excretions of plants and animals. Parsons (*33*) has estimated that for the surface waters of the North Atlantic the proportions (mg/liter) of these materials are ordinarily as follows: soluble organic matter, 2; detritus, 0.2; phytoplankton, 0.04; zooplankton, 0.004; fish, 0.00004.

In fresh waters, the content of soluble organic matter is ordinarily about 10 times the amount shown above, and there is not much reason to believe that the proportions of other classes of material are very different. The difference between fresh waters and the ocean is not due to primary productivity but rather to the more profound and direct contribution from terrestrial materials. According to the methods used for estimating these materials, bacteria would be included in the detritus fraction, and according to actual microbiological analyses, river waters have bacteria in much greater abundance than the ocean. Direct exchanges between mud and water are also obviously more important in small freshwater bodies than in the sea. Likewise, the category "fish" as used by Parsons is not comparable to the macro invertebrate and benthal fish populations of rivers and lakes. From the standpoint of organic matter assimilation, the important question is the extent to which the addition of soluble or particulate organic wastes can change the proportions of the Parsons pyramid of classes of material. It is well known that the floral and faunal components of the living classes of material are changed by organic pollution; the details of that phenomenon are discussed in Sec. IV.C.

D. Redox Reactions

Our discussion of redox reactions is limited to some chemical aspects of the subject as opposed to the measurement and interpretation of electrochemical potentials in water and sediments. The effect of redox potential and pH on the sorption and precipitation of iron and manganese in fresh waters has been clearly diagrammed by Morgan and Stumn (*34*). The implications of their diagram (Fig. 1) to the treatment and disposal of wastes containing heavy metals and acid are very clear. The behavior of other heavy metals such as zinc, cadmium, chromium, nickel, and molybdenum in the same redox and pH realms is an important subject for future study. Almost always, these heavy metals will be influenced (sorbed, precipitated, etc.) by iron and manganese, which are ordinarily present in excess of such elements as copper, cobalt, and molybdenum. Sugawara et al. (*35*) (Fig. 2) studied the behavior of molybdenum in a simple laboratory system of mud and water permitted to become anaerobic after aeration. Modern simple and reliable instrument methods of analysis for

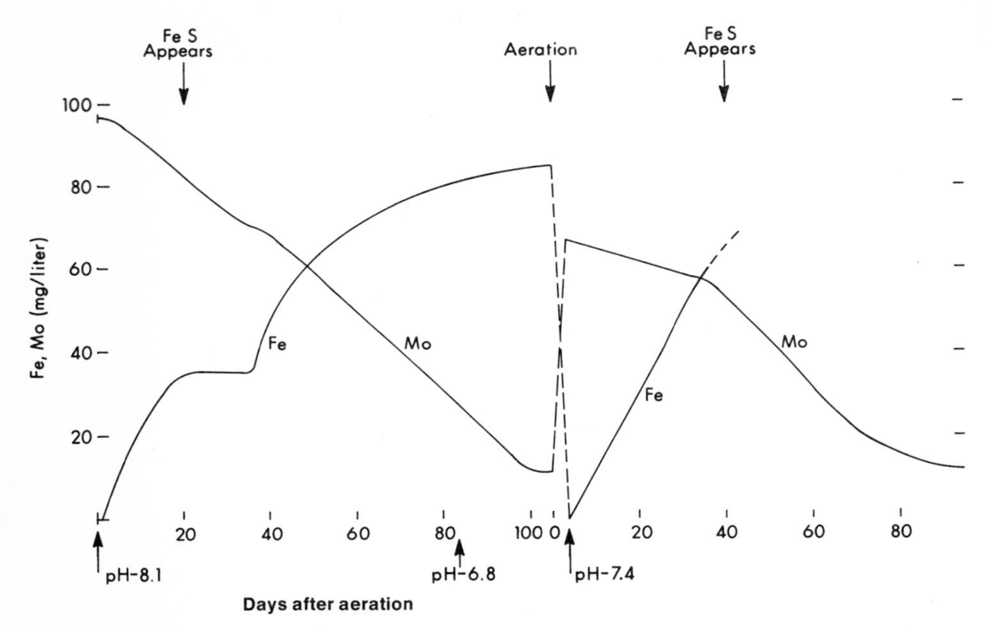

Fig. 2. Behavior of molybdenum in redox changes in water in the presence of mud. [After Ref. *35*, by courtesy of Nagoya University, Japan.]

heavy metals in solution should facilitate extensive studies of the same kind with other heavy metals.

E. Dissolved Oxygen Relations

As mentioned earlier, dissolved oxygen is a very commonly used water quality criterion; it is an important general index of quality albeit not all-pervasive. A comprehensive general review of the subject is given in Ref. (*36*). Loucks et al. (*37*) gave a general equation describing dissolved oxygen relations in a stream receiving oxygen-consuming waste, as follows:

$$\frac{dB}{dt} = -(K_1 + K_3)B + R \tag{9}$$

dB/dt is the rate of change of BOD with time, B is the BOD present, and R is the rate of BOD addition due to runoff and scour. K_1 and K_3 are rate constants for deoxygenation and sedimentation, respectively. A related expression, using dissolved oxygen deficit, D, rather than BOD, B, is the following:

$$\frac{dD}{dt} = K_1B - K_2D - A \tag{10}$$

where dD/dt is the rate of change of dissolved oxygen deficit with respect to time, D is the existing oxygen deficit (the difference between the existing dissolved oxygen concentration and the saturation value) A is the net rate of oxygen production due to photosynthesis and respiration of phytoplankton and/or waterweeds, and K_2 is the rate constant for reaeration.

Integration of the two equations yields:

$$B_t = \left(B_0 - \frac{R}{K_1 - K_3}\right)e^{-(K_1+K_3)t} + \frac{R}{K_1 + K_3} \tag{11}$$

and

$$D_t = \frac{K_1}{K_2 - K_1 - K_3}\left[\left(B_0 - \frac{R}{K_1 + K_3}\right)(e^{-(K_1+K_3)t} - e^{-K_2t})\right] + \frac{K_1}{K_2}\left[\left(\frac{R}{K_1 + K_3} - \frac{A}{K_1}\right)(1 - e^{-K_2t})\right] + D_0e^{-K_2t} \tag{12}$$

Equation (12) describes the typical oxygen sag curve as illustrated in Fig. 3. To the left of t_c in the figure, deoxygenation exceeds reaeration; to the right of t_c, the reverse is the case. The so-called critical time, t_c, is the moment at which dissolved oxygen reaches its lowest value. From

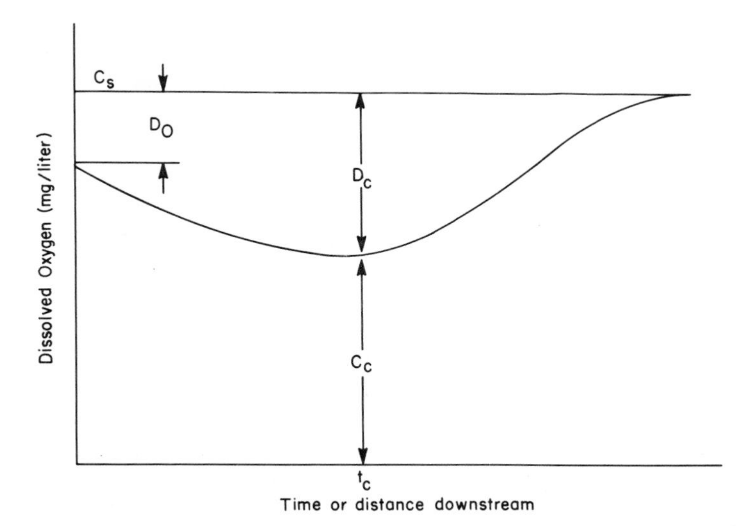

Fig. 3. Oxygen sag curve. C_s is dissolved oxygen (DO) saturation; D_0 is initial DO deficit; D_c is critical DO deficit; C_c is critical DO level (*37*).

stream flow information (velocity), the location of the critical point in the river course can be determined. It should be carefully noted that the application of the sag equation assumes that the parameters K_1, K_2, K_3, A, and R remain essentially constant and that actual values can be obtained or reasonably assumed. Equation (12) reduces to the classic Streeter–Phelps (*38*) equation if A, K_3, and R are taken to be zero.

The model of Loucks et al. (*37*) can be used to determine the minimum cost of meeting any set of dissolved oxygen standards (policy) for the various river reaches of a basin system; furthermore, the models determine the sensitivity of total cost to the allowed minimum standard for any particular reach. Each solution of the model is based on fixed (or design) flow conditions, but the sensitivity of cost and/or actual resultant water quality to departures from design flow conditions can be predicted from the model.

It should be noted that temperature does not enter the model of dissolved oxygen status as a variable, but it clearly has a profound effect on microbial and animal functions, on reaeration, on sedimentation through its effect on viscosity, and on the solubility of oxygen. Davidson and Bradshaw (*39*) list both pollutional and purification-enhancing effects of increased temperature and document specific cases studied. Davidson and Bradshaw provided a mathematical model showing the effect of temperature on dissolved oxygen status of BOD-loaded streams. The model demonstrates the existence of a theoretical optimum temperature profile downstream from a single source of BOD pollution. Even though the authors stress the hypothetical nature of the sample problems solved, one can look forward to the application of the refined model to policy decisions and river basin planning where options exist on the location of thermal discharges.

It is of critical importance, of course, that mathematical models be subject to empirical evaluation. Torpey, in very important recent papers (*40,41*) summarized the actual performance over many years of upper New York harbor and the Thames River (England) estuary with respect to summer dissolved oxygen levels and BOD loading. It is extremely fortunate that an extensive series of data has been gathered on those locales; these case histories should stand as examples of the value of planning for long-range water quality monitoring programs. Torpey interpreted the New York and Thames data as indicating the existence of a homeostatic plateau in dissolved oxygen between about 30 and 50‰ of saturation for loading rates corresponding to oxygen requirements between 20 and 135 lb/day per acre of water in the estuaries, which have average depths on the order of 30 ft. Figure 4 shows the response described by Torpey. He pointed out that the oxidation of nitrogeneous mat-

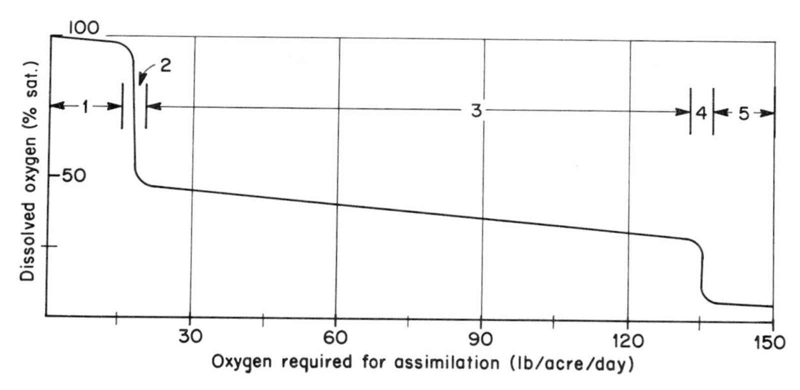

Fig. 4. Response of New York Harbor and Thames Estuary. [After Ref. *40*, by courtesy of the Water Pollution Control Federation.]

ter to nitrate was not discernible during the period of time covered by the plateau. Torpey attributed the plateau to a symbiotic relationship between algae (photosynthesis providing oxygen) and bacteria (mineralization providing nutrients for algae). Unfortunately, no direct evidence in the form of microbiological survey data exists for the period covered by the dissolved oxygen data. Torpey characterized zone 1 ecologically as supporting a biota including higher food chain forms such as crustacea, mollusks, and fish; zone 3, the homeostatic plateau, has a dominant microbial biota. It is surely not coincidental that the shad catch in the Hudson River during the transition from zone 1 (1901) to zone 3 (1904) dropped from 17 million to 5 million lb/year. Torpey warned that when receiving waters are being loaded at a rate falling toward the right of the homeostatic plateau, instituting sewage treatment will not necessarily result in dramatic increases in dissolved oxygen. At least for systems with hydraulic parameters similar to those of New York Harbor and the Thames Estuary, "the restoration of polluted waters to a condition where they can be used to support large fish populations usually requires a much higher degree of treatment to reduce the discharging load than would be indicated by oxygen-sag formulations." Although the data included in Torpey's study can be interpreted in other ways, there is no doubt that the relationships shown represent the overall response to all the complex parameters involved in estuarine chemical dynamics.

The contribution of bottom sludge or mud deposits to total oxygen demand in a river reach has been mentioned. Oldaker et al.(*42*) determined the oxygen demand of young and aged river sediments associated with sewage outfalls in a laboratory system consisting of 20-liter carboys and associated fittings. These authors were interested in the effect of depth of sediment on oxygen demand under quiescent conditions, but their data also provide insight into the maximum burden imposed by severe scouring (resuspension of sediment during turbulent flow conditions). The normal condition in the field, in which benthic animals work the sediments to some extent, would represent a condition of demand intermediate between fully suspended and settled sediment. It was earlier observed by Fair et al. (*43*) that the ultimate oxygen demand of muds was not approached until after a year. In the case of a quiescent laboratory system, the processes limiting the rate of oxygen consumption by mud are the diffusion of oxidizable matter from mud into water and the rate of diffusion of oxygen into the mud. The kinetics of the process thus is largely independent of microbial metabolic kinetics, but the laboratory tests of Oldaker et al. and Fair et al. are nevertheless of interest. Oldaker et al. (*42*) found the initial ultimate areal oxygen demand of aged Merrimac River sediments to be proportional to depth over the range studied (1.5–20 cm) at about 32 g/m^2 per centimeter of thickness. Recent sediment was studied at only one depth (1.5 cm) and it showed a demand of 120 g/m^2 for the depth tested or 80 g/m^2 per centimeter. The initial ultimate areal oxygen demand L_{d_0}, as used by Fair et al. and Oldaker et al. is conceptually comparable to the ultimate carbonaceous oxygen demand, BOD_u, defined by

$$BOD_t = BOD_u\,(1-10^{-kt}) \tag{13}$$

where no areal basis is involved and BOD_t is the value for time, t, and k, the rate constant. The total areal BOD of sediments is defined by

$$L_d = L_{d_0} \times 10^{-k_4 t} \tag{14}$$

where L_{d_0} is the initial ultimate areal demand as defined above. The rate constant, k_4, showed an exponential decrease with depth in the experiments with aged sediments, falling from 0.0033 at 1.5-cm depth to 0.0006 at 20 cm. With recent sediments, only a single depth (1.5 cm) was studied, and in that case k_4 had a value of 0.0051. From nitrate analyses on the influent and effluent waters, the demands attributed to nitrification were calculated. In the recent sediment (1.5 cm thick), nitrification rose to a peak at 70 days and then fell to very low values at 90 days. The highest proportion of the demand attributable to nitrification was about 50%. With aged sediment (1.5 cm thick) the time course of nitrification was more erratic, with a single value equal to about 90% of the total demand at 50 days, falling at 65 days to a low value of about 10%, then rising to about 40% at 80 days, and finally falling very low again. It would be unwise to press too far the significance of laboratory jar tests in relation to what can be found or expected in nature, but that problem is not peculiar to sedimentary metabolism. Laboratory studies must be used to determine at least the range of values of parameters not readily amenable to measurement in the field.

Velz(*20*), from extensive studies, suggested that the critical velocity for settling of organic solids is 0.6 fps (feet per second) (18 cm/sec) or less and that the velocity at which scouring of the deposits will occur is 1.0–1.5 fps (30–40 cm/sec). He developed an empirical function showing the course of development of equilibrium in a sludge deposit, wherein the rate of demand exerted is just equal to the deposition rate. It should be recalled that the critical value for settling was estimated to be 50 cm/sec by Jansa and Akerlindh(*21*).

F. Reaeration

The rate of absorption of oxygen by natural waters can be expressed by the function

$$dc/dt = k_2(c_s - c) \tag{15}$$

where c is the measured concentration of dissolved oxygen, c_s is the saturation value for the given temperature, t is time, and k_2 is the coefficient of reaeration, the second coefficient in the Streeter-Phelps formulation. Langbein and Durum(*44*) showed effects of hydraulic characteristics (velocity, v, and depth, H) on the value of k_2. They showed k_2 to be equal to $3.3v/H^{1.33}$ with laboratory and field data corrected to 68°F (20°C) by a factor of 0.01/1°F (0.55°C). For the Kansas-Missouri-Mississippi River system, they showed that the reaeration coefficient decreases in the downstream direction as the square root of the mean discharge (cubic feet per second). This is in conformance with the way in which velocity and stream height vary with discharge (velocity increases as the 0.10 power, whereas depth increases as the 0.40 power of discharge). Langbein and Durum showed that a regional difference exists in the relationship. For coastal plain rivers, k_2 varies as $15\sqrt{Q}$, whereas in the northern Rocky Mountain region, k_2 varies as $80\sqrt{Q}$, where Q is the mean discharge in cubic feet per second. That difference is attributed to slope of the stream bed, and the regions represented are extremes with regard to slope.

The general trend in the reaeration coefficient, k_2, is downward with increasing stage (height) of a river, but shallow reaches (riffles) behave differently from deeps (pools). Figure 5 shows that the coefficient in riffles falls rapidly as discharge increases, while in pools it rises only a

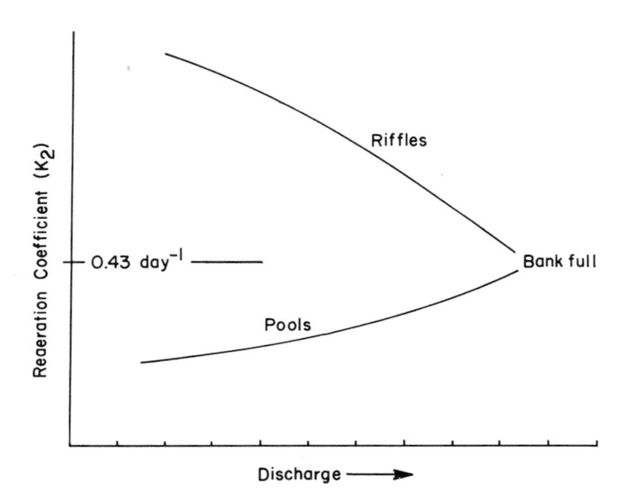

Fig. 5. Effect of discharge on reaeration. [After Ref. *44*.]

little. At some high stage the distinction between these stream features disappears. For a section of the Kansas River at Bonner Springs, Kansas, the coefficient had a value of 0.43 day^{-1} at the stage where riffles and pools were drowned out.

Langbein and Durum(*44*) estimated the total assimilatory capacity of eight size classes of streams on a nationwide basis. That information is not of much interest in the context of this chapter, but the data can be used to estimate the capacity per mile (per unit of deficiency in dissolved oxygen), that is to say, the number of tons of oxygen-consuming substances, as BOD, which can be assimilated per mile under conditions which would lower dissolved oxygen by one unit (mg/liter). The smallest class of streams (I) has an average discharge, Q, of 14 ft^3/sec, an average depth of 0.55 ft, an average velocity of 1.2 fps, and a coefficient of reaeration of 9.5 day^{-1}. For this class, the assimilatory capacity per mile per unit deficiency of dissolved oxygen is estimated to be 0.04 tons of BOD; the largest class III, represented by the Mississippi River, with a mean discharge of 700,000 ft^3/sec, an average depth of 45 ft, an average velocity of 5.0 fps, and a coefficient of 0.10 day^{-1}, has an estimated capacity of 5.2 tons of BOD/mile. An intermediate class, II, represented by the Allegheny, Kansas, and Rio Grande Rivers, has a capacity of about 1.0 tons/mile. Consideration of the assimilatory capacity per mile per foot of depth gives rise to an apparent paradox, as shown in Table 2. The capacity per foot of depth in all three classes is thus on the same order of magnitude in spite of the wide range of width in the three classes, which

TABLE 2
Relation of Assimilatory Capacity and Stream Size[a]

Stream class	Assimilatory capacity[b]	Average depth, ft	Assimilatory capacity per ft of depth	k_2 day^{-1}
I	0.04	0.55	0.07	9.5
II	1.0	5.0	0.2	—
III	5.2	45.0	0.12	0.1

[a]From Ref. *44*.
[b]Tons of BOD per mile per unit O_2 deficiency.

can be calculated from the data of Langbein and Durum to be 21, 560, and 3100 ft for classes I, II, and III, respectively. The apparent paradox is resolved by considering first the high reaeration coefficient of the first class, nearly 100 times the coefficient of the Mississippi River, and by considering that vertical mixing calls into play the whole depth of the river. The first two classes, with their shallow depth, are always virtually completely mixed, but the largest class, especially under velocities lower than the average, may show some vertical stratification (incomplete vertical mixing), a situation not to be overlooked in waste disposal policy considerations. Biochemical processes directly affecting dissolved oxygen relations in streams are discussed below.

IV. Biological Processes

Biological processes in relation to water pollution have two principal aspects. First, there is simply the activity of living plants and animals in the self-purification process, the transformation of deleterious substances into innocuous products, usually at the expense of dissolved oxygen and often at the expense of some species normally present in the diverse biota of unpolluted streams. Second, there are those changes in the biota itself in response to pollution; that is, the use of the biota as a complex quality criterion for assessing the impact of pollution on the environment. Recent comprehensive works on biological aspects of water waste management include those by Hynes(*45*), Doudoroff(*46*), Tarzwell(*47*), Bartsch and Ingram(*47a*), and Keup et al.(*48*). The last work is a collection of some classical papers not hitherto very readily available, including the important pioneer works of Kolkwitz and Marsson(*49,50*).

A. Normal Biological Processes

The biological processes of greatest importance in waste assimilation are: (1) feeding, digestion, and respiration of animals, including the protozoa and microscopic multicellular animals, and (2) photosynthesis and respiration of plants, including micro algae. Bacteria and fungi are classified as plants, of course. Although they are capable of absorbing and digesting organic substances from solution, they are commonly associated with surfaces of organic particles and thus an important part of their feeding must be mediated by extracellular enzymes. Furthermore, bacteria and fungi can live and thrive under conditions inimicable to plants and animals, and thus they dominate the biota in the most severely polluted reaches of streams.

Most protozoa live by ingesting food particles and digesting them in vacuoles within the cell body. Bacteria and microalgae are the normal food of most species of protozoa. They likewise make up a great part of the normal diet of filter-feeding invertebrate animals, both microscopic and macroscopic.

There is an old Chinese adage that says, "big fish eat little fish; little fish eat shrimp; shrimp eat mud." That statement expresses an important ecological principle in rudimentary form—the principle of the food chain or food web. This principle is discussed at length in Sec. IV. C. The adage overlooks the origin of the "mud," the base of the food web. Normally, the mud represents the cellular products of photosynthesis or the partically decomposed remains of plants, the remains of animals grazing on the plants, or the remains of animals preying upon other animals or feeding on bacteria, which attack organic matter in whatever form they find it. Photosynthesis and respiration are discussed as biochemical processes below.

Oxygenation in streams can be divided into the components from diffusion on the one hand, and the net excess of photosynthesis over plant respiration on the other. The latter component can be separated into contributions from the benthic flora (epiphyton or periphyton) and free-floating microscopic plants (phytoplankton). The net production of oxygen from photosynthetic metabolism is theoretically zero, according to the classic synoptic equation:

$$CO_2 + H_2O \underset{\text{respiration}}{\overset{\text{photosynthesis}}{\rightleftharpoons}} CH_2O + O_2 \tag{16}$$

In Eqs. (16) and (17) CH_2O symbolizes carbohydrate. Fats, proteins, and other cellular materials are synthesized from carbohydrate. For the synthesis of protein and nucleic acids, plants need a source of nitrogen, sulfur, and phosphate, of course. These synthetic activities require energy, and that energy comes from respiration of carbohydrate.

A positive net yield of oxygen thus can only result from transport of plant tissue out of the water zone of interest, either downstream or by sedimentation to the bottom where decomposition is arrested or proceeds anaerobically. There is a broad class of anaerobic microbial reactions that partition organic matter into carbon dioxide and molecular species more reduced than carbohydrate, e.g.,

$$2CH_2O \rightarrow CO_2 + CH_4 \text{ (methane)} \tag{17}$$

Redox reactions involving nitrogen and sulfur also play prominent roles in anaerobic microbial metabolism(*51*).

Stay et al.(*52*) described an apparatus and methods for measuring directly the components of oxygenation in streams—diffusion, respiration, periphyton photosynthesis, and phytoplankton photosynthesis. The commonly used method of light and dark bottles introduces an unnaturally high surface-to-volume ratio, and stagnant water conditions within the bottles prevent the normal interchange of nutrients and waste products. Furthermore, the contribution of benthos, of primary importance in shallow streams, is negated. Stay et al. used in their field studies three large plastic chambers: type 1—transparent and closed to the atmosphere; type 2—opaque and closed to the atmosphere; type 3—transparent and open to the atmosphere. The chambers can contain natural or artificial substrates. Paddle wheels, artificially driven or driven by external waterwheels, simulate natural flow velocities. Provision is made for sampling and analysis of gases leaving solution in the chamber and for taking water samples for dissolved oxygen, CO_2, and other analyses. In type 1, photosynthesis and respiration proceed simultaneously, but there is no physical exchange of oxygen or carbon dioxide between the water and the atmosphere (diffusion). In type 2, only respiration occurs. In type 3, photosynthesis, respiration, and diffusion occur. By using all three types of chambers, the contributions of the component processes in gas exchange in streams can be measured. For example, type 1 measures net photosynthesis, the excess of photosynthesis over respiration. Type 1 plus type 2 measures gross photosynthesis; type 3 minus type 1 measures gas exchange between air and water. The various possibilities are presented in Table 3.

By damming off the substrates from the upper chamber, the contribution of benthos versus phytoplankton can be measured. The preliminary data of Stay et al.(*52*) on the Blue River, Oklahoma, show clearly the

TABLE 3
Component Processes in Stream Metabolism from Chamber-Dissolved Gas Measurements[a]

Chamber type	Configuration	Processes occurring	Change in dissolved gases
1	Transparent, closed to atmosphere	Photosynthesis, respiration	$O_2 \uparrow$; $CO_2 \downarrow$ $O_2 \downarrow$; $CO_2 \uparrow$
2	Opaque, closed to atmosphere	Respiration	$O_2 \downarrow$; $CO_2 \uparrow$
3	Transparent, open to atmosphere	Photosynthesis, respiration, diffusion	$O_2 \uparrow$; $CO_2 \downarrow$ $O_2 \downarrow$; $CO_2 \uparrow$ $O_2 \updownarrow$; $CO_2 \updownarrow$

[a]From Ref. *52*.

predominant effects of benthos versus phytoplankton and other features of the system as described above. In addition, the effect of natural diurnal and day-to-day variation in light intensity on photosynthesis can be seen. The stability of the system can be seen in the relatively constant rate of respiration over the few 24-hr periods covered in the test runs reported. The details of the data are not presented here because of their preliminary nature; for meaningful generalization, synoptic data from several geographic regions will be required.

Gas exchange is only one of many manifestations of the activities of the living realm in streams. Other aspects are amenable to study by the relatively simple methods of physics and chemistry. The dynamics of the biota, the relationships between plant, animal, and microbial populations, however, cannot be measured in any simple way. Rather, one attempts to describe the biota as completely as possible and to devise a simplified model which describes the ecodynamics. Theoretical models of ecosystems are discussed next.

B. General Systems Model

Rivers, lakes, and estuaries are readily accessible to scientists, they are complex and fascinating environments, and they are of great value as water resources. For these reasons, they have been extensively studied and an enormous literature exists. That is not to say, however, that the information available on physical, geological, geochemical, and biological features of waters has been assembled, sorted, selected, analyzed, and synthesized into the meaningful generalities needed for prudent planning, regulations, and management of the resource. The sorting, analysis, and synthesis of the existing information should be done by means of modern techniques of operations analysis and information processing. General systems(*53*) offer ultimately a framework for the needed synthesis, but it is by no means certain that a useful general systems model of ecosystem dynamics (including social science aspects) can be devised within the present state of knowledge. Such a model should be a major long-range goal in any event, and for a beginning a crude conceptual general systems model should be devised. Klir(*54*) gave the characteristic traits defining a general system, and the operational submodels discussed in this chapter may someday be assembled into a general system model providing a very powerful management tool for predicting with known confidence the behavior of the system parts in response to perturbations.

The sub models which must eventually be synthesized include flow dynamics models (transport and mixing), dissolved oxygen status models, and geochemical equilibrium models already discussed, as well as trophic dynamic models which are discussed below. With regard to the general state of knowledge in relation to modeling, the situation is aptly summarized by Hem(*55*) in his review of *Equilibrium Concepts in Natural Water Systems*(*56*):

"The development of an equilibrium model for natural water systems generally must now be based on permissive evidence. That is, the conditions broadly observed in the systems do not contradict the equilibrium hypothesis. A much more rigorous set of observations is generally needed to give the proof of equilibrium conditions that would generally be demanded in other branches of solution chemistry.

"The general theme of a need for more and properly designed research is present throughout. Although the more elaborate models proposed in some of the papers are thought-provoking, the next symposium on this subject will, hopefully, provide more of the facts which are now missing, facts that must be learned before the more elaborate geochemical models can be fully accepted."

C. *Trophic Dynamic Model(s)*

The trophic dynamic ecosystem concept of Lindeman(*57*) is one of the most important generalizations in the science. Although specifically aimed at aquatic systems in the original version, the theory is just as applicable to land as to reverine, lacustrine, or other ecosystems [see, for example, Refs. (*58,59*)]. The Lindeman model states simply that there is a flow of energy (and materials) from one level of a food chain to the next, and that there are characteristic efficiencies in the transfer between levels. For example, a given amount of solar energy falls on a unit area in a lake, and a small fraction of this energy is utilized in a photosynthesis to produce a certain mass of algal cells with a certain energy (caloric) equivalent. These cells are grazed upon (eaten) by zooplankton and the organic material is assimilated with a certain efficiency. Herbivorous zooplankters are eaten by predators, e.g., small fish, resulting in the production of a certain amount of new fish flesh. In nature the efficiency of transfer between trophic levels is less than 10%.

Slobodkin(*60*) pointed out a logical error in Lindeman's classical model, but what is of great significance is that the model has always produced reasonable results and, even though formally incorrect, has enhanced understanding and further heuristic development. According to Hutchinson (*61*), the main deficiency in the classic model and many subsequent studies based upon it [for example, see Ref. *62*] is that the trophic dynamic approach considers the efficiency of chemical energy transfer between trophic levels of the food web and the quantities contained within each level (standing crops) without much regard to the actual species or kinds of organisms present in each level.

Hutchinson is probably not decrying the lack of taxonomically oriented studies but rather the lack of what he calls "biological reality." His notion of biological reality is aptly typified by a recent paper by Brooks and Dodson(*63*), who showed the effect of the alewife on the species composition of zooplankton in lakes where the fish was introduced (see also *63a*). The various ways in which different species exploit trophic level resources is precisely what Hutchinson meant by biological reality. Another special case of very timely importance is the passage of DDT (and other persistent pesticides) from one trophic level to the next. According to Odum(*64*), antarctic animals such as the Weddel seal contain significant quantities of DDT through a process he calls "biological magnification." In that process, metabolically persistent substances are passed undegraded and virtually undiminished through successive levels of the food chain or food web until secondary carnivores come to accumulate, mostly in fatty tissue, high concentrations of substances which exist at vanishingly low levels in the environment as a whole.

Hutchinson makes special note of the work of MacArthur(*65*), Margalef(*66*), and Patten(*67*), in which concepts from information theory are applied to descriptions of food webs and from which emerges the notion of stability in animal communities, largely the product of alternate paths in the trophic system. The important thing from the standpoint of a general system for describing the dynamics of aquatic ecology is the necessity for combining population dynamics and community dynamics into the trophic dynamic model. Population dynamics models take into account such factors as birth rate and fecundity, growth rate, life history, including the time span occupied in various stages of the life history, and death rate. These parameters are genetically determined in part but can be profoundly altered by environment (especially death rate). From a practical point of view as well as on heuristic grounds, it is important to know the sublethal effects of pollution on population growth parameters. Woelke(*68*), working in Puget Sound, reported the adverse effects of pulp mill wastes on growth, "condition," and reproduction (apart from mortality) of oysters. Given sufficient skill in devising or applying a population dynamics model, it would be possible to predict the effect of pollution on overall shellfish productivity.

Brezonik(*69*) discussed the application of mathematical models to the eutrophication process.

D. Biological Effects of Pollution

1. Indicator Species

In terms of effect on the biota, pollution acts mostly directly in the elimination of the most sensitive species. In a sense, these species act as so-called indicators or indicator species. Thus, mayfly larvae, salmonid fishes, and certain microalgae, especially diatoms, are characteristic only of pristine freshwater environments(*70,71*). A secondary effect of pollution is the prolific development of certain tolerant species or the presence of nothing but tolerant species at ordinary or low levels of population development. Thus, sludge worms, sewage fungus (*Sphaerotilus*), and certain blue-green algae are characteristic of polluted reaches of streams and have been assigned special significance as indicator species. It is of coincidental interest that the phrase "indicator species" is used in a different context in oceanography; that is, for species of plankton characteristically abundant in certain discrete oceanic water masses and presumably dependent for well-being upon some combination of ecological factors (niche) unique to those water masses. Jeffries(*72*), in a study of the copepods of Raritan Bay, New York, showed that the indicator species concept (in the oceanographic sense), usually regarded as a phenomenon of oceanic proportions, can be applied to estuarine circulation over a range of only a few miles. The most commonplace example of an indicator species in the broad sense is sargassum weed, which is the predominant or at least the most apparent member of the biota of the great Atlantic water mass called the Sargasso Sea.

2. Eutrophication and Algal Blooms

Eutrophication is a special case of alteration of biotic diversity involving the algal flora as the most apparent aspect. The phytoplankton of oligotrophic lakes is sparse in numbers of individuals but remarkably diverse in species. In eutrophic lakes the phytoplankton is more abundant, and certain species come to be strongly dominant. In hypereutrophic lakes certain species of colonial blue-green algae almost completely dominate the phytoplankton in late summer and fall. These include *Polycystis (Anacystis, Microcystis) aeruginosa*, *Anabaena circinalis*, and *Aphanizomenon flos-aquae*. Diatom species especially characteristic of the phytoplankton of hypereutrophic lakes in the spring, and in eutrophic lakes throughout the whole summer, are *Asterionella formosa*, *Fragillaria crotonensis*, *Tabellaria fenestrata*, and *Melosira granulata*.

A detailed discussion of eutrophication is beyond the scope of this essay, but a few additional elementary facts should be presented. Nuisance conditions in relation to recreational or aesthetic values of lakes exist when blue-green algae blooms occur (a hypereutrophic condition). The concentration of phytoplankton cells (or colonies of cells) that constitutes a bloom has been arbitrarily set at various numbers. Transparency (or optical density) measurements (*72a*) and chlorophyll (or mixed pigment) analyses (*72b*) have been made and correlated with cell concentrations. There is obviously no single best way to describe the size of microalgae populations in a lake, and the correlations between different measures are surely imperfect. Aesthetic criteria obviously depend upon how fastidious the judging party happens to be. It remains for regulatory agencies to set arbitrary standards on transparency, cell concentration, pigment concentration, etc., for various uses of lakes.

Certain biological characteristics of blue-green algae are of special significance to aesthetics. First, they are colonial in habit. They occur as filaments or clumps and the larger colonies can be seen with the naked eye as clumps or clots of greenish matter. Second, they form gas bubbles (pseudovacuoles) within their cells when they become moribund; this makes them buoyant and they tend to accumulate in the uppermost few inches of the water column and even as a scum on or in the water surface. Other phytoplankton sink when they become moribund. I have seen a very heavy blue-green bloom "disappear" overnight when a stable warm layer of surface water 6–12 in. thick was mixed down into the water to a depth of 6 ft or more by night winds. The bloom reappeared in just a few days of still weather which reestablished the stable warm surface layer into which the same crop of blue-green algae "settled." To show that these processes are important in the nuisance value of an algae bloom requires much more detailed measurements of temperature and plankton-sampling in the water column than are ordinarily carried out.

Kuentzel(*31*) discussed the role of bacterial carbon dioxide production in relation to algae blooms. High bacterial carbon dioxide production presupposes a large supply of organic matter. Kuentzel noted that bloom-forming species of blue-green algae have abundant sheath material which characteristically harbors bacteria. He cited experiments which showed that blue-green algal growth was stimulated in the light by sucrose, and he attributed the stimulation of photosynthetic growth to carbon dioxide production from respiration of sucrose by bacteria present in the cultures. He noted further that additions of phosphate to the same cultures had no effect, but he failed to note that laboratory cultures of microalgae need, and get from the standard growth media, much more phosphate than is ordinarily found in natural waters. Kuentzel calls carbon dioxide a limiting nutrient, and there is little doubt that at summer temperatures, and given an adequate supply of phosphate and light, blue-green algae will grow at a rate such that carbon dioxide supply will become limiting. With the rich media used for laboratory cultures, it is a common observation that cultures must be aerated or even aerated with carbon dioxide-enriched air to achieve maximum growth. We have deliberately neglected to mention nitrogen in this discussion because some blue-green species, e.g., *Aphanizomenon* and *Anabaena* can fix atmospheric nitrogen. *Microcystis* cannot, but *Microcystis* blooms often occur with *Anabaena* or following an *Anabaena* bloom.

In this discussion of chemical factors limiting algal growth, we hasten to point out that we mean rate of supply, not momentary concentration or standing crop. Unfortunately, we know of no satisfactory description of phytoplankton growth that is based on rates of supply of essential nutrients including carbon dioxide [but see Ref. *69*]. The momentary concentration of phosphate in solution during a bloom can be vanishingly low (less than 0.01 mg/liter according to Kuentzel and many other authors), but this says nothing about the concentration of phosphate in the bloom cells or the rate at which the element was supplied and absorbed by the cells during their growth. We(*72c*) have shown the nitrogen : phosphorous ratio in the phytoplankton of 20 nonblooming Connecticut lakes to be 20–30, but in three lakes that bloom the phytoplankton nitrogen : phosphorus ratio is less than 10. In other words, bloom cells have three or more times more phosphorus than nonblooming phytoplankton. This could be an inherent difference between blue-green algae and other types, but we interpret it to mean that blue-green blooms occur when phosphate is not limiting. A discussion of what limits algal growth when algae already exist in bloom numbers seems of no practical importance.

The distinction between standing crop and rate of growth of algae is also important. The species which cause nuisance blooms, insofar as relevant field measurements and laboratory growth experiments have been carried out, have their growth optimum between 20 and 25°C. At those temperatures, no alga has ever been shown to be capable of growth exceeding three or four doublings in 24 hr. Kuentzel(*31*) cited Mackenthun's(*73*) description of a blue-green algae bloom in Lake Sebasticook, Maine, which showed over 200,000 cells/ml, "most of which grew in a single August day." If there were 50,000 cells/ml present the previous day and these divided twice in 24 hr, it may be perfectly proper to say that most grew in a single day. It is also reasonable to expect from experience that one or two cell divisions involving doubling of the biomass could occur without any additional phosphate being absorbed.

One further point deserves discussion. Kuentzel(*31*) stresses the role of organic pollution in blue-green algae blooms, symbiotic bacteria supplying the carbon dioxide from respiration of the organic matter, and algae supplying the bacteria with oxygen via photosynthesis. First, we would point out that "organic pollution" also contains bound phosphate and this would be released through bacterial action about as fast as carbon dioxide would be released. Second, the sheath material of blue-green algae, on and in which symbiotic bacteria live, is part carbohydrate and as such represents a potential reservoir of carbon dioxide in itself. Third, blue-green algae blooms occur in lakes with no "organic pollution," although natural soluble or suspended organic matter has been cited by several authors(*74–76*) as an important ingredient in the etiology of blue-green algae blooms. There is no doubt that it takes carbon dioxide to make algae and that invasion of standing water with carbon dioxide from air is a slow process relative to algal growth rates; Kuentzel's thesis requires that oxygen and organic matter disappear in proportion to the amount of bacterial and algal growth occurring during a bloom pulse. These questions are amenable to observational and experimental inquiry, and I have no doubt that Kuentzel's paper will stimulate scientific inquiry.

The phenomenon of eutrophication has been primarily studied in lakes, but estuaries and even offshore ocean areas are well known to have been subject to biotic change due to eutrophication. The best-known consequence of eutrophication in the marine realm is "red tide"(*77*); in temperate zone shallow estuaries the growth of sea lettuce to the point of nuisance may occur rather than phytoplankton blooms(*78*). Very dramatic effects on fauna, e.g., mass mortalities, can occur as a result of phytoplankton blooms; the faunal effects of estuarine eutrophication short of mass mortalities are much less well known.

3. Zonation—the Saprobic System

Indicator species, in the sense of freshwater pollution biology, represent rather extreme conditions of pollution, generally speaking, rather than incipient danger to flora and fauna or borderline conditions. It is quite clear that public concern for the quality of the environment is now so intense that it demands control of water pollution before the fact, so to speak. For that reason it has been necessary to develop special knowledge of ecology and physiology of aquatic species (*47,79*) and to devise criteria for judging when biotic conditions are marginal or degraded. The criteria and the methodology most prominent in the field are concerned with biotic diversity (*80–82*). The measurement of such a complex quality as biotic diversity is not a simple matter; it involves gathering a large amount of data in the field and interpreting these data in terms of a mathematical model of biotic community dynamics.

Thus, a grossly simple concept of indicator species, as a touchstone for detecting pollution, cannot be used to promote an optimum program of water waste management. Perhaps the simplest, and probably the oldest, system for describing biological effects of pollution is that of Kolkwitz and Marsson (*49,50*). They introduced the concept that waters or stream reaches could be divided into three classes or zones: zone I, *polysaprobic*, heavily polluted, devoid of dissolved oxygen, and biotically dominated by bacteria; zone II, *mesosaprobic*, divided into "strong and weak" subclasses; and zone III, *oligosaprobic*, in which mineralization of organic matter is largely complete and in which the "biological organization is manifold" (see Table 4). The basis for their classification is biological, but some simple chemical or physical characteristics are also mentioned; some 300 plant species and 500 animal species are listed in assemblages representative of the various classes or zones.

The first zone, polysaprobic, characteristically contains bacteria and only a few species of algae, certain protozoa, sludge worms, and the rat-tailed maggot. The last species lives in the bud but breathes air by means of a long slender tube reaching to the surface. Bacterial numbers, by the then standard test (gelatin medium) as used by Kolkwitz and Marsson, are in the range of millions per milliliter. Zone II, subdivision alpha, characteristically has bacteria (at levels of thousands or hundreds of thousands per milliliter), filamentous blue-green algae, certain green algae including two filamentous species, and a few species of diatoms. The fauna is dominated by many protozoan species, several worm species in addition to the predominant sludge worms (*Tubificidae*), many rotifer species, a single mollusk species (*Sphaerium*), a single crustacean species (*Asellus*,

TABLE 4
Zonation in Polluted Streams[a]

Zone	Degree of pollution	DO status	Dominant bionts	Number of bacteria
I (polysaprobic)	Heavy	Zero or very low	Bacteria, sludge worms, rat-tailed maggot, algae (rarely)	> 10^6/ml
II (mesosaprobic)	Strong but diminishing	Reduced to full saturation	Bacteria, protozoa, worms, rotifers, midge larvae, diatoms, other algae, carp	10^4–10^5/ml
III (oligosaprobic)	Weak or absent	Saturated	A diverse flora and fauna including diatoms, mayfly larvae, game fish	< 10^3/ml

[a]From Refs. *49,50*.

sow bugs or hog slaters), and larvae of a few insect species, especially midges (*Chironomidae*). In zone II, subdivision beta, diatoms and green algal species become much more common, in the sense that many more species are present without regard to the numbers of invididuals, filaments, or colonies of the various species that may be found. Waterweeds such as duckweeds, pondweed (*Potamogeton*), and pond lily-like species make their appearance. As to animals, several new mollusk species are found; many crustacea, including water fleas and copepods, a few insects, including mosquito larvae, and about half a dozen species of fish (especially carp-like forms) are characteristically present. Zone III, oligosaprobic, contains a characteristic diverse assemblage of diatoms, green algae, dinoflagellates (diagnostically), and waterweeds. The bacterial fauna is diverse and includes obligate aerobic species; numbers of bacteria in the open water are usually less than 1000/ml. The fauna includes diagnostically a diverse assemblage of protozoa, rotifers, and other invertebrates; mayfly larvae (Ephemeroptera) and water beetles are especially diagnostic. Fish species now include those more sensitive than carp-like fishes, such as bass, perch, or salmonid fishes.

It must be pointed out that I have grossly simplified the saprobic classification system, which is actually based on species lists. Unfortunately, some of the most useful groups of organisms, such as bacteria, algae, protozoa, and rotifers, cannot be identified except by specialists. Furthermore, the situation in nature is a dynamic continuum with no clear boundaries between classes; both the length of each zone and the location of zones within river reaches change in time with varying flow conditions and seasonally. The general relationship between the saprobic zones and the zonation shown by oxygen sag curves has been illustrated by Bartsch and Ingram (*70*) and others, but it should be carefully noted that physical and chemical conditions are more transient than the biotic characteristics, which tend to integrate environmental conditions over time.

Fjeringstad (*83*) subdivided the zones further into a total of nine, and he gave a simple mathematical expression for obtaining a quantitative indication of the saprobic state of a stream reach based on a fairly extensive collection of microbes, plants, and animals. He expressed very little confidence in the quantitative index; I agree with him and therefore do not repeat the details of the computation. Of genuine interest and great value, on the other hand, are his diagrams (see Fig. 6) showing the relationship of some chemical parameters—BOD, H_2S, total N, NH_4-N, and NO_3-N—to the nine zones, biologically defined. The range of values for the chemical parameters in the various zones indicates that a few analyses of a single chemical parameter can only provide a very uncertain measure of ecological conditions in a stream station. It should be specially noted in Fig. 6 that the average value of the chemical parameters falls almost always in the middle of the range shown on a log scale. This means, of course, that the frequency distribution of a population of analyses of a chemical parameter taken repeatedly at a single station tends to be lognormal, and not normal as is commonly assumed.

4. Biotic Diversity

The use of folk taxonomies rather than formal biological nomenclatural methods has been proposed for diagnosis of pollution (*84*). Folk taxonomies are simple systems of classification of plants and animals based on ordinary or superficial characteristics. For example, one might collect plants and animals at a stream station and divide them into groups according to their appearance—large worms, small red worms, caterpillar-like forms, small white clams, large clams, etc. In that way, an estimate of the diversity of the biota could be obtained, and one could note the abundance of the various types or categories in the folk taxonomy devised.

It is a lamentable fact that formal biological nomenclature and taxonomic procedures have resulted in a temporal instability in the system of

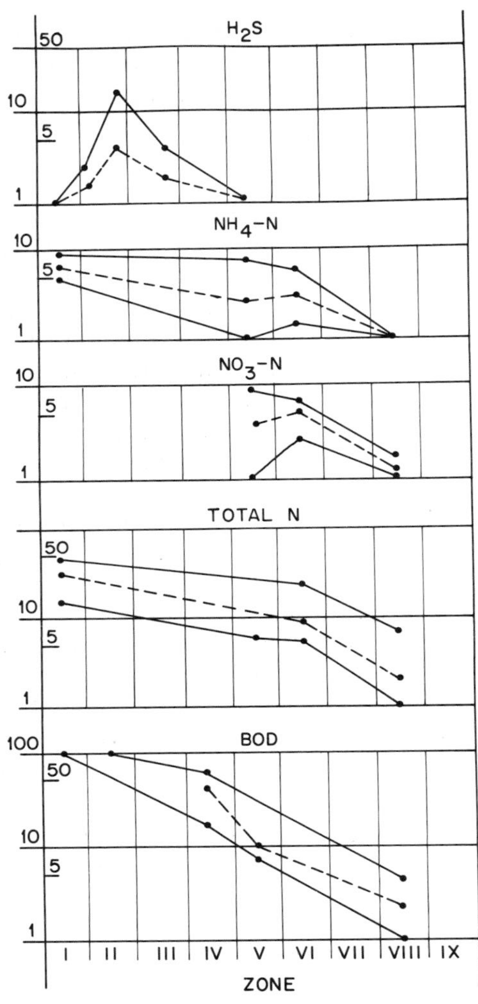

Fig. 6. Relationship between biotic zones in polluted streams and certain chemical parameters. Ordinal scales are logarithmic; units in mg/liter. Average values are shown as dashed lines between maximum and minimum curves. [After Ref. *83*.]

biological classification, instituted in the first place to replace the geographically variable folk taxonomies. The scientifically proper names of many common animals have been changed according to the taxonomic rules of priority and nomenclature at the same time that local common names or names widespread in a single language have remained the same. For the solution of practical problems, the science should be made to serve the art, but both biologists and engineers should continue to work together in resource management. The scope of the survey made to evaluate the state or quality of the environment will usually be dictated by nonscientific considerations.

Cairns et al. (*84*) described a sequential sampling method for estimating biological diversity in stream pollution studies, based not upon the correct scientific identification of plants and animals or their taxonomic status but merely on one's ability to recognize simple differences in shape, structure, color, or size. The method was shown to be capable of detecting differences in biotic diversity and environmental quality confirmed by more exhaustive and more sophisticated biological survey procedures. For the simple sequential procedure to come into more widespread use or for it to become accepted as a standard procedure, it will be necessary to accumulate an extensive body of correlation data for several habitat types over a broad geographic range.

McIntire et al. (*85*) compared the fatty acid spectra of six laboratory streams operated under differing regimes of light and flow. Some streams were dominated by blue-green algae and others by diatoms, and these biotic differences were reflected in the fatty acid spectra. Using the diversity index concept of MacArthur (*86*) and McIntosh (*87*), McIntire et al. showed an inverse relationship between species diversity and fatty acid "redundancy" in the six streams. That work is a most interesting example of how chemists can measure biotic parameters, but the final usefulness of the technique will depend upon the accumulation of a large amount of correlation data.

5. The Biological Basis For Public Policy

The ecosystem is, in the long run, a complex quality criterion in its own right. The salient characteristics of ecosystems are the species contained, their fluctuating abundances, the niches they occupy, and the functional relationships between the species comprising the biota, whether producer, consumer, predator, or prey. We use the term niche in the abstract sense of a hyperspace bounded by the ranges of various ecological factors relevant to the survival and welfare of each species. The response of species to fluctuations in environmental parameters is not simply all or none. One simple concept of the general form of the response was given by Carpenter (*88*). His concept, shown in simplified form in Fig. 7, is important to the establishment of allowed or permitted levels of injurious substances. We advise against the use of the terms "allowable" or "permissible" because those words imply that it has been determined that the levels allowed or permitted do not cause any deleterious effect, which is

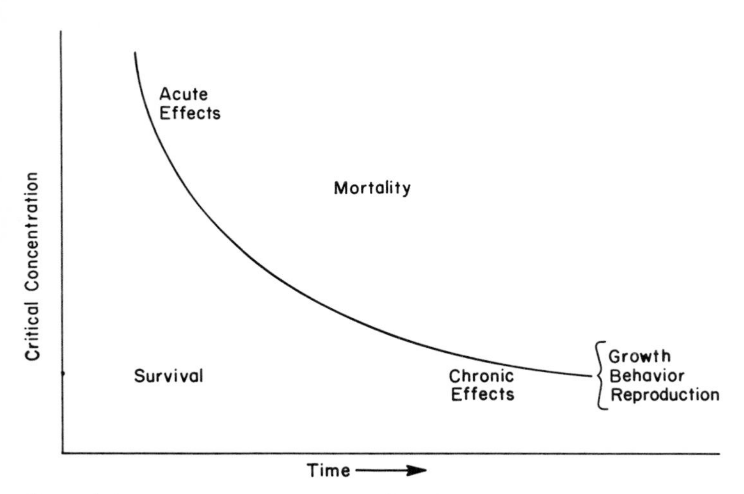

Fig. 7. General dose-exposure-response relationship. [After Ref. *88*, by courtesy of the American Chemical Society.]

not usually the case. The abscissa of Fig. 7 is time of exposure and the ordinate is the concentration or level of the injurious substance which will result in some particular response in exposed organisms, for example, the concentration which will kill 90% of the subject species for an exposure of the given time duration. Allowed levels or quality standards must be set not only on the basis of protection of the environment, but in relation to costs to achieve various degrees of protection. The biologist, using knowledge from laboratory toxicity bioassays and field survey data, stands between the conservationist and the parties with wastes to be disposed of, such as industry and public bodies.

REFERENCES

1. P. Doudoroff, B. G. Anderson, G. E. Burdick, P. S. Galtsoff, W. B. Hart, R. Patrick, E. R. Strong, E. W. Surber, and V. M. Van Horn, *Sewage Ind. Wastes*, **23**, 1380 (1951).

2. *Standard Methods for the Examination of Water and Wastewater*, 12th ed., American Public Health Assoc., New York, 1965.

3. R. J. Benoit, J. Cairns, and C. W. Reimer, *Am. Fisheries Soc. Symp. Reservoir Fisheries Resources*, 1968, pp. 69–99.

4. R. Patrick, M. H. Hohn, and J. R. Wallace, *Notulae Naturae, Acad. Nat. Sci. Publ. No. 259*, 1954.

5. T. Laevastu, T. Thompson, and T. G. Thompson, *J. Conseil Exploration Mer*, **21**, 125 (1956).

6. G. F. Lee, paper presented at National Symposium on Water Quality Standards for Natural Water, July, 1966, Ann Arbor, Mich.

7. S. M. Kenton and G. A. Rohlich, *Eutrophication—A Review, Publ. No. 34*, State Water Quality Board, The Resources Agency, State of California, Sacramento, 1967.

8. G. A. Rohlich, paper presented at 41st Annual Convention of the Soap and Detergent Assoc., New York, January 25, 1968.

9. U. M. Cowgill, *Proc. Natl. Acad. Sci.*, **57**, 198 (1967).

10. W. S. Von Arx, *Introduction to Physical Oceanography*, Addison-Wesley, Reading, Mass., 1962.

11. T. R. Folsom and A. C. Vine, in *Publ. No. 551*, National Academy of Sciences, National Research Council, Washington, D.C., 1957, Chap. 12.

12. H. Craig, in *Publ. No. 551*, National Academy of Sciences, National Research Council, Washington, D.C., 1957, pp. 1–25.

13. R. Revelle and M. B. Schaefer, in *Publ. No. 551*, National Academy of Sciences, National Research Council, Washington, D.C., 1957, pp. 1–25.

13a. V. T. Bowen, personal communication, 1968.

14. C. Eckart, *J. Marine Res.*, **VII**, 265 (1948).

15. T. Von Karman, *J. Marine Res.*, **VII**, 252 (1948).

16. T. Ichiye, in *Proc. Symp. Diffusion in Oceans and Fresh Waters, August 31–September 2, 1964* (T. Ichiye, ed.), Lamont Geological Observatory of Columbia Univ., Palisades, N.Y., 1965.

17. C. C. Kisiel, in *Proc. Symp. Diffusion in Oceans and Fresh Waters, August 31–September 2, 1964* (T. Ichiye, ed.), Lamont Geological Observatory of Columbia Univ., Palisades, N.Y., 1965, pp. 28–41.

18. W. Stumm, discussion of a paper by G. A. Rohlich, in *Proc. Intern Conf. Water Pollution Res., 1st, London, 1962*, Pergamon, New York, 1962, pp. 216–229.

19. R. H. Hsu, *Soil Sci.*, **99**, 398 (1965).

20. C. J. Velz, in *Robert A. Taft Sanit. Eng. Center Tech. Rept. W58-2*, U.S. Dept. Health, Education, and Welfare, Public Health Service, Cincinnati, 1958, pp. 47–62.

21. V. Jansa and G. Akerlindh, *Sewage Works J.*, **13**, 551 (1941).

22. H. C. Bramer and R. D. Hoak, *Ind. Eng. Chem.*, **1**, 185 (1962).

23. R. D. Hoak and H. C. Bramer, paper presented at Chicago Regional Technical Meeting of American Iron and Steel Institute, September 25, 1963, The Mellon Institute, Pittsburgh.

24. H. C. Bramer and R. D. Hoak, *Ind. Eng. Chem.*, **3**, 46 (1964).

25. H. C. Bramer and R. D. Hoak, *Ind. Eng. Chem.*, **5**, 316 (1966).

26. R. J. Benoit, in *Eutrophication: Causes, Consequences, and Correctives*, National Academy of Sciences, Washington, D.C., 1969, pp. 614–630.

27. J. R. Kramer, in *Great Lakes Res. Div. Publ. No. 11*, Univ. Michigan, Ann Arbor, 1964, pp. 147–160.

28. J. C. Sutherland, J. R. Kramer, L. Nichols, and T. D. Kurtz, in *Great Lakes Res. Div. Publ. No. 15*, Univ. Michigan, Ann Arbor, 1966, pp. 439–445.

29. D. A. Livingstone and J. C. Boykin, *Limnol. Oceanog.*, **7**, 57 (1962).

30. G. E. Hutchinson, *A Treatise on Limnology*, Vol. I, Wiley, New York, 1957, Chap. 10.

31. L. E. Kuentzel, *J. Water Pollution Control Federation*, **41**, 1737 (1969).

31a. P. K. Park, L. I. Gordon, S. W. Hager, and M. C. Cissell, *Science*, **166**, 867 (1969).

32. G. F. Lee and A. W. Hoadley, in *Equilibrium Concepts in Natural Water Systems* (R. F. Gould, ed.), American Chemical Society, Washington, D.C., 1967, Chap. 16.

33. T. R. Parsons, in *Progress in Oceanography* (M. Sears, ed.), Vol. 1, Pergamon, New York, 1963, pp. 205–239.

34. J. J. Morgan and W. Stumm, in *Proc. Intern. Conf. Water Pollution Res., 2nd, Tokyo, 1964*, Pergamon, New York, 1965, pp. 103–131.

35. K. Sugawara, S. Okabe, and M. Tanaka, *J. Earth Sci. (Nagoya Univ.)*, **9**, 114 (1961).

36. *Robert A. Taft Sanit. Eng. Center Tech. Rept. W58-2*, U.S. Dept. Health, Education, and Welfare, Public Health Service, Cincinnati, 1958, p. 194.

37. D. P. Loucks, C. P. ReVelle, and W. R. Lynn, *J. Inst. Management Sci.*, **14**, B166–B181 (1967).
38. H. W. Streeter and E. B. Phelps, *Public Health Serv. Bull. No. 146*, U.S. Dept. Health, Education, and Welfare, Washington, D.C., 1925.
39. B. Davidson and R. W. Bradshaw, *Environ. Sci. Technol.*, **1**, 618 (1967).
40. W. N. Torpey, *J. Water Pollution Control Federation*, **39**, 1797 (1967).
41. W. N. Torpey, *Water Sewage Works*, **115**, 295 (1968).
42. W. H. Oldaker, A. A. Burgum, and H. A. Pahren, *J. Water Pollution Control Federation*, **40**, 1688 (1968).
43. G. M. Fair, E. W. Moore, and H. A. Thomas, Jr., *Sewage Works J.*, **13**, 270, 756 (1941).
44. W. B. Langbein and H. W. Durum, *U.S. Geol. Surv. Circ. No. 542*, U.S. Dept. Interior, Washington, D.C., 1967.
45. H. B. N. Hynes, *The Biology of Polluted Water*, Liverpool Univ. Press, Liverpool, England, 1960.
46. P. Doudoroff, in *The Physiology of Fishes* (M. E. Brown, ed.), Vol. 2, Academic, New York, 1957, Chap. IX.
47. C. M. Tarzwell, ed., *Public Health Serv. Publ. No. 999-WP-25*, U.S. Dept. Health, Education, and Welfare, Cincinnati, 1965.
47a. A. F. Bartsch and W. M. Ingram, *Verhandl. Intern. Ver. Limnol.*, **16**, 786 (1966).
48. L. E. Keup, W. M. Ingram, and K. M. Mackenthun, eds., *Biology of Water Pollution*, U.S. Dept. Interior, Federal Water Pollution Control Admin., Cincinnati, 1967.
49. R. Kolkwitz and M. Marsson, in *Biology of Water Pollution* (L. E. Keup, W. M. Ingram, and K. M. Mackenthun, eds.), U.S. Dept. Interior, Federal Water Pollution Control Admin., Cincinnati, 1967, pp. 47–51.
50. *Ibid.*, pp. 85–95.
51. K. R. Butlin and J. R. Postgate, in *Autotrophic Microorganisms* (B. A. Fry and J. L. Peel, eds.), General Microbiology, London, 1954, pp. 271–305.
52. F. S. Stay, W. S. Duffer, B. L. DePrater, and J. W. Keeley, *The Components of Oxygenation in Flowing Streams*, U.S. Dept. Interior, Federal Water Pollution Control Admin., Robert S. Kerr Water Res. Center, Ada, Okla., 1967.
53. L. von Bertalanffy and A. Rapoport, eds., *General Systems, Vol. 10, Yearbook of the Society for General Systems Research*, Ann Arbor, Mich., 1965, pp. vii, 211.
54. J. Klir, *ibid.*, pp. 29–42.
55. J. D. Hem, *Environ. Sci. Technol.*, **1**, 1017 (1967).
56. R. F. Gould, ed., *Equilibrium Concepts in Natural Water Systems, Advan. Chem. Ser. No. 67*, American Chemical Society, Washington, D.C., 1967.
57. R. L. Lindeman, *Ecology*, **23**, 399 (1942).
58. R. M. Darnell, *Ecology*, **42**, 553 (1961).
59. J. M. Teal, *Ecology*, **43**, 614 (1962).
60. L. B. Slobodkin, in *Marine Biology* (G. A. Riley, ed.), Vol. I, American Institute of Biological Sciences, Washington, D.C., 1963, pp. 235–238.
61. G. E. Hutchinson, in *Limnology in North America* (D. J. Frey, ed.), Univ. Wisconsin Press, Madison, 1963, pp. 683–690.
62. H. T. Odum, *Ecology*, **37**, 592 (1956).
63. J. L. Brooks and S. I. Dodson, *Science*, **150**, 28 (1965).
63a. J. L. Brooks, *Systematic Zool.*, **17**(3), 272 (1968).
64. W. E. Odum, *Sea Frontiers*, **14**, 234 (1968).
65. R. MacArthur, *Ecology*, **36**, 533 (1955).
66. R. Margalef, *Limnol. Oceanog.*, **6**, 124 (1961).
67. B. Patten, *Limnol. Oceanog.*, **5**, 26 (1961).

68. C. E. Woelke, in *Public Health Serv. Publ. No. 999-WP-25* (C. M. Tarzwell, ed.), U.S. Dept. Health, Education, and Welfare, Cincinnati, 1965, pp. 67–76.

69. P. L. Brezonik, *Proc. Conf. Great Lakes Res., 11th*, 1968, pp. 16–30.

70. A. F. Bartch and W. M. Ingram, *Public Works*, **90**, 104 (1959).

71. K. M. Mackenthun and W. M. Ingram, *Biological Associated Problems in Fresh Water Environments—Their Identification, Investigation, and Control*, U.S. Dept. Interior, Federal Water Pollution Control Admin., Washington, D.C., 1967.

72. H. P. Jeffries, *Ecology*, **43**, 730 (1962).

72a. J. H. Ryther and C. S. Yentsch, *Limnol. Oceanog.*, **2**, 281 (1957).

72b. C. S. Yentsch, in *Primary Productivity in Aquatic Environments* (C. R. Goldman, ed.), Univ. California Press, Berkeley, 1966, pp. 325–346.

72c. R. J. Benoit, unpublished data.

73. K. A. Mackenthun, L. E. Keup, and R. K. Stewart, *J. Water Pollution Control Federation*, **40**, R72–R81 (1968).

74. W. H. Pearsall, *J. Ecology*, **20**, 241 (1932).

75. U. T. Hammer, *Proc. Intern. Assoc. Theoret. Appl. Limnol.*, **15**, 829 (1964).

76. B. Komorovsky, *Bull. Res. Council Israel*, **II**, 379 (1953).

77. M. Bongersma-Sanders, in *Treatise on Marine Ecology and Paleo-ecology* (J. W. Hedgpath, ed.), Memoirs Geological Society of America, 1957, pp. 941–1010.

78. C. N. Sawyer, *J. Water Pollution Control Federation*, **37**, 1122 (1965).

79. E. L. Cooper, *Am. Fisheries Soc. Publ. No. 4*, 1967.

80. R. Patrick, *Proc. Acad. Nat. Sci. Philadelphia*, **101**, 277 (1949).

81. R. Patrick, in *Algae and Man* (D. F. Jackson, ed.), Plenum Press, New York, 1962, pp. 182–185.

82. H. B. N. Hynes, in *Public Health Serv. Publ. No. 999-WP-25*, (C. M. Tarzwell, ed.), U.S. Dept. Health, Education, and Welfare, Cincinnati, 1965, pp. 235–243.

83. E. Fjeringstad, *ibid.*, pp. 232–235.

84. J. Cairns, D. W. Albaugh, F. Busey, and M. D. Charay, *J. Water Pollution Control Federation*, **40**, 1607 (1968).

85. C. D. McIntire, I. J. Tinsley, and R. R. Lowry, *J. Phycology*, **5**, 26 (1969).

86. R. H. MacArthur, *Biol. Rev.*, **40**, 510 (1965).

87. R. P. McIntosh, *Ecology*, **48**, 392 (1967).

88. R. A. Carpenter, *Environ. Sci. Technol.*, **2**, 518 (1968).

Chapter 5 Use of Computer Technology to Develop Mathematical Models for Natural Bodies

A. R. LeFeuvre
CANADA CENTRE FOR INLAND WATERS
DEPARTMENT OF ENERGY, MINES AND RESOURCES
BURLINGTON, ONTARIO

I. Introduction

Modern computer technology has had an almost unbelievable impact on practically every phase of man's existence. Important applications range from the placing of airline reservations to the calculation of the trajectory of space capsules. One promising use of this technology is in the simulation or modeling of natural bodies of water. This chapter discusses these modeling techniques.

The rapid growth and rate of obsolescence in this field of study obviate detailed description of particular programs. Rather, the subject is approached from the direction of functional application. The details of methodology must be tailored to the particular equipment at hand.

A brief discussion of mathematical models and computer technology is followed by a description of the various natural bodies to be simulated. Finally, the techniques of optimization are introduced to point the way toward future applications.

II. Mathematical Models

A. General Comments

A mathematical model is a description of a natural relationship. As with any description, it is imperfect and incomplete. Being stated in the precise language of mathematics, however, it has the advantage of being unambiguous. Unfortunately, this precision can be achieved only by the use of simplifying assumptions.

Some mathematical models can be represented in very simple forms which are easily applied. Other models are so complex, either in statement or in repetitive manipulation, that they require the speed and vast memory of a digital computer. This chapter deals primarily with the latter type of mathematical model.

Mathematical models may be classified in a variety of ways. They may be grouped according to the type of phenomenon being modeled, i.e., hydrology, chemistry, etc. Another classification could be in terms of the nature of the process, such as deterministic or stochastic, steady-state or dynamic. The type of model chosen will depend on the nature of the process involved and on the degree of approximation permitted by the end use of the output.

B. *Biochemical Models*

A biochemical model, as used in a study of natural bodies, describes the interrelationships between biological organisms and their chemical environment, occurring in a confined space and over a relatively short time period. The classic model for this interrelationship is that due to Streeter and Phelps(*1*). It assumes two compensating reactions occurring simultaneously: deoxygenation due to the biochemical oxygen demand of the organic wastes, and reaeration due to absorption of atmospheric oxygen at the free surface. The rate of change of the oxygen deficit is given as:

$$\frac{d}{dt}D = K_1L - K_2D \tag{1}$$

in which D is the oxygen saturation deficit in mg/liter, L is the biochemical oxygen demand in mg/liter, K_1 is the coefficient defining the rate of deoxygenation, K_2 is the coefficient defining the rate of reaeration, and t is the time of reaction in days.

This classic relationship, or one of the many modifications of it, forms the basis of almost all simulations of the oxygen profile in flowing streams.

C. *Hydrologic Models*

Hydrology is the study of the origin and distribution of the waters of the earth. Many mechanisms play a role in this overall, cyclic phenomenon, Precipitation depends on a complex combination of meteorological conditions. Runoff depends on the condition of the receiving surface and the pattern of precipitation. Stream flow depends on the physical characteristics of the watershed and the runoff distribution. Each of these mechanisms can be approximated by mathematical relationships.

Most hydrologic models are extensive rather than intensive. That is, the model attempts to simulate the behavior of a large and complex system covering many square miles. Because of this, it is necessary to recognize local mechanisms in terms of average conditions. For instance, the infiltration capacity of the ground varies with soil characteristics and vegetal cover. When considering a watershed of several hundred square miles, therefore, it is meaningless to refer to specific local conditions. The Stanford watershed model(*2*) recognizes this by assuming a linear variation of infiltration rate as precipitation continues. This mechanism assumes that ground surface conditions vary throughout the watershed and that as time progresses, various elements of the watershed area will control the average infiltration rate.

D. Economic Models

Economic models deal with yet another level of abstraction. In this case an attempt is made to represent the complex interrelationships of the entire economic community in terms of mathematical relationships. One of the more difficult problems which must be faced in establishing a mathematical model of an economic system is defining the objective function. This and other facets of economic models are discussed in detail in a later section of the chapter.

E. Deterministic Models

Deterministic mathematical models are the most widely used form of mathematical relationship. For mechanism modeled in this case it is assumed that for a given set of input conditions there is but one uniquely determined solution. This is the basic assumption of most engineering calculations. For example, the well-known Rippl(*3*) method for determining reservoir capacity in a water supply reservoir assumes that an analysis of the historic records of flow at that site will yield a single deterministic answer for the storage required. In this case there is an inherent implication that the hydrology which the structure will experience in the future will be the same as the hydrology reported in the relatively short historic trace that was analyzed.

F. Stochastic Models

Stochastic models are mathematical relationships which recognize that some of the inputs are random and unpredictable in nature. This randomness is of such magnitude that the solution must be described in statistical terms rather than as a unique solution. Stochastic relationships attempt to incorporate such things as the capriciousness of natural occurrence, e.g., the occurrence of rainfall. This can be illustrated by returning to the example of the required storage capacity in a reservoir. The Rippl method analyzes the historic trace of stream flow records at the site and determines the amount of storage which would provide for the design-sustained flow from the structure on the assumption that future events will be a mere image of past history. Unfortunately, the project life is usually considerably longer than the period of historic record. An alternative to the Rippl method would be a statistical analysis of the historic record to estimate the statistical parameters of the population of stream flows from which the historic record is but a small sample. These statistical parameters could then be incorporated into a stochastic model along with a randomness generator to generate a much longer sequence of flows. This sequence could then be routed through the proposed reservoir, and the percentage of time that the design was satisfactory could be determined. A series of such runs for a variety of reservoir capacities would permit the selection of an optimal design on the basis of the economic cost of the failure of the structure. In this case, failure would be represented not by the collapse of the structure but by the inability of the storage to meet the design draft.

Stochastic models find application in other areas, such as management and economics, in which the unpredictability of human behavior is simulated.

G. *Steady-State Models*

Many mathematical relationships assume that the process being modeled is a steady-state process. That is to say, the various components or inputs to the system are invariate with time. In most cases, this assumption is invoked in order to bring to the simulation a greater degree of mathematical simplification. Very few natural phenomena are, in fact, steady-state. In many cases, however, the assumption of steady-state conditions introduces only a minor error in the answer. Before the advent of high speed digital computers, the assumption of dynamic conditions made many engineering calculations prohibitively complex and lengthy. Even with the availability of high speed computers, it is often unnecessary to include the dynamic effects caused by unsteady conditions.

H. *Dynamic Models*

The dynamic nature of some physical phenomena makes it imperative to incorporate the unsteady effect in any mathematical simulation. For example, it would be quite unrealistic to attempt to model the water quality of an estuary without recognizing the dynamic effects of the tidal variation that acts upon and, in large measure, controls the flow in an estuary. The ability to incorporate these dynamic effects into mathematical models has been greatly enhanced by the advent of the high speed digital computer. Finite difference techniques(*4*) and other numerical methods, when applied with the ultrahigh speed of the modern computer, make possible many engineering determinations which heretofore were economically infeasible.

In summary, it should be pointed out that every mathematical model of a natural process is to some extent an approximation. The fidelity with which the model can represent the natural process is a function of the level of our understanding of the basic mechanisms operative in the phenomenon in question and the amount of detail which can be justified in a given case. In many cases a simple slide rule computation of a deterministic nature is still perfectly adequate for our needs. In other cases we are justified in applying sophisticated numerical techniques employing the very latest in high speed digital equipment.

III. Computer Technology

A. Digital Computer

A modern digital computer is basically a very fast and essentially automatic desk calculator and filing system. Mathematical relationships are solved arithmetically. The power of this machine lies in its ability to call from storage the required input data, operate on these data arithmetically, and store the result for future printout. This entire process takes place in several millionths of a second. Programming a succession of such operations permits the solution of many complex problems.

An essential adjunct to digital computer technology is the mathematical discipline of numerical analysis. Solutions to functional relationships are found by transforming the problem to a form which can be solved or approximated by a succession of arithmetic operations. The digital computer is ideally suited to solution by successive approximations or any iterative processes.

The speed of the digital computer permits evaluation of a range of solutions to a given problem so as to approach an optimal solution. The time required for hand calculation would limit an investigator to calculating a small number of alternative solutions.

B. Analog Computer

An analog computer is a device which permits the analysis of a physical system by measuring comparable effects on a second physical system which is analogous to the prototype. The basis for many analog solutions is the fact that the differential equations describing two different physical systems may be essentially identical from a mathematical point of view. For example, a hydraulic network consisting of pipes and pumps with water as the fluid medium can be shown to be analogous in many respects to an electric circuit composed of resistances and potential differences in which electric current is the fluid medium. The behavior of the hydraulic system may be predicted by measurements of current, voltage, and resistance which are made on the electric system. In this case the hardware comprising the electric system would be a single-purpose analog computer. Some physical systems are of sufficient importance to warrant such a single-purpose computer. The flood routing analog(*5*) of the Missouri River, which is used by the Kansas City office of the Corps of Engineers, is an example of this kind of computer installation. Many analog computers are designed for general use by providing plug-in circuit components that can be assembled in a variety of configurations.

The analog computer is ideally suited to the solution of dynamic problems. The time variability of the output can be accommodated by the use of oscilloscope readout or *xy* plotting equipment. The nature of the data output format is at once a convenience and a limitation of this kind of equipment. Many unsteady phenomena are best reported in graphical form. On the other hand, graphical readout has very limited accuracy. In many cases this limitation of accuracy is not a serious disadvantage, in other cases the necessary accuracy can be achieved by multirange selection of scale.

An important property of analog computation is the ability to vary input signals continuously over a wide range of values. This permits the operator to adjust the system inputs so as to approach an optimum solution. This capability becomes less useful as the number of input variables increases.

Analog computers hold bright promise for the solution of many water pollution-oriented analyses. Very likely, their greatest utility will be achieved as they are combined with digital computers into hybrid digital-analog systems.

C. Data Input and Manipulation

1. Form of Raw Data

All digital computer programs require the input of certain basic data which are stored internally for recall during the execution of the program. In the field of water resources, these background data frequently are large amounts of field data. The form of the data can be convenient or inconvenient for use in a digital computer. With this fact in mind, serious consideration should be given to the way in which basic field data are assembled and recorded.

Field data which are the result of direct observation by a human observer are most usually reported in digital form on observation sheets. This information must be transcribed to punched cards before it is suitable for input into the computer. In some cases field data are gathered continuously and automatically in graphical form on recording strip charts of one kind or another. Such data must first be interpreted by office personnel and translated into digital form, then either recorded in tabular form or punched directly onto punched cards. More recent field observation equipment has been designed to provide the basic field measurements in some form of punched code which can be fed directly to the computer. This requires some method of digitizing the output signal from the sensing element. Such a digitized signal is desirable if telemetering of the field observation to some central processing point is required. Every effort should be made to provide field data in as convenient a form as practical.

2. Machine Reduction

One of the main problems associated with the handling of field data is the sheer bulk of the data. Usually it is desirable to reduce this bulk by some systematic means. The computer can aid in this reduction. In some cases only the maximum or extreme value of the continuous signal is of significance. In other cases the mean, the standard deviation, or some other statistical parameter is desired. In some instances a reordering of the data is called for. All of these manipulations can be handled automatically and rapidly by the digital computer.

Another possible function of the computer would be to scan a record and determine if a value observed in the field was above or below an established standard or critical value. In this way it would be possible for the computer to report the number of occurrences and duration of the violation of such a standard.

3. Correlation Analysis

One of the important operations performed on many kinds of field or experimental data is correlation analysis. The computer can assist us in this task.

First, the computer can perform a variety of mathematical operations on each unit of data so as to bring it into a form which makes the relationship with other data most meaningful. The data can be plotted graphically so as to show by visual observation whether or not an obvious correlation exists. A mathematical refinement of this procedure would be to perform a least squares analysis and least squares fit of the regression equation relating the two sets of data.

In some natural phenomena we must deal with more than a two-dimensional correlation. Multivariate analysis can handle this situation. There is practically no limit to the variety of statistical procedures which can be devised to aid in this type of analysis.

4. Output Format

The form of the output from a computer operation is just as important as the format of the input to that program. One danger in the use of high speed computers is the possibility that the amount of material produced in the output will be such that the operator will not be able to comprehend the significance of the data because of its sheer bulk. Thus, it is very important to plan carefully the output that is required.

Output which is of a terminal nature is most usually presented in numerical, tabular format. In some cases it is more convenient to have the information displayed in graphical form. The digital computer is capable of producing the information in either or both forms, as required.

In many instances the output from one computer program may become input to a subsequent program at some other time. If this is the case, it may be desirable to have the output in the form of punched cards or magnetic tape. This format greatly facilitates the input of this information into a subsequent program.

The modern digital computer is capable of producing output in almost any format required. The cardinal rule to remember is that output should be limited to essentials to reduce the amount of material to be scanned by the user.

5. Sensitivity Analysis

Another important application of high speed digital computation is in the field of sensitivity analysis. By this we mean the repetitive computation of a mathematical relationship in which a single variable or input is varied systematically over a wide range of values in order to determine the effect of this variable on the final result. This procedure permits the investigator to determine those input variables which are most sensitive to the solution and those which are least sensitive. Such information can be valuable in deciding upon the necessity for sophisticated measuring techniques in field observations. Analyses of this kind can aid in the planning of future studies.

IV. Simulation of Natural Bodies

A. Simulation

Natural bodies may be simulated in many different forms. Such simulations are commonly called "models" and may be physical, pictorial, or mathematical. The river and harbor models constructed by the Corps of Engineers at the Vicksburg laboratory simulate the flow in large river systems by the flow of water in very small replicas of these systems. Models of this kind, therefore, are physical models. The geodetic maps provided by the U.S. Geological Survey are examples of pictorial models. Some of these pictorial models may be three-dimensional, e.g., relief maps. In other cases the natural body may be simulated by a mathematical model. Our discussion is limited to this last kind of model.

Any mathematical expression which describes a physical cause and effect relationship can be thought of as a model. In the context of this section, however, we restrict the use of the term "simulation" to those mathematical relationships which simulate a rather complex natural system. The following discussion considers such natural bodies as rivers, estuaries, basins, near ocean waters, and groundwater. Each of these bodies has characteristic peculiarities which distinguish it from the others. But there is one common denominator which joins them: the combination of water and potential pollution of that water. These models will have to simulate both quantity and quality of water.

B. Rivers

1. General Comments

The most common body of water is the stream or river. A system of streams and rivers forms the natural connection between the occurrence of rainfall over the watershed and the formation of larger bodies of water, such as lakes, estuaries, and oceans. The systems of major rivers throughout the United States formed the original lines of commerce and transport in the early days of this nation's history. As commerce and industry grew along the shores of these rivers, the waters were used for municipal and industrial water supply and later for agricultural irrigation. Waterwheels converted the energy of these flowing streams into mechanical power for mills of all kinds. Finally, the flowing water was used as a means of disposing the sewage produced by municipalities and industries.

2. River Hydraulics

A complete simulation of a river system requires consideration of both quantity and quality of the water flowing in the system. The quality of the water usually is a function of the quantity of water flowing at that point. The converse, however, is not usually true. That is, the quantity of water or the hydraulics of the system is essentially unaffected by quality considerations. For this reason it is customary to establish the hydraulics of the river system and then to superimpose on this the quality considerations of interest.

The flow diagram for the Kansas River model is shown in Fig. 1. In this model the discharge in a given reach is established by the addition of the inflow from upstream reaches and the local runoff, *QADD*. The average velocity and average hydraulic depths are computed as functions of the discharge. Time of flow through the reach is related to the average velocity.

The reaeration coefficient, K_2, in the Streeter–Phelps equation is computed as a function of the average velocity, the average hydraulic depth, and the slope of the bed. Several forms of this functional relationship are reported in the literature (*6*).

3. HYDROLOGY

Stream flow is the result of surface runoff plus base flow contributed from groundwater; it varies with time and location because runoff is related directly to rainfall patterns. Thus, consideration of the hydrology of the river system requires the modeling of a process with uncertainty as one of its components (a stochastic process).

Absolute deterministic values have little significance in the analysis of a river system. When striving for a low flow design value for water quality calculations, it is necessary to state this flow in statistical terms. The Federal Water Pollution Control Administration has adopted the low flow of 7-day duration, which would occur once in 10 years, as the design low flow for stream analysis purposes. That this particular statistic is completely arbitrary is undeniable. However, the low flow of record would be just as arbitrary and not nearly as indicative.

The simulation of an entire river system must recognize the special variation and complexity of many branches and junctions. Thus, the simulation technique must include some logical procedure for routing flows through such a system. Because flows downstream from a junction are determined by the flows coming from upstream branches, it usually is desirable to start with assigned flows in the uppermost reaches of the river system. In this way the flows can be accumulated throughout the river system as the outflow from one reach becomes the inflow to the reach downstream. The Willamet model by Worley (*7*) and the Kansas River model by Smith et al. (*8*) utilize such a scheme in their simulations; e.g., the reaches are numbered in ascending order from the mouth. In actual simulation, however, the program starts at the uppermost or

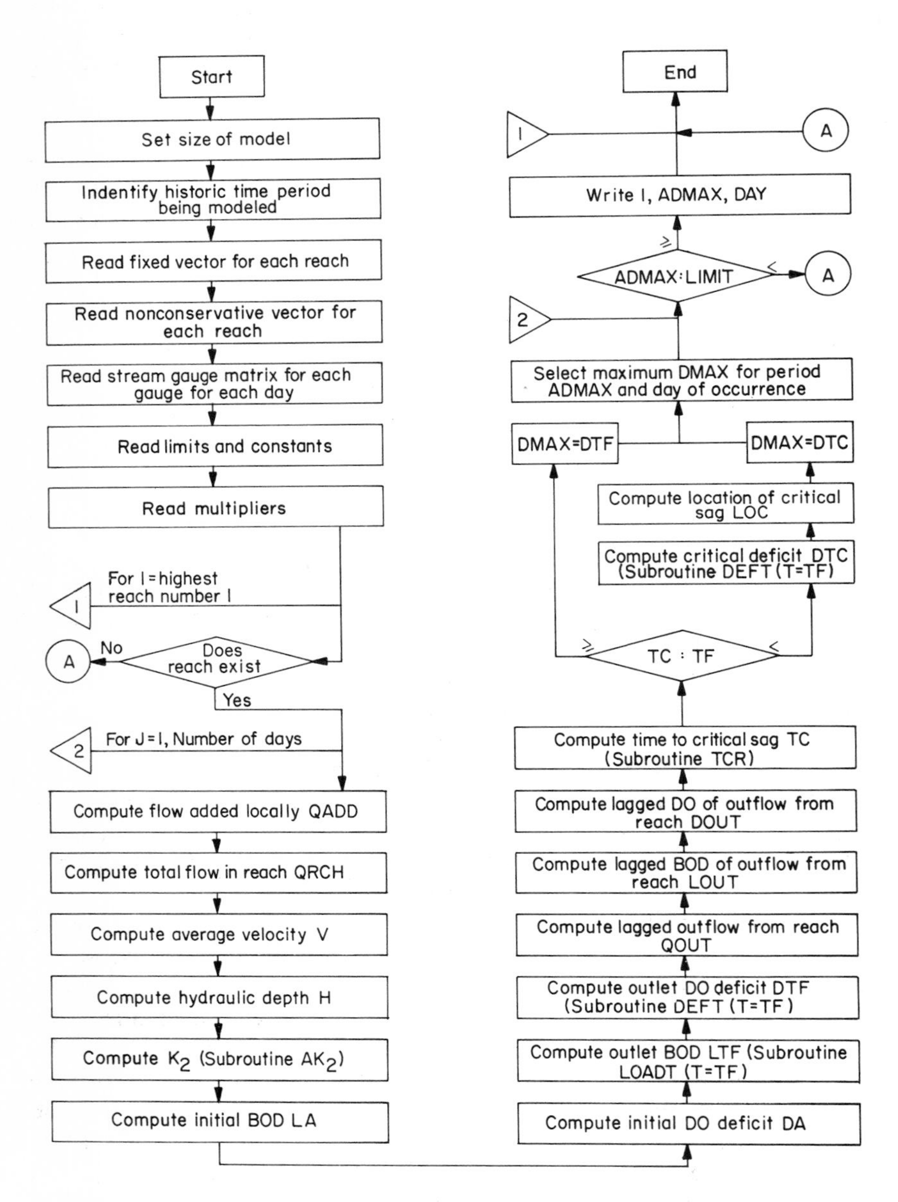

Fig. 1. Flow diagram for a digital computer model of the Kansas River. [After Ref. *8*.]

highest-numbered reach and proceeds in the direction of flow. At each junction point the reach numbering transfers the simulation to the uppermost reach of the tributary.

The integers actually assigned to the reaches proceed in steps of five so as to leave space for future revision of some reaches. In this way it is possible to subdivide a reach into five parts without altering the remainder of the identification system. This scheme is illustrated in Fig. 2.

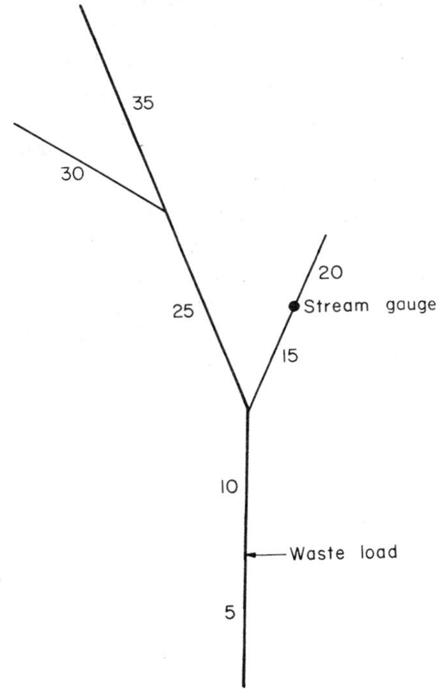

Fig. 2. Reach identification scheme for a digital computer model of a river system. [After Ref. *8*.]

Many stream simulations are concerned with only a relatively small section of the total river system. In such a case it is necessary to decide whether or not local additions of flow have a significant effect upon the end result of the study. Worley's model(*7*) assumed that the flow contributed locally was a very small proportion of the total flow in the reaches under consideration. On the other hand, the Kansas River model(*8*) dealt with a section of that river system which gained a large portion of its total flow from local addition. One common method for handling this problem is to assume that the yield in cubic feet per second per square mile is constant over a considerable area of the watershed. In this way the local contribution to an ungauged reach can be estimated from the corresponding flow in a gauged reach of the system.

Water quality models very often are concerned with low flow conditions as contrasted with flood flow analysis. In such cases the base flow in a stream system and its relationship to groundwater are important. River flow contributed from groundwater sources is very difficult to predict in a mathematical model. Field measurement of this contribution can be made only by noting the difference between main stem gauges. A further difficulty is the unknown mechanism which controls the rate at which this groundwater contribution will take place. The level of the groundwater relative to the stream level is significant. This relationship can be determined only by observations in a number of test wells along the river bank. Such information is rarely available. In the case of the Kansas River model, it was assumed that the groundwater contribution in the lower reaches of the main stem could be approximated on a constant basis by assuming 1 ft^3/sec per mile of river length as the contribution in that area. A more accurate mechanism would recognize the effect of river stage, reducing the contribution during high stage and adding to it at low stage, such a refinement is seldom justified.

4. Conservative Pollutants

Pollutants which are not degraded over time are termed "conservative pollutants." Many soluble materials are included in this classification. One of the most common of these is the salt that is accumulated in the watershed, either by natural erosion and solution of the crust of the earth or by the action of oil well operation, in which brines are drawn to the surface and disposed of in order to produce the oil. Acid mine drainage also causes this kind of pollutant. Because this kind of pollutant cannot be degraded by bacteriological action or by some natural, built-in decay mechanism, it can be controlled only by dilution or by some method of chemical removal.

The simulation of a conservative pollutant within the river system is quite straightforward. The distribution throughout the river system can be accounted for by a simple mass transport analysis. The concentration of this pollutant at any point in the system can be determined from the known quantity of the material and the flow at that point. The major difficulty in handling this kind of pollutant is the fact that it is seldom introduced into the river system as a point source. More commonly, this kind of pollutant occurs as a wash load from the general watershed area. The amount of pollutant contributed by wash load from a given portion of the watershed is greatly affected by the geology of the area. For example, in the Kansas River model it was found that a very large portion of the total sodium chloride load came from the wash load of one fairly small section of the total watershed. This could be inferred only from rather extensive field observations which were correlated with the flows. The relationship between addition of conservative pollutants and the local wash load indicates that the mechanism for this contribution must be connected to the addition of water from local sources. Such pollution loads can be incorporated into functions which predict the amount of local flow contribution. In many cases these functions are complicated by the temporal and spatial variation of local stream flow contribution throughout the watershed.

In the Kansas River model(*8*) for instance, a significant portion of the inorganic load carried by the stream network originates within the model area. This is illustrated in Fig. 3, which shows the average chloride load in tons per day for the 12 longer-term stations and the average increase in chloride load between adjacent stations. These values were computed by integrating the discharge versus chloride concentration curve with the flow duration curve for each water quality station.

5. Nonconservative Pollutants

Any pollutant which is reduced in quantity over time is considered a nonconservative pollutant. This reduction in pollution load can be due to a variety of mechanisms. The most common nonconservative pollutant, municipal sewage, is reduced by biodegradation. In this case the microorganisms in the water consume the pollutant along with the oxygen in the stream. Another example of a nonconservative pollutant is a radioactive substance which decays according to the normal decay of radioactive isotopes. Temporarily suspended sediment loads, when considered a pollutant, are another form of nonconservative pollutant.

The most common form of nonconservative pollutant is that associated with the effluent from sewage treatment plants. The natural decay of this pollutant and the interrelationships of temperature and reaeration capacity have been the subject of investigation for many years. The classic model used to describe this process is that attributed to Streeter and Phelps(*1*). This model was illustrated in Sec. II.B of this chapter.

6. Impoundments

The effect on water quality of impounding (storing) a stretch of river involves a very complex relationship. The entire hydraulic regime of this reach of the river is changed by the flooding out of the original turbulence

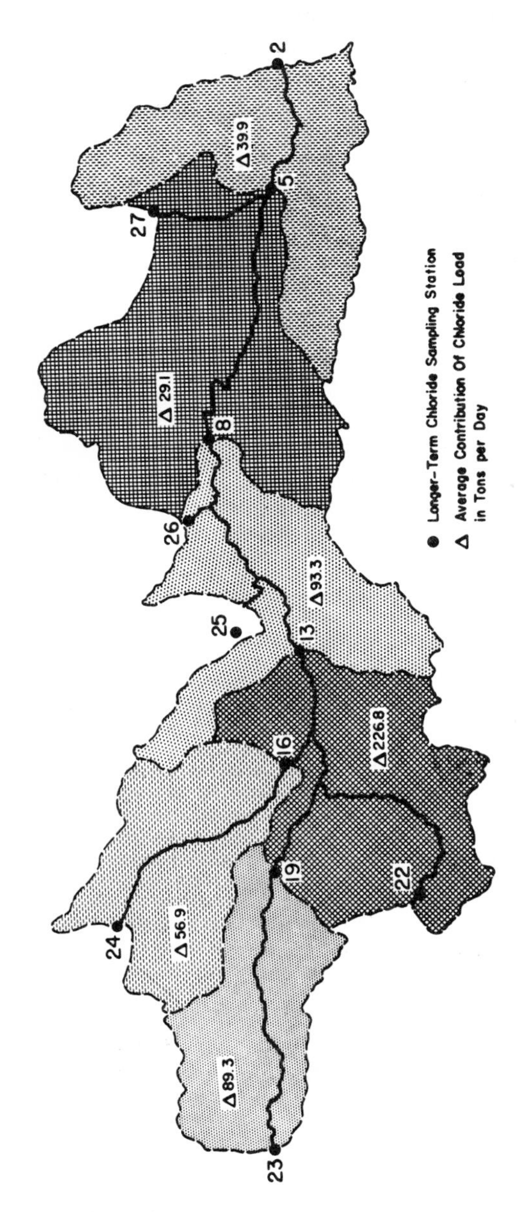

Fig. 3. Source areas for chlorides contributed to the lower Kansas River. [After Ref. 8.]

and velocity patterns. The reaeration coefficient is greatly affected by this change. Other complications, such as thermal stratification and density flows, are created. In most cases the construction of a reservoir serves to complicate the analysis and to some extent degrade water quality. In many cases the reservoir design is such that the water drawn from the reservoir is drawn from the bottom through low penstocks for hydroelectric development. This water in the lower level of the reservoir, known as the hypolimium, is often totally devoid of dissolved oxygen because of thermal stratification which inhibits vertical mixing within a reservoir. This particular problem can be alleviated somewhat by changing the location of the intake structure or by building a skimming weir inside the reservoir which would force the drawing of surface waters into the intake structure. An alternative to this would be a multilevel intake structure which would permit the operator to select the level within the reservoir from which the water would be drawn. High level withdrawal of water from the reservoir is not an ideal solution because the poor quality water or water that is low in dissolved oxygen would still accumulate in the bottom of the reservoir. In cases of high flow from the reservoir, the zone of influence of the withdrawal would increase so that periodic slugs of poor quality water would be carried from the reservoir, which could result in fish kills because of very low dissolved oxygen levels (see Chap. 6).

One technique being investigated at the present time which holds considerable promise as a solution to this problem is that of reservoir mixing. Under this scheme some mechanical device would be used to mix the reservoir in a vertical direction so as to break up the thermal stratification and provide what amounts to a homogeneous reservoir. One method of achieving this mixing is to mechanically pump water from the lower reaches of the reservoir to the surface. Another alternative is to provide this lifting motion by injection air at the bottom of the reservoir. The artificially aerated water, being lighter than the surrounding water, rises to the surface. An additional benefit associated with the air injection method is that there is some contribution of oxygen to the reservoir waters.

Mathematical models of impoundments(*9*) incorporate the diffusion mechanism and also recognize the density gradients which are responsible for thermal stratification. Present techniques account for solar energy inputs as well as advective heat sources.

Some shallow reservoirs have little stratification. In such cases the chief difficulty is to estimate the reaeration coefficient. As a first approximation this coefficient often is set equal to zero, which yields a so-called mass balance solution. In this case the oxygen resource is assumed to enter the impoundment with the inflow of water from upstream.

7. Pathogens

A water pollution parameter of great concern to public health authorities is the concentration of pathogens—bacteria which are harmful to man. Pathogens can produce a variety of diseases. Microbiology has provided us with a number of standard tests for certain indicator microbes. The presence and concentration of these indicators serve as a measure of the suitability of water for drinking and water sports. In estuarine waters, concentrations of these indicators also are singificant with regard to the suitability of certain areas for shellfish harvesting.

The presence of coliform bacteria in the water traditionally has been accepted as an indication of pathogenic content. In recent years, more selective indicators have been used. Currently, the fecal coliform content is recommended as the indicator of pathogenic content (see Chap. 13). In the future we look to specific identification of pathogens such as salmonella as one of the indicators. In each case the number of samples taken at a given location is so small relative to the variability of occurrence of these organisms that the data are reported in terms of the most probable number of coliform existing in the sample. The Public Health Service has established allowable limits on the most probable number of coliform and fecal coliform bacteria for specific water used(*9a*). Pathogens are living organisms, and it is therefore difficult to develop models to simulate their behavior. Recent studies in growth kinetics(*10*) hold some promise for the future, but present knowledge permits only very crude estimates of expected behavior. Pathogens represent a highly nonconservative type of pollution. The assumption usually made is that the bacterial pollution will "die off" in a certain number of hours.

8. Storet

An important problem in the handling of water resource and water quality data is the sheer bulk of field data which has been collected. In order that this information may be available when required for analysis, some systematic scheme of identification, storage, and retrieval must be available. The availability of high speed digital computers with very large storage capacities makes such a scheme entirely feasible. One such scheme currently employed by the Federal Water Pollution Control Administration is called STORET, the code name for a storage and retrieval system. This scheme provides for the storage of a large number of individual parameters for a given observation station. The observation station is identified by a systematic coding arrangement which relates the station identification to the location of the station within the river systems of the United States. Under this scheme it is possible to call for the retrieval of data from all of the observation stations between two points on a river or alternatively to call for data from all stations above a given point on a stream system. The latter scheme would provide information on all of the waters flowing to that point from any stream in the watershed. Details of this scheme are contained in publications of the Federal Water Pollution Control Administration(*11*).

C. Basins

1. General Comments

A river basin is the geographic area drained by a given river system. Water-oriented activities within the basin always should be planned and evaluated on a basin-wide approach. This approach is necessary because of the interrelationship between various parts of the basin. For example, water-polluting activities in an upstream region of the basin will have impact on potential water uses throughout many other parts of the basin. Conversely, water stored in an upstream reservoir for release as low flow augmentation can have beneficial effects throughout the entire length of the stream channel from that point down.

Mathematical models which simulate the river system of the basin can be applied to the problem of optimizing water quality management in the basin. River models such as the Kansas River model(*8*) predict the effect on the entire system of any change in boundary conditions. Thus, it is possible to investigate the effect of various management schemes.

The ability to test a number of alternative schemes leads to the determination of an optimum scheme. The techniques for systematically approaching an optimum solution are discussed in a later section of the chapter.

2. Polluter versus User

If the entire watershed were owned and operated by a single entity, the economics of profit maximization would require that different units of the system located in different parts of the watershed would be operated in such a manner that the total effort would be maximized or optimized. For example, if a single corporation operated an industrial plant on an upstream tributary and another plant some miles downstream, the decision as to the treatment of the wastes from the upstream plant versus treatment of water supply for the downstream plant would be an internal decision for that corporation. The savings gained by dumping untreated waste into the river upstream would show up as increased treatment costs for the downstream plant. In this situation the natural laws of economics would dictate the optimum arrangement of effluent treatment versus water supply treatment. Unfortunately, this is not usually the situation in reality. The savings on the part of the upstream polluter represent a treatment cost or use foregone by some *other* individual or corporation. This situation is called an external diseconomy. The normal rules of economics are no longer operative. In many instances in the past, the self-purifying properties of the flowing stream between the upstream and downstream plants have been able to rejuvenate the water and to a large extent alleviate the problem. But as industrial development proceeds and river systems become more fully utilized, we soon reach the point where the self-purification capacity of the stream is exceeded and there is a direct cost to downstream users because of the activity of upstream polluters.

The lack of economic incentives to induce an equitable division of responsibility between upstream and downstream water users has resulted in the need for political action in this area. State and federal governments have recognized a responsibility for the actions of both upstream and downstream water users and have passed laws to control the pollution of the streams within their jurisdiction. Unfortunately, we frequently find that river basin areas do not coincide with normal political boundaries. Because of this, there is a need for regional jurisdiction, i.e., intermunicipal or interstate authority to manage these areas. Because of the logic of managing the river basin as a geographic unit, it is becoming popular to devise new institutional arrangements which have political responsibilities over areas which coincide with the river basin itself. The Delaware River Basin Commission is an example of this kind of arrangement.

3. Basin Management Concept

The concept of basin management(*12*) of water resources and water quality goes beyond the mere formation of a political institution which coincides with the river basin as a geographic entity. Basin management is the ultimate in planning and control of the interrelated structures and physical features of the basin so as to optimize the use and development of the water resource of that area for the good of the people living within it. Basin management implies comprehensive planning in both space and time. A comprehensive plan must look at the spatial interrelationship of the inputs and withdrawals within the region and the interaction of polluter and water user. In addition, it must ensure that present structures and facilities will be compatible with future needs and structures. Finally, a comprehensive plan must look beyond the construction of facilities to the detailed management schemes which will be necessary in order to operate the system effectively.

The power to control is always accompanied by the responsibility of deciding how that control will be exercised. One very powerful tool of management is the mathematical simulation of the river system of the basin. If this model is capable of simulating both the quantity and quality distribution in place and in time, then this model permits management decisions to be based on objective analysis. That is to say, any proposed change in the system can be evaluated in terms of the overall effect that this change will have on the system. For example, if an industry requests permission to locate a plant in a particular place with a particular proposed pollution load, the effect of the new plant can be evaluated beforehand by including this new input in the model and thereby simulating the overall effect that such an addition would have. Going one step further, as discussed in a later section of this chapter, it is possible with such models to arrive at optimal combinations of treatment method to give the most economic operation of the system for a given set of inputs.

4. Competitive Uses of Water Resources

The water resources of a basin may be used for a wide variety of purposes. Many of these purposes are competitive, that is, a given resource can be used for one purpose or another, but not for both. In some instances the competition for the water resource is not clear-cut, because one use may aggravate but not eliminate entirely a second use of this resource.

One example of competition for a water resource is that between water quality control and navigation. On the one hand, navigation requires a slack water route through the river system which is maintained by a series of low head dams and locks in the main stem. On the other hand, the use of the water resource for the maintenance of water quality in the system requires storage of water in tributary reservoirs to supply low flow augmentation in the main stem, and a main stem which is maintained as a free-flowing river such that a large portion of the waste load contributed to it could be assimilated by the natural purification capability of this flowing stream. Another competition for the use of water is that between reservoir-based recreation and low flow augmentation for water quality control. For recreational uses a stable water surface elevation is needed to maintain beach areas and docks. Low flow augmentation requires a varying water surface elevation in order to utilize the storage capacity of the reservoir.

Mathematical models which simulate these differing requirements could assist in the determination of the optimum relationship between possible uses of the water resource.

Hydroelectric power plants designed and operated for peaking power are a serious problem in water quality control. These operations consume no water but cause a time discontinuity which can seriously aggravate the pollution problem in a given stream. Under normal operating conditions these plants discharge all of the available flow for a given day in a period of approximately 3–4 hr which coincides with the peak power demand on the electric system. This means that the plant discharges no water during the remaining 20 hr of the day. The river system is particularly aggravated by weekend operation in which there is no peaking from Friday until Monday. During this prolonged time the flow from the plant would be zero under optimum operating conditions from the standpoint of hydroelectric peaking power development. In some cases the Federal Power Commission has granted licenses which permit the peaking power plant to literally shut off the flow of the stream completely for these prolonged periods. In other instances, however, it has been necessary to require that a minimum base flow be maintained during off-peak periods. Mathematical models which attempt to simulate the operation of these peaking power plants must, of necessity, be dynamic models. The time parameter becomes all important in this case. An added complication in such an operation is the fact that the rapid rise and fall of the flow in the stream results in a momentum wave which does not flush the system as might be expected. Dye studies(*13*) on the Chattahoochee River near Atlanta have indicated that this short duration, high flow wave passes down the river system, leaving the water behind. A mathematical model is needed which can simulate this momentum wave and the effect on water quality of this intermittent operation.

The total river basin and the river system draining it are the logical units upon which to apply mathematical simulations. These models hold promise for our ultimate ability to manage the water resource within a basin in such a way that the optimal good for the various uses will result.

D. Estuaries

1. General Comments

As the bridge between ocean transport and inland transport, the estuary has become the focal point for a great deal of commerce, which is both a benefit and a problem (see Chap. 2). Associated with the many transshipment facilities is the pollution that comes from inadvertent or intentional spills of product and waste.

Estuarine areas are also important for the growth of shellfish. These shellfish harvesting areas are very susceptible to pollution and many are now closed because of water quality problems.

A third and increasingly important use of our estuaries is for recreation. The proximity of these waters to large concentrations of population make them desirable for water sports of all kinds. It is a deplorable situation when the masses of people in great metropolitan areas are prohibited from using the many thousands of acres of estuarine area surrounding them because of water pollution.

2. Distribution of Pollutants

The distribution of pollutants to estuarine areas follows many complex patterns in space and time. The geometry of an estuary is usually somewhat more complex than that of a flowing river, and the level of complexity varies from one area to another. The upper reaches of some estuaries such as the Delaware(*15*), for example, are more like a wide river. Hillsborough Bay near Tampa, Florida(*17*), on the other hand, has an extremely complex geometry with very poorly defined current patterns.

The current patterns in an estuary are the result of three different driving forces. The first is that of fresh water inflow or advection. This force is highly variable, as it is dependent upon the hydrology of the watershed. Its effect diminishes with nearness to the ocean.

The second driving force in an estuary is the tidal effect, which also is time dependent but of a more uniform and predictable nature. The current patterns generated by tides are, in many cases, very complex and dependent upon the geometry of the estuary. In some instances, the pattern for a flood tide is altogether different from the pattern of an ebb tide, even though they traverse essentially the same channels. In most estuary studies, the tidal effect is the dominant driving force.

The third force in producing current patterns is wind. Wind setup has an effect similar to a tidal effect. A strong wind blowing up into an estuary can add considerably to the normal tidal levels. The currents associated with these tides are likewise affected.

An important mechanism which affects the distribution of pollutants in all bodies of water is diffusion. This mechanism is dependent upon one or several of the previously mentioned forces. It is, in the final analysis, the mechanism by which the pollutant is conveyed to the various parts of the estuary. In portions of the estuary where current velocities are significantly high, the diffusion is related to this velocity. In open waters with very low tidal currents, the diffusion effect is most likely connected with wind-generated wave action and surface mixing.

As can be seen from the above discussion, the distribution of pollutants within estuarine waters can be extremely complex, and any model must be an abstraction of the real situation with all the attendant simplifications and assumptions.

3. Estuarine Models

Estuarine models have been developed for a number of the more important estuaries in the country. The form of each model was dictated by the peculiarities of the local problem and the local estuary.

O'Connor(*14*) has developed the general differential equations which describe estuarine flow. His model incorporates concepts of advection, tidal effect, diffusion, and decay of nonconservative pollutants. In applying this theoretical model to actual estuaries, it is necessary to make some simplifying assumptions, such as one-dimensional flow. That is to say, variations in velocity and all other parameters are assumed to vary only in the direction of flow. This assumes a homogeneous cross section.

An applied model of the one-dimensional type is that of Thomann (*15*), which was produced for the Delaware Estuary. Thomann's model is a segmented estuary model in that it uses the concept of finite differences and finite lengths of channel to replace the differential equations of the O'Connor model. The Delaware Estuary model has been verified by field measurements and has been useful in predicting water quality conditions in that estuary.

Some estuaries, because of their geometry, demand a two-dimensional model. One example of this requirement is the San Francisco Bay delta area. In this case the estuary consists of a complex pattern of interconnected channels. The model(*16*) which was developed to simulate this region was similar to a pipe network analysis model.

A second example of a two-dimensional estuary model is the Hillsborough Bay study(*17*) near Tampa, Florida. The geometry of this embayment requires a two-dimensional approach, but in this case an arbitrary grid system was established and the flow patterns within this area will be patterned after field determinations. The Hillsborough Bay study(*17*) proposes to follow the general method of Thomann(*15*) in developing a segmented estuary model.

The Thomann model(*15*) avoids the complication of time-dependent solution by looking at conditions as they exist on consecutive occurrences of the same point in the tidal cycle. In this way the diffusive effect of the tidal motion is recognized but the transport of water throughout the system by this mechanism is ignored.

The peculiar characteristics of each estuary require modifications of these general methods for each specific application. The digital computer is the tool most often used for this kind of analysis, but in some cases the analog computer also is helpful.

E. Near Ocean Waters

Ocean waters near a coastline present special problems in analysis and modeling. Myriad bays and inlets with their complex geometry and even more complex current patterns make it very difficult to represent these waters by mathematical models. The development of pollution loads along the coastline, however, requires analysis of pollution effects in these waters.

At the outset of such a discussion it is necessary to define the limits of the area under consideration. This in itself is a difficult task because of the great expanse of the ocean. For the most part, analyses must confine themselves to bays and inlets or at least to those areas on the continental shelf or within a barrier reef. Indeed, some waste disposal systems avoid the problem by carrying the outfall to the edge of the continental shelf and assuming that the tremendous size of the pollution sink at that point will easily accommodate the waste flows supplied to it.

As with other bodies of water, it is necessary to define the motion of the fluid medium and then to superimpose upon this motion the effect of the pollutant. Water currents in these near ocean waters are extremely complex and of considerable magnitude. The driving force producing these currents is a combination wind, tide, and density. Although an estuary can be approximated by a one-dimensional model, it is completely impractical to apply such approximation to near ocean waters. As the geometry of the body of water broadens toward the sea, the tide has a less pronounced effect upon the current, and wave action and the consequent littoral drift become more significant. There is no clearly defined interface between estuary waters and bays and open seacoast.

The generally accepted analysis of waste disposal along the open coast assumes that dilution by the ocean is so great that tracing of the pollution becomes inconsequential. This view is challenged by concern with the ever-increasing quantity of material being disposed of in this way. Cohesive substances such as crude oil have been known to travel many miles over the surface of the sea to pollute beaches. It is possible that other forms of pollution not so obvious also could be transported great distances in the ocean currents.

Mathematical models of these systems must recognize the mechanisms of dilution and dispersion and also the probability concept. Fortunately, there is a considerable similarity between the action of water movement in the near ocean areas and the motion of the water in large freshwater bodies such as the Great Lakes. Current field sampling programs (*18*) in the Great Lakes hold promise of a greatly increased understanding of the fate of pollutants in these waters and the currents that distribute them. At this point in time, there are very few mathematical models of near ocean waters, but the growing concern for their pollution will require more rigorous analyses and simulations in the future.

F. Groundwater

1. General Comments

The largest supply of fresh water available for man's use is under the ground (see Chap. 1). From ancient times this source of drinking water has been the basis for the location of towns and cities. Today, large industries and great tracts of irrigation land depend on this source of supply. More recently, the vast reservoirs located beneath the ground have been used as disposal areas for polluted water. Increased pumping of groundwater near the coastline has resulted in saltwater intrusion (*19*) into the freshwater aquifers. Both of these pollution problems have resulted in growing concern for the preservation of the quality of our groundwaters.

2. Principles of Groundwater Movement

With the exception of some areas with subterranean solution channels, the motion of water through the ground is very slow and produces large quantities only because of the extremely large cross-sectional area experiencing this flow. This low velocity coupled with the confining pore spaces between soil particles produces a flow which can be analyzed by relatively simple hydraulic equations (*20*).

a. Darcy's Law

The law governing the movement of viscous liquids through porous materials under saturated conditions was discovered by Darcy (*20a*) in 1856. It has come to be called Darcy's law. Darcy's law states that:

$$Q = KA \frac{h_1 - h_2}{L} \tag{2}$$

where Q = discharge rate
A = cross-sectional area
K = hydraulic conductivity
L = length of flow
h_1 = inflow head
h_2 = outflow head

Rearranging Eq. (2) we get

$$K = \frac{Q}{A} \frac{L}{h_1 - h_2} \tag{3}$$

Thus K has the units of velocity. A variety of units are used, but for simplicity only one set of units is used here. If Q is in gallons per day, L, h_1, and h_2 are in feet, and A is in square feet, K has the units of gallons per day per square foot (gpd/ft^2). This gives a better conceptual feeling for K than velocity, although it should be remembered that K has velocity units. It should also be pointed out that in a strict sense the value of K varies with temperature. However, the effect of temperature change on K is small in most cases and is generally ignored.

As an example, consider a stratum of sand 10 ft thick and with hydraulic conductivity of 10^3 gpd/ft^2. We assume that the loss of head is approximately 2 ft/mile and that we are interested in the flow or discharge rate (Q) through a 1 ft width of aquifer extending from the bottom to the top of the aquifer. Then, substituting the following values into Eq. (2)

$$\frac{h_1 - h_2}{L} = \frac{2}{5280}$$

$$A = 1(10) = 10 \text{ ft}^2$$

$$K = 10^3 \text{ gpd/ft}^2$$

the discharge rate, Q, is easily calculated:

$$Q = (10^3)\frac{(10)(2)}{5280} = 3.79 \text{ gpd}$$

b. An Analogy

An electric analog model of a groundwater system (*20b*) consists of a network (usually a square lattice) of resistors selected in such a manner that the flow of electrical current and the distribution of electrical potential in the network is analogous to the flow of water and the distribution of hydraulic head in the groundwater system.

If nonsteady-state problems are to be analyzed on the model, capacitance must be added to simulate the storage characteristics of the groundwater system.

The elements that are analogous between the two systems are as follows:

MODEL	GROUNDWATER SYSTEM
Electric potential	Hydraulic head
Conductivity	Transmissibility
Capacitance	Storage
Time (milliseconds)	Time (decades)
Electric current	Flow rate

c. *Information Needed to Construct a Model*

To select the correct values for the electrical components of the model the following information must be available:

1. Well logs – establish aquifer thickness and extent
2. Pump tests – determination of transmissibility and storage characteristics
3. Water level data – establishment of a known distribution of hydraulic head to test the mode.
4. Pumpage data – establish current demands on the system for testing purposes.

3. Applications of Analog Models

Once a model has been constructed and tested it can be used to examine a number of problems. One of the more obvious uses would be to evaluate the effect of pumpage in the future. With the rapidity of computerized solutions available, the effect of a number of different pumping schemes can be analyzed rapidly. The effectiveness of artificial recharge could also be studied.

The potential of a model of this nature for the study of the economics of groundwater mining and overdraft also are great. Since the model closely duplicates the actual response of the aquifer to changes, the effects of nearly any contemplated groundwater development can be analyzed.

Having defined the flow of the fluid medium in the ground, the next problem is to develop methods of modeling the pollution of these flow fields. One form of pollution which has been analyzed is that of saltwater intrusion along the coast. Charmonman(*21*) developed mathematical models using the digital computer to predict the movement of the saltwater-freshwater interface along the coastline under the action of withdrawal or injection of water into the groundwater aquifer. These models have been applied to the problem of saltwater intrusion in the Bangkok area of Thailand.

The tracing of conservative pollutants in the groundwater is possible with present technology. A more complex problem is that of nonconservative wastes, such as BOD (biochemical oxygen demand) or nuclear energy wastes.

Surface waters have the ability to rejuvenate themselves and/or flush themselves to the sea in a short time period. Groundwaters, on the other hand, because of their slow movement, remain polluted for very long periods of time. Because of this, responsible officials are becoming increasingly concerned about the potential of groundwater pollution.

V. Optimizing Techniques

A. Introduction

Introduction of the high speed digital computer has presented engineers with the possibility of finding optimal solutions to many problems that heretofore could be approached only from the standpoint of an adequate solution. In almost all cases, optimization must be approached by a repetitive (iterative) calculation technique. This kind of approach is ideally suited to the abilities of the digital computer.

Computerization of a problem requires very close attention to the definition of the system under study. The boundaries of the system must be clearly defined and the important interrelationships between components of the system must be stated in mathematical terms.

Optimization implies that one is working toward a given objective. This objective is stated mathematically in some form of objective function. In the case of large water resource projects, it often is necessary to look into the areas of welfare and resource economics and politics in order to properly assess the objective of the project.

B. System Definition

A water resource system is usually bounded by the natural watershed of the river basin under study. Exceptions to this rule are such circumstances as the transwatershed diversion of water(*21a*) either by man-made activities or by groundwater flows, and the economic impact from neighboring watersheds. The river system in the basin is the basis for most of the important interrelationships that must be modeled. Upstream activities, either beneficial or detrimental, have effects on all portions of the system downstream. This is the basic relationship that must be investigated within a river system. The interrelationship is best expressed in terms of a simulation program that models the behavior of the entire river system.

C. Objective Functions

One of the more difficult tasks in an optimization scheme is that of defining the objective of the system. In the realm of federally sponsored water resource development schemes, it is possible to have a wide variety of ultimate objectives. One legitimate objective of such a program would be the redistribution of the national income in a particular direction. The Appalachian program(*21b*) currently under way is an example of this kind of objective. Under these circumstances, considerations such as economic efficiency must be subservient to the broad concept of income redistribution presently espoused by Congress. But within the framework of this general policy decision, it is still desirable to achieve this distribution with a high level of economic efficiency. Having decided that development should be encouraged in a particular locality, the next question is how to accomplish this development most efficiently. Once again, it is necessary to state the objective function in mathematical form so that it can be handled by the digital computer. This becomes very difficult in many cases because the ultimate decision has been rather subjective.

Economic efficiency for resource development is usually stated in terms of a benefit-cost(*22*) analysis. This form of analysis attempts to monetize all the variables in the system so as to measure all of the benefits as well as the costs of the system in dollar value. A benefit-cost analysis can be used in two different ways. First of all, the analysis can be used to determine the optimal scale of development of a given project. In this case it is customary to increase the size of the development until the net benefits are maximized.

A second use of the benefit-cost analysis is to rank various projects in order of their economic efficiency. For this purpose it is customary to determine a benefit-cost ratio. The larger this ratio, the more desirable that particular project becomes. When the choice is between several projects which will contribute to the same basic policy of resource development, it is customary to choose the project which has the highest benefit-cost ratio. This decision, however, if often shaded and influenced by the political expediency that is attendant in any public works program.

D. Operations Research

Operations research (*23*) is a mathematical method which can be used for assigning scarce resources in such a way as to optimize a given objective function. As such, it is ideally suited to the problem of resource economics. This mathematical process may take on a wide variety of forms, depending upon the circumstances involved.

One example of the application of operations research is the question of the level of treatment that should be applied to waste sources before they are injected into a flowing stream. The natural flowing stream has a large capacity for self-purification after pollution. The question arises as to the extent to which we can rely upon this self-purifying capacity, because of the effect of even a slight degradation of quality upon downstream uses.

Conservationists would prefer complete treatment of waste at the source with essentially zero degradation of the flowing stream. This approach disregards any self-purification capacity that the flowing stream might possess. The opposite viewpoint is to permit the maximum allowable degradation of the stream so as to fully utilize its self-purification capacity. Unfortunately, there is no general agreement on the level of degradation which is permissible in terms of uses in the downstream area.

Pollution parameters such as dissolved oxygen and fecal coliform concentration are used as indicators of pollution level. The states have set standards for these parameters which must be maintained in certain classified streams.

Given a minimum value for dissolved oxygen concentration, it is possible, by means of operations research, to determine the amount of treatment that should be applied at various plants along a river. Some consideration must be given to the future needs of the basin so that existing plants are not permitted to utilize all of the self-purification capacity of the river and leave no additional capacity for future growth. Computer operations require that such decisions be made in order to provide a firm mathematical relationship.

Strict economic efficiency might dictate a varying level of treatment for the various plants and outfalls along a water course. Such a scheme could be developed. However, present institutional arrangements are not adequate to this task. The new techniques available by use of computers and simulations, if they are to be used effectively, will require new forms of jurisdiction and responsibility and the development of new institutional arrangements.

VI. Summary

Computerized solutions of mathematical models have been demonstrated for a number of natural bodies of water. The discussions in the chapter are in no way exhaustive. Indeed, the growth of this technology is such that any text must be obsolete before the ink is dry. For this reason the author has attempted to point out a number of current applications and indicate some of the tremendous potential for future development of these techniques.

This chapter is in no way a handbook. Detailed methodology will vary from model to model and from one computer to another. It is hoped, however, that this chapter has introduced the reader to basic concepts of computer models so that he can recognize those areas of water quality study which may be amenable to solution by this methodology.

REFERENCES

1. H. W. Streeter and E. B. Phelps, *Public Health Bull. No. 146,* U. S. Public Health Serv., Washington, D.C., 1925.
2. N. H. Crawford and R. K. Linsley, *Tech. Rept. No. 12,* Dept. Civil Eng., Stanford Univ., July, 1962.
3. W. Rippl. *Proc. Inst. Civil Engrs.,* **71**, 270 (1883).
4. R. G. Stanton, *Numerical Methods for Science and Engineering,* Prentice-Hall, Englewood Cliffs, N.J., 1961.
5. *Electric Flood Control Model for the Lower Kansas Basin,* Dept. of the Army, Kansas City District, Corps of Engineers, 1964.
6. W. E. Dobbins, *Sanit. Eng. Div., Am. Soc. Civil Engrs., 90, SA3, Proc. Paper 3949,* 53 (1964).
7. J. L. Worley, Div. Water Supply and Pollution Control, U.S. Public Health Serv., Region IX, Portland, Ore., 1963.
8. R. L. Smith, W. J. O'Brien, A. R. LeFeuvre, and E. C. Pogge, Federal Water Pollution Control Admin., Missouri-Souris-Red River Basins Comprehensive Water Pollution Control Project, 1967.
9. G. T. Orlob and L. G. Selma, Paper presented at Conf. on Current Research into the Effects of Reservoirs on Water Quality, Portland, Ore., January, 1968.
9a. *Public Health Serv. Publ. No. 1195,* U.S. Dept. of Health, Education and Welfare, Washington, D.C., 1965.
10. J. T. Marlar, M.S. Thesis, Georgia Institute of Technology, Atlanta, Ga., 1967.
11. R. S. Green, *Public Health Serv. Publ. No. 1263,* U.S. Dept. of Health, Education and Welfare, Washington, D.C., 1964.
12. J. V. Krutilla and O. Eckstein, *Multiple Purpose River Development,* Johns Hopkins Press, Baltimore, Md., 1958.
13. Dye Study of Chattahoochee, unpublished study of the Chattahoochee River near Atlanta, Ga., Southeast Region, Federal Water Pollution Control Admin., Atlanta, Ga., 1967.
14. D. J. O'Conner, *Trans. Am. Soc. Civil Engrs.,* Part III, **126** (1961).

15. R. V. Thomann, *J., Sanit. Eng. Div., Am. Soc. Civil Engrs.*, **89** (October, 1963).

16. R. P. Shubinski, J. C. McCarty, and M. R. Lindorf, *J. Hydraulics Div., Am. Soc. Civil Engrs.*, **91** (September, 1965).

17. J. E. Hagen and T. P. Gallagher, unpublished report on the Hillsborough Bay Technical Assistance Project, Southeast Region, Federal Water Pollution Control Admin., Atlanta, Ga., October, 1967.

18. *Lake Erie Surveillance Data Summary 1967–1968*, Great Lakes Region, Federal Water Pollution Control Admin., Cleveland Program Office, 1968.

19. R. K. Linsley and J. B. Franzini, *Water-Resources Engineering*, McGraw-Hill, New York, 1964, p. 104.

20. D. D. Franz, private communication, 1966.

20a. H. Darcy, *Les Fontaines Publiques de la Ville de Dijon*, Dalmont, Paris, 1856.

20b. J. D. Winslow and C. E. Nuzman, *Special Res. Publ. No. 29*, State Geological Survey of Kansas, Lawrence, Kans., 1966.

21. S. Charmonman, *Proc. Am. Soc. Civil Engrs.* **93**, 13 (1967).

21a. E. A. Ackerman and G. O. G. Lof, *Technology in American Water Development*, Johns Hopkins Press, Baltimore, Md., 1959, p. 507.

21b. *The Appalachian Regional Development Act of 1965*, Public Law 89-4, U.S. Government Printing Office, Washington, D.C.

22. O. Eckstein, *Water Resources Development*, Harvard University Press, Cambridge, Mass., 1961.

23. A. Kaufman, *Methods and Models of Operations Research*, Prentice-Hall, Englewood Cliffs, N.J., 1963.

Chapter 6 The Effects of Pollution upon Aquatic Life

Max Katz
COLLEGE OF FISHERIES
UNIVERSITY OF WASHINGTON
SEATTLE, WASHINGTON

I. Introduction

The wastes which are discharged into our waterways can be classified into four general types: domestic, industrial, agricultural, and radioactive. In this chapter, the effects of the first three types of wastes upon aquatic organisms, particularly fish, are briefly discussed; major emphasis is given to industrial waste components. Radioactive wastes have been adequately and voluminously discussed by others, in a readily available literature.

Two other subjects, of timely importance, connected with water pollution are taste and odor problems and thermal pollution. The tainting of seafood with "off" tastes and odors is a serious defilement of a commodity so essential in a world of expanding population. Temperature changes have a profound effect on metabolic processes and thereby are a consequential influence when concomitantly applied to existing pollution problems.

II. Types of Wastes

A. Domestic Wastes

The earliest interest in water pollution was stimulated by the discovery by microbiologists that the waterborne diseases of humans are to a large extent caused by pathogenic bacteria. The early studies on water pollution were motivated primarily by public health considerations and were largely bacteriological investigations. These studies led to the development of methods of water treatment and disinfection that proved to be very effective—as a result, waterborne diseases such as typhoid and cholera are now a rarity in the United States, Europe, and other sections of the world where the relatively simple technologies necessary to provide safe water for domestic uses are employed.

The intensive studies that led to the development of water treatment technologies also gave the biologists and chemists who were interested in the sanitary problems of water a good insight into the chemical, physical, and biological changes in waterways that received domestic wastes. We now have a good working knowledge of the biological as well as the chemical and physical changes that result in a body of water receiving domestic wastes.

We know that the biological effects of the introduction of domestic wastes into streams are a reflection of the introduction of large amounts of nutrients into a limited volume of water. When the area that is now the United States was largely a wilderness, the wastes of a small community of a few hundred or even thousands of people could be disposed of readily, and without harm to the aquatic biota, into the closest stream. These wastes were rapidly diluted and the nutrients available in the waste were utilized to good advantage by the aquatic community.

But as the population increased and the small farming and trading communities grew into large cities with populations in the hundred thousands and millions, the streams no longer had the capacity to dilute the huge volumes of untreated or partially treated wastes they received. The nutrients of the wastes, in the process of being utilized by the aquatic communities, caused marked ecological changes in the streams. The resulting changes in the chemistry and physical conditions of the stream and of the aquatic biota were typical and characteristic and usually undesirable. These alterations in the ecology and uses of the receiving stream are referred to as "water pollution."

The chemical, physical, microbiological, and biological changes that occur in unpolluted streams and streams polluted by domestic waste are adequately, if not completely, understood. Various aspects of these changes have been studied extensively by many investigators and, with the exception of the biological changes, are usually discussed in textbooks. The most complete knowledge of the biological changes that occur in a stream receiving domestic wastes was obtained by Tarzwell and his co-workers (*1–3*) at the R. A. Taft Sanitary Engineering Center of the Public Health Service in Cincinnati. A series of interesting and authoritative studies was conducted on Lytle Creek, which is located near Cincinnati, and the data obtained have been published in the professional literature.

With adequate sewage treatment, a large percentage of the nutrients in domestic wastes can be removed. Adequate secondary sewage treatment removes most of the carbon compounds that are utilized by bacteria, and the resulting effluent is high in nutrient salts, primarily nitrates, nitrites, and phosphates, which stimulate algal growth in receiving waters. These algae may and often do grow in such great numbers that they are a nuisance in themselves. The problem of excessive algal growth has become acute in certain areas, particularly in lakes adjacent to large cities, for example, Lake Erie. This condition, commonly referred to as eutrophication (see Chap. 4), has stimulated the development of the so-called tertiary waste treatment procedures which are designed to remove the nutrient salts (see Chap. 8). Some of the current research in tertiary treatment procedures is directed toward the removal of phosphates which many authorities believe is the important limiting factor for algal growth in the freshwater environment.

B. Industrial Wastes

The principal area of concern today is the need for an understanding of the biological effects of the various industrial wastes and, of course, the development of effective methods for adequately and economically treating these wastes so that they have minimal deleterious effects on the biological resources of receiving waters. Industrial wastes can be characterized in the following manner:

(1) wastes with a high BOD (biological or biochemical oxygen demand: fruit and vegetable canneries, sugar refineries;
(2) wastes with a high BOD and a significant toxicity: kraft pulping mills, petroleum refineries;
(3) wastes with low or no BOD and high toxicity: metal refineries, chemical industries, acid mine wastes;
(4) thermal wastes: electric power plants, nuclear power plants, steel mills.

1. Biological Effects of Industrial Wastes with a High BOD

Among the industries that produce wastes with an extremely high BOD are fruit and vegetable canneries, sugarcane and sugar beet refineries, and some wood pulping operations. In these plants, large amounts of liquids are discharged containing large amounts of various sugars which are leached out from the fruits and vegetables or are released from the wood by the pulping process. Because of these carbohydrates, these plants produce wastes which are very high in BOD. A moderately sized wood pulping mill can produce waste with the population equivalent of a city of several million people.

These carbohydrates, when discharged into the stream, supply an abundant food source for bacteria and fungi. The metabolic activities of these organisms in consuming these wastes deplete the dissolved oxygen resources of the stream, and the stream quickly becomes anaerobic. Any organisms, including fish, which require dissolved oxygen, are forced our or are killed, and the only organisms remaining are, of course, the anaerobic bacteria and fungi and the few invertebrates which are adapted to get their oxygen from the atmosphere. In turn, a predator population quickly develops which feeds upon the slurry of bacteria in the water. These predators are usually stalked and free-swimming ciliate protozoa(*4*). In addition, in areas along the edges of the stream, the bottom may be blanketed with red mats of annelid worms or insect larvae which feed upon the bacteria, the protozoa, and other nutrients in the bottom deposits. Anaerobic or near anaerobic conditions persist in the stream until most of the nutrient is utilized. At this time the bacterial population decreases, and aquatic invertebrates which can survive in the reduced amounts of dissolved oxygen characteristic of such areas become the dominant organisms(*5*).

The only other forms of life which may be observed in the anaerobic or septic portions of the stream are some invertebrates which have modifications for obtaining oxygen from the atmosphere. These are usually the larvae of certain flies, *Eristalis,* which have a long, telescoping tail which can be thrust to the surface of the water to obtain air, certain beetles which can capture bubbles of air on the ventral surface of their abdomen, and pulmonate snails.

The only areas in these polluted streams where fish may be observed are at the mouth of unpolluted tributary streams, where clouds of minnows often will be observed feeding at the interface of the unpolluted tributary and the main stream.

Some interesting observations were made by the author regarding the behavior of fish in a stream receiving fruit cannery waste. The stream to which the wastes of this cannery were discharged served, for all practical purposes, as a drainage ditch which contained large amounts of water only during the spring and summer irrigation seasons and during the period of cannery operation. Before the cannery started operations in the early summer, there were good fish populations throughout the stream wherever there was sufficient water and in the lateral ditches which carried irrigation runoff water into the stream. When the cannery started operation, the volume of the stream increased markedly as a result of the waste discharged from the cannery. The dissolved oxygen in the stream dropped to zero because of the oxygen-depleting wastes, and no fish were to be found in the main stream although there were considerable numbers of small minnows feeding actively in the mouths of the irrigation lateral ditches.

The cannery usually closed down for the weekend, and by Saturday afternoon the dissolved oxygen content of the water increased to about 5–6 ppm and the areas that were without fish during the working days were observed to contain large populations of fish. When the cannery reopened on Monday, the dissolved oxygen content started to decrease as the effects of the wastes became evident. When the dissolved oxygen of the water decreased to 0.5 ppm or less, large schools of fish were observed drifting downstream with their noses at the surface. When these schools of fish reached a larger tributary in which there was adequate dissolved oxygen, they immediately sounded into the deeper water.

Although the dissolved oxygen in the stream receiving the wastes went down to zero during the operation of the cannery, only a few fish kills, involving small numbers of fish, were observed. Obviously, the fish in this stream and its tributaries were utilizing as food the large amount of living animal matter (oligochaetes) that was the end result of the nutrients discharged by the cannery. In fact, the populations in the cannery stream were much larger than the populations in another tributary which did not receive any wastes.

Of course, the fish in the stream receiving the wastes were species that are included among the "rough fish" species, primarily minnows. The game species that might be resident in this stream if it were not polluted were missing. Bass and bluegills are characterized by the habit of competing for and maintaining a home territory. These species probably refuse to leave their home area even when oxygen conditions deteriorate to the point where they suffocate. The population structure of the normal stream, which is made up of the desirable predator species such as bass and bluegills, and the forage minnows, is destroyed by polluting wastes. The final population consists entirely of the undesirable forage minnows and other species of no economic interest to man.

It is commonly, even if uncritically, accepted by most biologists that the end product of pollution caused by wastes high in nutrients is always an undesirable or useless species or complex of species. There are now, however, some indications that the food chain which is sustained by nutrient wastes may result in an increase of certain desirable food species. English (*6*) observed that the English sole populations living in Port Gardner, Washington, seem to be on the increase. Port Gardner receives the untreated wastes of three pulp mills and the partially treated domestic wastes of Everett, Washington. Statistics indicate that the catch of sole is increasing, and a rapidly growing population with all size and age groups of fish has been observed. Although these situations have not been studied in detail, there is no good theoretical reason why nutrient wastes cannot result in increased populations of fish useful to man, under the proper conditions.

2. Biological Effects of Wastes with Toxicants and a High BOD

Some industrial wastes not only have a high BOD but may contain either inorganic or organic toxicants or may have high or low pH. An example is wood pulping waste produced by the kraft process. Kraft pulping wastes have a high BOD derived from the wood sugars and the other soluble compounds created by the pulping process, and they also contain compounds toxic to fish, such as sodium hydroxide, sodium sulfide, methylmercaptan, formaldehyde, and the sodium salts of resin and fatty acids (*7*). The alkalies present in the kraft pulping wastes may increase the pH of streams to a level too high for fish to survive (*7*).

In England, there is a good deal of concern about the toxic effluents produced by the distillation of coal for the production of gas, coke, and organic chemical intermediates. The waste is called ammoniacal gas liquor and contains free ammonia, ammonium salts, cyanides, sulfides, and organic substances which include pyridine, phenols, cresols, xylenols, and some hydrocarbons (*8*). These substances are not only toxic to fish but also contribute a substantial BOD. The biological effects of both the inorganic and organic components of the ammoniacal gas liquor are detrimental to fish. Near the point of discharge, the toxicants are lethal to both fish and aquatic invertebrates that may be present. Actually, in the case of well-established industries, the areas below the effluents have long been fishless. By the time the waste moves downstream to the point where many of the toxicants have been diluted or removed by any one of several imperfectly known mechanisms, the bacterial populations may have become large enough so that their activities have reduced the dissolved oxygen to a level where the stream is not suitable for fish. If no other wastes are introduced, a point is reached in the stream where the nutrients have been utilized and oxygen is again present. Then fish and desirable aquatic invertebrates would be able to exist if it were not for the toxicants that may still persist.

It is well-established that most substances are more lethal to fish in water of reduced oxygen content. Lloyd(*9*), in a thorough study of the variables that affect the toxicity of zinc sulfate to trout, found a significant decrease of tolerance to the zinc ion in waters of reduced dissolved oxygen. Trout survived about 180 min in water containing 8 ppm of zinc with a dissolved oxygen content of 3.8 ppm, while they survived in the same concentration of zinc for about 290 min in water of 8.9 ppm. Lloyd(*10*) observed that the toxicity threshold of rainbow trout to ammonia was significantly reduced in water with reduced oxygen content. Lloyd and Herbert(*11*) reported that lead, copper, and phenols, as well as zinc, were more toxic in water of reduced oxygen content. This apparent increase of the toxicity of substances to fish in water of reduced dissolved oxygen content has been reported by Lloyd(*12*) for several toxicants.

The increased toxicity of substances in waters with reduced dissolved oxygen has been explained by the fact that fish compensate for reduced dissolved oxygen content by increasing their rate of respiration(*13*). If the oxygen in the water is reduced to the level where the fish compensate by increasing their respiratory rate and the water contains some toxicant, the gill surfaces may be exposed to a greater amount of toxicant in a given period of time than they would be in water with sufficient dissolved oxygen. Unfortunately, with the expansion of population and the resultant industrial growth, our watercourses are subjected to heavy loads of wastes, some of which have a significant oxygen demand and others of which are toxic. The fish population in the recipient streams might be able to survive (but probably not flourish) in either the waste with the high BOD or the toxic waste but not in the mixture.

The toxicity bioassays aimed at helping industry to design waste treatment procedures and to determine dilution factors help perpetuate this situation. The standard bioassay procedure(*13*) recommends that toxicity tests be carried out in water with adequate dissolved oxygen. In addition, these procedures further recommend that data from tests that were conducted in water with reduced oxygen content be discarded. It would be more representative of the true situation if it were specified that toxicity bioassay tests always be carried out in water of reduced oxygen content, because this is the condition that often prevails in our heavily utilized streams.

3. Toxic Wastes

The effect of toxic wastes upon fish and other aquatic forms has stimulated perhaps more interest and research in recent years than any other subfield of water pollution biology. This research has resulted in an extensive scientific literature which has given a quite useful, if not complete, understanding of the effect of toxic substances on fish.

The necessity of determining the toxicity of wastes to fish and other aquatic forms has resulted in the development of the toxicity bioassay into a useful and accepted, although far from perfected, tool in water quality management. The usefulness of the data obtained in toxicity bioassays and the fact that these data can be expressed in numbers and tables have made the compilations of toxicity data "best sellers." These titles include Ellis's publication(*14*), the reviews of Doudoroff and Katz(*15, 16*), and the very valuable but sadly abused *Water Quality Criteria* of McKee and Wolf(*17*).

There are some who feel, however, that the ease of getting quantitative data which can be readily applied to certain water pollution problems has led to an overemphasis upon toxicology and a neglect of the ecological changes which are of equal, if not greater, importance but which are much harder to measure. Hynes(*4*) has expressed this admirably in his discussion of the biological effects of poisons. He states: "much of this work is of very high quality and of great theoretical interest, but for many reasons I feel that a disproportionate amount of the effort spent in all countries on the problems of pollution has been directed into this narrow field of research. Admittedly, it has told us much about the actions of poison and their interactions between themselves and with the environment, but the final answer as to how much of any particular poison can be tolerated *by a population of fishes living a natural life* in a river still remains unanswered."

Hynes's criticisms can be applied to several studies in the United States in which the author has been involved. Heavy penalties have been placed upon certain industries, where the basis of data used to support this action was derived largely from studies which indicated that the waste effluents were somewhat toxic under laboratory and semilaboratory field tests. The regulatory agencies, in their eagerness to present convincing evidence of damage, were reluctant to look at the populations of valuable fish which were thriving in the areas of concern. In these cases, the use of bioassay test has obscured, if not defeated, the purpose for which the test was designed.

Most of the current research work in fish toxicology is of excellent quality, and the data obtained are useful. Experimental conditions are usually clearly defined by the experimenters, but the people who apply these data to their particular problems seldom read anything else except the summary tables. Toxicity data for a particular substance obtained with small Atlantic salmon in soft waters, at 12°C in a constant flow system, have been known to be applied directly to a problem involving adult bluegill sunfish, in alkaline waters at 25°C. There is little that can be done to correct the abuse of good data except to hope that water resource managers will finally come to the realization that aquatic animals are more specialized in their ecological requirements than are humans.

a. *Factors That Affect the Toxicity of Wastes*

Some of the factors that affect the toxicity of substances to fish are water temperature, oxygen content of the water, pH, and the dissolved salts. The effects of these variables upon the toxicity of substances are reported in many studies and are summarized and discussed in detail by Hynes (*4*), Doudoroff (*18*), and Jones (*13*).

(1). Temperature and Toxicity

Temperature has a marked effect on toxicity (*16*); in general, the toxicity of a substance increases with the water temperature. In experiments with the insecticide, endrin, for example, the toxicity as expressed by the 96-hr TLm was about 33 times greater in experiments conducted at 30°C than was the toxicity in experiments conducted below 5°C (*19*). These data suggest that concentrations of a toxicant which might be tolerated by fish during the winter may result in significant mortalities during the warm summer months.

TLm is the median tolerance limit. This is the concentration of a toxicant which will kill 50% of the fish exposed within a fixed time. In this case, the time is 96 hr.

(2). Reduced Dissolved Oxygen and Toxicity

As has been discussed previously, many substances are more toxic when the dissolved oxygen of the water is reduced. The ability of water to retain dissolved gas is greater in cold waters than in warm waters. Thus, in waters receiving both toxic and oxygen-depleting wastes, there can be a significant increase in the effective toxicity due not only to the toxicant but also to the reduced oxygen content and the increased respiratory activity of the fish which is associated with the reduced dissolved oxygen.

(3). pH and Toxicity

The chemistry of the water, particularly the pH, has a marked effect upon the toxicity of substances commonly present in industrial wastes.

The comprehensive literature review of Doudoroff and Katz(*15*) indicated that well-developed fish could tolerate unpolluted waters between pH 5.0 and 9.0 and certain tolerant warm water species could tolerate a pH above 10 for at least a day. But, even within this range, pH can have a marked effect upon the toxicity of certain compounds. The toxicity of ammonium salts, in particular, is markedly affected by pH(*20*). When the concentration of un-ionized ammonia (NH_3) was between 2 and 3 ppm (in solutions with total ammonium concentrations of 10.1, 38.4, and 167 ppm at pH 8.84, 8.16, and 7.52, respectively), the experimental fish lost equilibrium in 170–330 min(*20*). When the ammonia content was increased to 16–24 ppm, either by increasing the ammonium salt content of the solutions without changing the pH, or by raising the pH, the overturning time was reduced to about 26 min(*20*).

This observation can be of significance in a river receiving several effluents. A plant could discharge a waste containing ammonium salts into a river with water near pH 7.0 without any significant harm occurring to fish populations. A plant downstream, discharging a nontoxic waste which changes the pH of the water to above pH 8.0, could cause fish kills in the areas where the two wastes mix.

On the other hand, Brinley(*21*) and Wuhrmann and Woker(*22*) observed that the toxicity of HCN decreased markedly as the pH of the water increased.

Perhaps the most dramatic increase in toxicity with change in pH was reported by Doudoroff(*23*) and Doudoroff et al.(*24*), who studied the toxicity of metal cyanide complexes to fish. When combined with nickel to form a nickel cyanide complex, 1.0 ppm of cyanide (as CN^-) at pH 6.5 was more toxic to minnows than 1000 ppm at pH 8.0. In another experiment, they found that 10–13 times as much nickel cyanide was required to kill half the test fish at pH 7.8 than was required at pH 7.5.

Sunlight, although fundamental to the biological activities of most aquatic ecosystems, is not usually regarded as an important factor in affecting the toxicity of substances to fish. Burdick and Lipschuetz (*25*), however, reported that harmless potassium ferrocyanide and ferricyanide solutions became highly toxic to fish on sunny days because of the liberation of cyanide through the photodecomposition of the otherwise stable complexes.

(4). Dissolved Mineral Content and Toxicity

The dissolved mineral content of the water into which heavy metal salts are discharged can markedly affect the toxicity of these salts to fish. Most natural waters contain various amounts of sodium, calcium, potassium, and magnesium ions, as well as bicarbonate and sulfate ions. Concentrations of metal salts which are harmless to fish in dilute seawater or hard fresh waters can be extremely toxic in soft fresh waters. The toxicity of the heavy metals such as copper and zinc is counteracted or antagonized particularly by calcium ion. The toxicity of the salts of heavy metals and the antagonistic action of certain ions on the toxicity of these metals are discussed in thorough detail by Doudoroff and Katz(*16*).

Yet, it has been observed that an increase in dissolved salt content does not automatically afford protection to heavy metal ion toxicity. Herbert and Wakeford(*26*) observed that the resistance of yearling rainbow trout and Atlantic salmon smolts to zinc sulfate increases with salinity up to 30–40% seawater. At this concentration, these fish could tolerate 13–15 times as much zinc sulfate as in seawater. A further increase in salinity to 72% seawater reduced the tolerance for the zinc salt. Proper regard for these data would suggest that industries with a heavy metal waste disposal problem should consider the desirability of locating on an estuary with rapid dilution.

b. *The Effects of Heavy Metals on Fish*

Among the earliest systematic water pollution studies concerned primarily with fish was that regarding the wastes from the lead and zinc mines in Wales. Jones(*13*) mentioned documents published in 1878 regarding the effects of pollution from the Cardiganshire, Wales, lead and zinc mines upon the resident fish species. The intensive biological research dating back to 1919 in this area has resulted in a considerable understanding of the biological effects of heavy metals upon aquatic life, especially fish. Carpenter(*34*) noted that in lead solutions the test fish that died exhibited the symptoms of acute respiratory distress and that the bodies of affected fish were covered with a veil-like film which looked like coagulated mucus. The gills were covered with a thick film of mucus which upon chemical analysis was found to contain lead. Further studies by other workers indicated that salts of zinc, copper, cadmium, and mercury caused similar precipitates, and the fish killed by relatively high contents of these heavy metals died displaying the symptoms of acute respiratory distress. This syndrome was termed coagulation film anoxia(*35*).

The heavy metals attack the respiratory apparatus in three ways. First, the spaces between the gill filaments become filled with the mucus-metal precipitate which prevents the water from reaching the gill filaments and supplying its dissolved oxygen to the organ. Then the spaces between the gill lamellae become so filled with the precipitate that movement becomes impossible and circulation of the blood through the gill capillaries is retarded. The stasis of the blood circulation through the gills results in a partial heart block. Schweiger(*36*) demonstrated the destruction of the gill epithelium by heavy metal salts. Skidmore(*37*), in a thorough review of the toxicity of zinc compounds to fish, mentioned several studies which described the histopathological effects of zinc on the gills, gut, and liver.

Donald Mount of the Federal Water Pollution Administration outlined autopsy procedures used in fish kills attributed to zinc or to cadmium (*38, 39*). The zinc autopsy technique is based on the fact that in nonlethal concentrations of zinc salts, the opercular bone accumulates zinc at about the same rate as the gill tissue, but in acutely lethal concentrations, the opercular bone does not accumulate zinc as rapidly as does the gill tissue. The ratio of zinc in the gill tissue to that in the bone is increased 100-fold in lethal exposures. The autopsy technique for cadmium is based on the observation that living fish never accumulated more than 130 μg of cadmium per gram of gill tissue based on dry weight, but in fish that died of acute cadmium poisoning, the accumulation of cadmium was never less than 150 μg/g.

c. *Other Toxic Wastes—Mixed Wastes*

There is no good reason to expect that industrial wastes will contain only one toxic component. Many industrial wastes prove to be complicated mixtures of metals, organic substances, greases and oils derived from the lubricants used for the plant machinery, and any other materials most easily disposed of through a floor drain. Even if the waste is a simple one, it will be discharged into a waterway containing substantial amounts of toxicants or oxygen-reducing substances derived from other industries or municipal waste sources. An industrial toxicity problem is for all practical purposes the study of the toxicological effects of mixed wastes.

Some of the early studies regarding the toxicity of mixed heavy metal salts, as reviewed by Doudoroff and Katz(*16*), indicated that mixtures of some of the heavy metals were synergistic. Synergism implies, in effect, that mixed solutions of two heavy metals are more toxic than the sum of the toxicities of the individual solutions of single metals of corresponding concentrations. Bandt(*27*) reported that the toxic effects of the sulfates of zinc and cadmium and of nickel and cobalt in mixed solutions of the two salts were additive. The toxicity of mixtures of nickel and zinc, copper and zinc, and copper and cadmium, however, were greater than would be expected if the toxicity were additive. Doudoroff(*28*), using mixtures of cupric and zinc sulfate, found a pronounced synergism in tests lasting for about 8 hr.

Lloyd(*29*), however, found that synergism is exhibited only at high concentrations of copper and zinc in hard waters. Under other conditions, the combined action of the two metals is additive.

Herbert and Shurben(*30*) and Herbert and Van Dyke(*31*) studied the toxicity to rainbow trout of mixtures of ammonium and zinc salts, ammonium and copper salts, and zinc salts and phenols. They found that the toxicity of these mixtures to fish followed the empirical rule that:

$$P_s/P_t + Q_s/Q_t = 1$$

where P_t is the threshold concentration of toxicity of toxicant P, P_s is the concentration of toxicant P to which fish are exposed, Q_s is the concentration of toxicant Q to which fish are exposed, Q_t is the threshold concentration of toxicity of toxicant Q.

With the background of the above studies and the knowledge of the toxicity of the substances (ammonia, phenols, heavy metals, and cyanide) known to be present in the sewage of English cities, Lloyd and Jordan(*32,33*) made a study in which they tried to predict the toxicity of sewage effluents to rainbow trout. For about two-thirds of the effluents, they were able to predict the toxicity with an accuracy of $\pm 30\%$. Lloyd and Jordan emphasized that bioassays were still necessary, but it is obvious that enough information has been gathered regarding the toxicity of substances to fish and the effects of the environmental factors such as water temperature and reduced oxygen upon toxicity so that many laboratories can monitor their wastes by chemical analyses. Of course, sufficient preliminary bioassay studies are needed to ensure this course of action.

Space and the state of knowledge does not allow a detailed or even a brief discussion of the biological effects of the hundreds or perhaps thousands of toxic substances present in industrial wastes that are discharged into our waterways. In fact, it has been a source of concern to the author that many plant operators seldom have more than a vague idea of what they are discharging in their waste discharges. Actually, until the recent federal actions insisting that the various states impose water quality standards, there was seldom any reason that a mill operator subject himself to the onerous task of locating his waste discharges and determining what substances were being discharged, and in what quantities. In many cases, these wastes are so complex in mixture and so variable in concentration that the only effective analytical tool that the plant operator has is the toxicity bioassay with fish or other aquatic organisms. The toxicity bioassay with aquatic organisms is discussed in some detail in Chap. 14.

Perhaps the best single source of information on the toxicity of the substances that may be present in industrial and other wastes is the splendid compilation by McKee and Wolf(*17*) which was published under the auspices of the Water Pollution Control Agency of the State of California. The toxicities and other biological effects of several hundred compounds are listed in this compilation, but it must be emphasized that this compilation is by no means a critical review. Some compounds listed in the compilation have been studied by competent investigators, and McKee and Wolf were able to analyze and interpret the data so that a biologist or waste treatment engineer can apply the information to his particular problem without any further study. With other compounds, and these are in the great majority, it is advisable that the person concerned go to the original studies quoted by McKee and Wolf and make his own judgment as to whether the work is sufficiently authoritative to warrant the use of the data, either in planning his own waste treatment procedures or as an authority to be used in litigation.

C. Agricultural Wastes

Many conservationists have expressed the belief that the wastes from agriculture pose a greater threat than do industrial wastes. The chlorinated hydrocarbon insecticides and the organic phosphorous insecticides include the most toxic compounds to fish of any known. Prior to World War II, the main types of insecticides were arsenic compounds, lime-sulfur mixtures, petroleum derivatives, nicotine, substances derived from plants, such as pyrethrum and rotenone, and some organic compounds. Although these compounds were known to be toxic to fish (in fact, rotenone is used in fisheries management procedures to kill unwanted species of fish), they were not regarded as major threats to aquatic life (see Chaps. 1, 3, 4, and 23).

1. Chlorinated Hydrocarbon Insecticides

The chlorinated hydrocarbon insecticides, principally DDT, were introduced at the time of World War II. Since that time, DDT and the other chlorinated hydrocarbon insecticides have largely superseded the conventional insecticides because of their toxicity to a wide spectrum of insects, their persistence, their ease of application, and their low cost. Because of these virtues, they have been produced and applied by the thousands of tons for use in agriculture and to control forest pests and arthropod disease vectors.

Concurrently with the beginning of the widespread use of these insecticides in mosquito abatement control, C. M. Tarzwell of the United States Public Health Service became aware of their great toxicity to fish. As the use of chlorinated hydrocarbon insecticides became more widespread [51 million lb/year in the United States in 1964(*40*)] so did the awareness of their great toxicity to fish. But the protests of the conservationists were buried under the many proofs offered by the chemical industry, entomologists, agriculturists, and public health officials of the great benefits to mankind that these compounds provided. It was not until Rachel Carson's book, *Silent Spring*(*41*), was published that sufficient public opinion was aroused to force authorities to look into the detrimental effects of chlorinated hydrocarbon insecticides. Part of this belated willingness to evaluate the detrimental effects of insecticides was no doubt due to the fact that many insect species were developing a tolerance to the chlorinated hydrocarbon insecticides, and to the fact that entomologists and the chemical industry were already searching for and finding other types of insecticides.

The massive fish kills in the Mississippi River which were documented by Mount and Putnicki(*42*) brought to the attention of the general public the great toxicity of the chlorinated hydrocarbon insecticides to fish. Carson's book and the Mississippi River kills have stimulated a large volume of productive research which is still continuing. Typical of these studies is that of Katz(*43*), which indicated the great toxicity of some of the chlorinated hydrocarbon compounds to salmonids. One of these compounds, endrin, is the most toxic of all compounds known to fish. The 96-hr TLm of endrin to coho salmon and to rainbow trout is less than 0.5 ppb. Katz's study also indicated that the salinity of the water in which the fish were tested had little effect on the toxicity of the chlorinated hydrocarbon insecticides. Three-spine stickleback were tested in waters of 5 and 25 parts per thousand salinity and the toxicity was about the same in both waters.

Although the spectacular fish kills recorded by Mount and Putnicki (*42*) and others have implicated the synthetic insecticides among the pollutants, recent studies emphasize that these compounds are perhaps the most important single group of environmental poisons to which fish and other aquatic life are exposed. The Federal Water Pollution Control Administration stream surveillance studies have found DDT, endrin, and dieldrin to be persistent in significant quantities in all of the major watersheds of the country(*43a*). Bailey and Hannum(*44*) found the average concentrations of some chlorinated hydrocarbon insecticides in the San Joaquin basin of California to be as high as 0.231 ppb. Although these substances are usually only found in concentrations of hundredths of a part per billion, these concentrations may have a profound effect upon resident fish populations. Johnson(*45*) subjected the medaka, the Japanese rice fish, to long exposures (up to 30 days) to small concentrations of endrin. At concentrations of about 0.03 ppb, the fish fed well, grew at a normal rate, mated, and produced viable eggs which often hatched. But even at these low concentrations, 70% of the larvae which hatched from these eggs were grossly deformed and died soon after hatching. The concentrations used in these experiments were in the same range as the concentrations regularly found in many of our major rivers.

If the data which were found in the laboratory with medaka can be applied to our native species [and the data reported by Katz(*43*) indicate that many of our fish species have about the same tolerance to insecticides], then we can expect losses of our fish resources in the streams draining many of our agricultural areas. It is probable that in these streams there would not be the typical spectacular fish kills–the populations of fish would gradually disappear because of the failure to produce sufficient young to maintain the population structure.

With the exception of Minimata disease of Japan and the tastes produced in fish flesh by some phenolic wastes, it is believed that most industrial wastes do not make surviving aquatic organisms undesirable to humans. The chlorinated hydrocarbon insecticides may, however, affect the health of humans who eat contaminated fish or oysters. The chlorinated hydrocarbons are readily absorbed by fish from water and these compounds are stored by fish in their fats. Johnson(*45*) observed that the living medaka had concentrations of insecticides in their tissues 20,000–26,000 times the concentrations in the water in which they were swimming. Ferguson et al.(*46*) have found populations of fish in the cotton-growing areas of the Southern states containing over 200 ppm of endrin in their tissues. Some authorities(*46a, 46b*) believe that many of the rural residents of the cotton-growing areas, who regard fish as one of their favorite protein sources, may ingest enough of the insecticide to affect their health.

No one can discount the usefulness of these chlorinated hydrocarbons to our agriculture and to our entire standard of living, but their usefulness is blemished by our recognition of the harmful effect that they are having upon our birds, wildlife, and fish, as well as their yet undetermined effects upon humans. It is obvious that these compounds must be used with greater care than in the past. The chemical industry must search for substances that are more specific than the chlorinated hydrocarbon compounds and must also look for substances that can be decomposed rapidly in the soil and in water.

2. Organic Phosphorous Insecticides

The other major group of insecticides in common use are the organic phosphorous compounds, which include parathion, malathion, guthion, dipterex, coral, and others. These compounds, while very toxic to fish, are significantly less toxic than the chlorinated hydrocarbon insecticides. Weiss *(47)* demonstrated that the organic phosphorous compounds inactivate the acetylcholinesterase of the nervous system of fish. This enzyme inactivation inhibits the transmission of nerve impulses. The fish, after removal from exposure to the insecticide, have the ability to regenerate the enzyme, but the regenerative period may be as long as 40 days. It is obvious that a significant inhibition of the activities of fish will reduce their ability to carry on their life functions. No doubt the ability to collect food will be inhibited and the fish will lack the coordination necessary to evade predators. By the end of a 40-day period, a large part of an exposed population may disappear due to predation or starvation.

In general, the organic phosphorous compounds present a minor problem as compared to the chlorinated hydrocarbon insecticides, not only because of their lesser toxicity but because they decompose rapidly in water into nontoxic compounds.

III. Related Factors

A. Thermal Pollution

The greatest volume of water used by industry is for cooling, but until very recently the discharge by industry of large quantities of warm water was not regarded as a degradation of water quality because the change was physical, not chemical. With the increasing necessity for the reuse of water in our industrial areas, and the changes in the temperature regimen in many of our rivers caused by the construction of reservoirs, the biological effects of the warmed waters have become a matter of acute concern. The present and projected nuclear power generating plants require large amounts of water for cooling, and it is known that the contribution of heat by a large nuclear power complex can be great. It is estimated that the Columbia River, which has a mean flow of 100,000 ft^3/sec, is warmed by the nuclear plants of the Hanford Project in Washington State from 3 to 6°F over ambient temperatures.

Thermal pollution is potentially one of the most critical of all water pollution problems. With the increase in our population, the reuse of our water to an ever-increasing degree is inevitable. Even if the wastes are well treated to remove BOD and toxic components, each reuse undergone by water will usually result in some increase in water temperature. To satisfy our demands for goods and for power, the plants constructed in the future will be ever larger. To preserve the biological resources of our waters, it is obvious that in any river system where water is employed several times, cooling will be one of the waste treatment necessities.

1. General Effects on Fish and Aquatic Organisms

Because physiological processes of aquatic organisms are regulated closely by the temperature of their medium, physiologists have made many studies of the effects of temperature upon the physiology of fish and other aquatic organisms. There is a great background of knowledge available which can be applied toward the understanding of the undesirable effects of thermal pollution in the aquatic environment (*48, 49*).

The current interest regarding the deleterious effects of heated discharges upon aquatic organisms has stimulated a great deal of directed research, and the compilation of some excellent bibliographies has made much of this material quite accessible; see, for example, Kennedy and Mihursky (*50*), Raney and Menzel (*51*), and The Committee on Thermal Pollution, American Society of Civil Engineers (*52*).

Of particular interest, because of the impending necessity to dispose of heated effluents into the estuarine and marine environment, particularly by the many proposed nuclear power plants, is the review by Naylor (*53*) regarding the effects of heated effluents upon the aquatic organisms in these areas.

Fish, in common with most living organisms, can tolerate a relatively narrow range of temperatures. The upper and lower lethal temperatures have been determined in the laboratory for some species. The lethal temperatures for some common species of fish are listed by Altman and Dittmer (*49*) and Brett (*54*).

It is known that water temperature is of prime importance in determining the distribution of fish species. For example, even those with little biological training know that trout and pike are cold water species and that the sunfish and bass are warm water species. The distribution of fish species is reflected by their upper lethal temperatures; those with low upper lethal temperatures are usually the cold water species and those with high upper lethal temperatures are the warm water fish.

The upper lethal limits for salmonids are between 23.7 and 25.9°C (*54*), and for bass and catfish about 36.4°C. Hence, it is obvious that if a trout stream is subjected to thermal pollution episodes that exceed the upper lethal temperature limits for trout, this stream which was once characterized by a salmonid population will be dominated by bass and catfish or by other warm water species. Ingram and Towne(*55*) cite instances in which the temperature of the water in the Ohio River below Youngstown rises to over 110°F. They also observed water temperatures in the Mahoning River as high as 130°F. This is above the lethal temperatures of most American species of fish. The temperatures observed by Ingram and Towne were recorded in the late summer, after most of the young fish were well developed. However, there is a very good chance that high temperatures well below the lethal temperatures for well-developed adult fish might have deleterious, if not lethal, effects upon fish eggs and very young fish, if these temperatures prevail during the spawning and egg development periods.

For freshwater organisms in general, the normal population structure is maintained to about 32°C, and above that temperature the population loses many of its typical organisms. Cairns(*56*) felt that to maintain normal fish populations in most parts of the United States, stream temperatures should not exceed 30°C for long periods.

The effects of thermal addition are not restricted to fish. Increase in water temperature has a marked and measurable effect upon the other organisms in the stream, including the aquatic bacteria, the phytoplankton and zooplankton, and the macroinvertebrates. Changes in the population composition of the flora and the invertebrates which form the food chain of fish will ultimately adversely affect the desirable fish populations even if the temperature does not approach the lethal point for the important species of fish.

Fish, as well as other aquatic organisms, depend upon the dissolved oxygen content for survival. In water the saturation solubility for oxygen is 9.1 ppm at 20°C and 7.5 ppm at 25°C. Thus, at warmer temperatures less oxygen is available for the activities of fish, yet the dissolved oxygen requirement for fish is greater at increased temperatures. In addition, it is not unusual for thermal power plants to be located on rivers that are receiving other wastes which may contain significant amounts of putrescible organic material. The increased temperature will increase bacterial activity, which will result in an accelerated utilization of the already depressed oxygen content in the water.

2. Effects on Aquatic Plants

In a stream with a normal, mixed population of algae, it was observed by Cairns(*57*) that diatoms are dominant at 20°C, green algae at 30°C, and blue-green algae at 35–40°C. An increase in water temperature will shift the algal population from species that do not present any unusual problems to man, to species of blue-green algae. Some blue-green algae species have been implicated in causing undesirable tastes in domestic water supplies, while others are known to be toxic to domestic animals. Trembley(*58*) observed that in Martin's Creek, a stream receiving a heated effluent, the blue-green algae increased in abundance in the heated area while the green algal species sharply decreased in number.

The rooted aquatic plants are often encouraged by temperature increases. Prolonged increases in water temperatures might encourage the spread of nuisance aquatic plants like the water hyacinths which could interfere with stream flow and navigation, as they have in many areas of the southern states. Trembley(*58*) observed that the rooted plants *Elodea* and *Potamogeton* have acclimated to the increased water temperatures. In Martin's Creek, the weed beds were greatly reduced near the effluent, but they have reestablished themselves in most areas in which they were normally found before the introduction of the effluent.

3. Effects on Macroinvertebrates

Trembley(*58*) observed that in Martin's Creek the macroinvertebrates as a group were most affected by water temperature increases. There is fierce competition for food and space among the many organisms that compose the macroinvertebrate fauna of a stream. Trembley observed that species acclimated to low temperatures can be eliminated by a slight temperature increase which warms the water to a temperature optimal for other species. Although the increased temperature is not lethal to the first species, the second species can eliminate it by taking over its living space.

Coutant(*59*) observed a drastic reduction in the numbers, species diversity, and biomass of the macroinvertebrate riffle fauna in the 1600 ft of river below the discharge point of the thermal wastes at Martin's Creek. The temperature increment was about 20–25°F. Coutant observed, however, that some insect larvae and a snail species seemed to tolerate the heated effluent. However, even if fish were not affected adversely by the increased water temperatures, they would leave the area because they could not find the food organisms to which they were accustomed. Thus, these emigrant fish would soon perish after leaving the area with increased water temperatures, because other areas were already supporting the maximum biomass of fish and would not provide sufficient food for the added population.

A final point is the effect of higher winter temperatures of the water upon the life cycle of fish and other organisms. Completion of the life cycle of some organisms is dependent upon cold water temperatures during the winter. The eggs of *Daphnia*, an inportant fish food organism, must be chilled before they will hatch. Seasonal water temperature fluctuations are important to aquatic insect cycles as well as to zooplankton succession and reproduction. Disruption or dislocation of the timing of the life cycles can destroy macroinvertebrate populations upon which desirable fish species must feed.

4. Effects on Fish Reproduction

The reproduction of many species of fish is temperature dependent. Alewives first enter their home stream when the temperature at the stream mouth reaches 50°F(*60*). Artificially warmed waters may induce the fish to move into the river before other conditions suitable for spawning prevail. Brett(*54*) cites several workers who observed that increases in water temperature caused several species of fish to become sexually mature. A temperature of 17°C was found sufficient to induce the completion of spermatogenesis in the stickleback, *Gasterosteus*. Some species, e.g., the pike (*Esox lucius*) and the mosquito fish (*Gambusia*), will spawn only on rising temperatures.

Increase in water temperature may induce some species to spawn prematurely. Even if the eggs develop normally, the fry will emerge from the eggs much earlier than normal. The young fish may migrate to their usual nursery areas to find that the organisms that they normally feed upon are not yet present. These young fish will then starve. Bass spawn in the spring at about 60°F. If thermal wastes induce the bass to spawn early, normal food of the fry may not be available. If the waste comes from a plant which may shut down for brief periods of time, the adult fish may desert the nest when the water cools, leaving the fry exposed to predation and disease.

5. Fish Diseases

Fish, in common with all other organisms, are afflicted with a wide variety of bacterial and viral diseases. Although epidemics are usually problems of fish cultural establishments where crowded conditions and inadequate diets favor the transmission of diseases, epizootics have been reported in nature. Brett(*54*) reported the observations of F. F. Fish, who noted that during the summer of 1941 the temperature of the Columbia River at Bonneville averaged 68.5°F. On July 20, 1941, the Columbia reached an all-time high temperature of 74.5°F. This is very close to the thermal lethal level. The elevated temperature favored the activity of the myxobacterium, *Chrondrococcus columnaris*, which was particularly lethal to the sockeye salmon found in the river at that time. The high temperatures along with the *columnaris* disease caused an almost complete loss of the 1941 run of sockeye in the Columbia River.

The continued decline of the Columbia River from its rank as the leading salmon river of North America is due in large part to the warming of its waters. A series of hydroelectric dams has changed this river from a cool, deep, fast-flowing river to a system of warm lakes. Although the fish have had serious and as yet unresolved problems surmounting the dams on their way to their spawning grounds, and returning to the ocean as fry, the increased water temperatures and the resulting endemic diseases are the final blows which have reduced the Columbia from the most important salmon-producing river system of North America (if not the world) to a minor salmon producer.

6. Effect on the Toxicity of Substances

There are many records (*13, 19*) in the literature detailing the increase in the toxicity of waste substances in waters of high temperature. This can be readily explained by the fact that the basic metabolism of the fish increases as the water temperature increases. Increased amounts of dissolved oxygen are required by the fish to sustain its metabolism, and the fish must increase the amount of water that passes over its gills in order to secure this oxygen. Thus, the fish exposes its gills to an increased amount of whatever substances may be present in the water, including toxicants. In addition, warmer water contains less dissolved oxygen than cooler water, and the fish must further increase its respiratory rate to compensate for the reduced dissolved oxygen, thus again increasing its exposure to toxicants.

7. Thermal Pollution in the Marine and Estuarine Environment

The electric power industry is by far the largest user of cooling water, and the demands for power will require a doubling of facilities each 15 years (*54*). The rivers will be unable to supply sufficient amounts of cooling water, and thermal and nuclear power plants will have to be located on estuaries to obtain sufficient cooling capacity. Nuclear power plants generate about twice the amount of heat generated by the conventional power plant. Naylor (*53*) cites a Scottish Marine Biological Association report which states that the Hunterston Nuclear Generating Station in Scotland was expected to discharge up to 20 million gal of water hourly at about 10°C above ambient sea temperatures.

Concern for the biological effects of heated effluents in the marine and estuarine environments has made the review by Naylor(*53*) of great value. Naylor observes that in the marine environment, as is the case with freshwater ecosystems, heat death is significant. Heat death is of greater concern in the tropical environment, for many of the tropical species often live at temperatures very near their thermal death points, whereas some arctic forms may live in water as much as 13–16°C below their death temperatures. In addition, the tolerance of different organisms in the same area may vary greatly. Intertidal organisms usually have a greater tolerance to high temperatures the higher up on the shore they are found and the longer they are normally exposed to the air when the tide recedes. The sublittoral organisms are, as a rule, less tolerant than the intertidal organisms.

Naylor also cites workers who have observed that estuarine species may be expected to have a greater range of temperature tolerance than sublittoral marine species. The estuarine species have adapted to the greater temperature fluctuations characteristic of estuaries.

One of the concerns regarding the effects of heated effluents is that the reproductive range of many species is significantly narrower than the temperature ranges for growth and survival. Many organisms that may grow well in heated areas may find their reproduction inhibited. Some species may disappear because of reproductive failure, unless young of the same species are carried in by current from areas in which reproduction is not inhibited.

Naylor(*53*) observed in artificially heated docks that a cold–temperate species, *Carcinus maenas,* was abundant but apparently was not able to breed there. In fact, this species grew and molted throughout the winter months when growth and molting normally cease and when the females carry their eggs. Breeding occurred during unheated periods when the water temperature in the docks dropped near that of the ambient seawater temperatures. Naylor observed in the heated docks the presence of organisms usually found only in the lower latitudes. They were observed to grow well and breed there.

In certain waters, a slight increase in water temperature has increased the abundance and spawning success of certain organisms. Naylor(*53*) cites the increase in density of barnacles and mussels at Southampton, England. In the harbor at Swansea, Wales, heating has increased the depredations of the borer, *Teredo*, and it has been necessary to replace wooden jettys with structures of reinforced concrete. (Naylor presents a table of 20 species typical of subtropical waters which are now found in heated marine and estuarine locales in the British Isles.)

8. Beneficial Effects of Warmed Waters

There are some benefits to be gained from heated effluents. The thermal control of fouling organisms at power stations has been accomplished by periodic discharge of the heated effluents into the intake channels to kill these organisms. In certain stations, periodic discharge of the hot effluent through alternative intake conduits has been sufficient to prevent the establishment of mussels.

Certain fish, e.g., carp, rainbow trout, and salmon, have been known to grow rapidly in an estuarine dock used as a water-cooling facility by a power station, and it is suggested that heated effluents be used in conjunction with proposed marine fish farming operations (*61*). A suggestion has been made to utilize heated effluents to prevent the severe losses which are sometimes caused by very cold winters in clam and oyster beds.

Another bright spot in the picture in regard to the warming of waters is that at least during the winter some of the fish tolerant to warm temperatures tend to congregate in the areas receiving warm effluents, and often splendid fishing can be had during the winter in these vicinities. In addition, some oyster biologists (*62*) in the Western states feel that the judicious location of warmed effluents in estuaries of Puget Sound and other oyster-producing areas may help solve certain problems of the oyster industry in that area. The oyster culture of the West Coast depends upon the Japanese oyster, which does not spawn or set regularly in the cool waters of Puget Sound. The oyster industry is forced to import seed oyster annually from Japan to replenish the oyster beds. Sufficient and judicious warming of oyster waters may allow the local development of sufficient oyster seed to supply the needs of the West Coast oyster industry.

B. Taste and Odor Problems

Pollutants that are introduced into our water supplies can cause severe economic losses to our fisheries resources without killing the fish or mollusks concerned. One of the important losses that can be suffered by a fishery is the tainting by wastes of the flesh of fish or mollusks that live in the receiving waterways. It is common knowledge to most anglers that at certain times of the year, fish in certain lakes and streams have a musty or earthy odor which can be related to the decomposition of various species of algae. The fish that are caught are either discarded by the fishermen or must be processed by certain culinary rituals before they can be prepared for table use. Fish and mollusks that are resident in waters receiving petroleum wastes and wastes containing phenolics, in particular, are tainted to such a degree that they are no longer marketable. Thus, severe economic losses are suffered by fishermen, oyster growers, and resort operators.

Hawkes(*63*) briefly describes the situation in Narragansett Bay, near Providence, Rhode Island, in which there are quahog (*Mercenaria mercenaria*) beds: "They also assure me that each and every one of these quahogs has a taste and odor which would make the strongest stomach turn." Hawkes also states that tainting of European mussel beds with oil has caused serious economic problems. And he quotes the Regional Director of the Fish and Wildlife Service, Atlanta, Georgia, who stated: "Fishermen on the lower St. Mary's and Savannah Rivers in Georgia have reported that the fish caught in these reaches are unfit for human consumption. Pumping of bilges is thought to be responsible for the taste in the lower Savannah River."

Most of the tainting of the flesh of fish and oysters has been correlated with the activities of the petroleum industry(*64–66*) and the chemical industries(*67*).

Mann(*65*) points out that the species of fish most noticeably tainted by petroleum wastes are those with high contents of body fat, such as salmonids and eels. These species are the most valuable of the commercial species in northern Europe. In addition, Mann states, in agreement with Boetius(*67*), that the chlorinated phenols derived from chemical industries can impart a phenolic taste to eels and oysters when the concentration of phenols in the water is as low as 0.001 ppm. The fish and oysters can absorb and store these phenols in their tissues, and the phenolic odor will persist in the fish for as long as a month after disappearance of the pollutant from the water. Bandt(*68*), as part of an extensive study of pollution by phenols, presented a table of several substances which impart tastes to fish and the threshold concentrations at which or above which the tastes are imparted (Table 1).

Mann also stated that not only do fish derive an oily taste from the oils dissolved in water, but they can develop oily flavors from feeding on invertebrates living on the bottom in areas which are polluted by oil. Oils, after being spilled, will often sink to the bottom and the invertebrates, particularly tubificids, become permeated with the oils. Fish feeding on the tubificids develop an oily taste.

Schulze(*69*) conducted a study to determine which tissues in fish had the greatest affinity for phenols. He kept carp in water of 10 ppm of phenols for 5 days and analyzed various tissues for phenols. As Table 2 indicates, the greatest concentrations of phenol were in the liver and gills.

In Japan, Nitta and co-workers(*66*) studied the oily odors in fish in

TABLE 1
Threshold Concentrations of Various Substances that Taint Fish Flesh[a]

Compound	Concentration, mg/liter
p-Quinone	0.5
p-Toluidine	20.0
Pyridine	5.0
Quinoline	0.5–0.1
Naphthalene	1.0
α-Naphthol	0.5
β-Naphthol	1.0
α-Naphthylamine	3.0
Xylenol	1.0–5.0
Pyrocatechol	2.5
Resorcinol	30.0
Pyrogallol	20.0–30.0
Phloroglucinol	10.0

[a]From Ref. *68*, by courtesy of Academia Verlag, Berlin.

Yokkaichi Harbor, adjacent to which there are many petroleum refineries. They determined in laboratory experiments that fish can become tainted in concentrations of oil of 0.01 ppm in water and 0.2% (by ether extraction) in bottom muds. The oily taint was obvious after 20 hr of exposure. It was further observed that fish would not avoid water masses polluted by oil. In fact, in the winter months, especially large

TABLE 2.
Concentration of Phenol in Various Fish Tissues[a]

Tissue	Concentration of phenol, mg/kg of tissue
Muscles	10
Kidneys	12
Testicles	9
Intestine	7
Gills	17
Liver	19

[a]From Ref. *69*. Fish were exposed to 10 mg/liter of phenol for 5 days.

numbers of fish congregated in Yokkaichi Harbor, whose waters were contaminated by oil.

Boetius (*67*) conducted an experimental study on the tainting of fish and oysters from the chlorophenols derived from effluent of a Danish plant making herbicides and insecticides. Analysis of the wastes indicated the presence of paranitrophenol and either chlorophenols or herbicide intermediates. Laboratory experiments indicated that these compounds were responsible for the tainting. Paranitrophenol imparted a taint after 10 days in a 10 ppm solution, while chlorophenols imparted a taste to fish flesh at 0.001 ppm. A similar problem was investigated at the same time in regard to tainting of oysters by chlorophenols in Limfjord. Oysters were found to be tainted by chlorophenols at a concentration of 0.0001 ppm within 4 days at 16°C. Boetius felt that the oysters and the eels concentrated the phenols and stored them in their fatty tissue.

Tainting in spring Chinook salmon has been reported by Shumway and co-workers (*70*) in the Willamette River near Portland, Oregon. The tainting substances were derived from a plant manufacturing herbicides.

IV. Water Quality Standards

The Federal Water Quality Act of 1965, Public Law 89-234, amended to provide for establishment of interstate standards, requires the states to establish water quality standards for their interstate waters. It is of interest to review the standards which apply to fish and fisheries of the State of Washington (*70a*) which are typical and meet the criteria established by the Federal Water Pollution Control Administration.

Standards

No sewage or industrial waste shall be discharged into any of the water of the state that will cause:

(a) The reduction of the dissolved oxygen content to less than five parts per million (5 ppm).
(b) The hydrogen-ion concentration (pH) to be outside of the range of 6.5 to 8.5.
(c) The liberation of dissolved gases such as carbon dioxide, hydrogen sulfide or any other gases in sufficient quantities to be deleterious to fishes or related forms.
(d) The development of fungi or other growth having a deleterious effect on stream bottoms, fishes and related forms, or be injurious to health, recreation or industry.

(e) Toxic conditions that are deleterious to fishes and related forms or affect the potability of drinking water.
(f) The formation of organic or inorganic deposits detrimental to fishes and related forms, or be injurious to health, recreation or industry.
(g) Discoloration, turbidity, scum, oil slick, floating solids, or coat the aquatic life with oily films, or be injurious to health, recreation or industry.
(h) The temperature to be raised above the limit of tolerance of fishes and related forms.

These approved water quality standards, which are typical, are valuable steps in the direction of protecting our water supplies, but they fall far short of the powerful instruments they could and should be. The people who drafted these standards and those who approved these standards were unaware of the depth of knowledge in regard to aquatic ecology that was readily available and could be applied to make these standards precise and powerful documents that would fulfill their stated goals.

The discussion in the following paragraphs illustrates some of these defects.

Standard (a) – Regarding dissolved oxygen is adequate for warm water fish such as catfish, carp, perch. The standard on dissolved oxygen is clearly inadequate to protect salmonid fish which have been shown to require about 7 to 8 ppm DO for them to carry out all of their necessary activities. The definitive experiments on the dissolved oxygen requirements of salmonids and warm water species were conducted by Fry and co-workers of Toronto (*71–73*), and Peter Doudoroff and his associates at Oregon State University.

Standard (b) – The allowed pH range of 6.5 to 8.5 is most likely based upon the experimental data in the Doudoroff and Katz (*15*) review.

Standard (c) – The standard for dissolved gases neglects chlorine and does not give specific amounts of carbon dioxide and hydrogen sulfide. In any case, concentrations of carbon dioxide that would affect fish are so high (*15*) that a standard for it is largely academic. Chlorine, which is present almost universally in sewage treatment plant effluents and which is toxic at the levels used for disinfection, is not specified unfortunately.

Standard (d) – The proscription against discharges causing the development of fungi or other growth probably reflects the administrator's concern over so-called sewage fungus, *Sphaerotilus natans*, which is actually a filamentous bacterium. There is nothing fundamentally bad about this standard except that it may be regarded as specifying tertiary waste treatment in specific instances.

Standard (e) – The requirement that waters be nontoxic to fish or related forms is quite confusing. What is meant by "related forms"? Did those who wrote this requirement mean oysters, clams, and crustaceans and other animal species of interest because of their commercial or recreational importance? The only *biologically* related forms are other vertebrates that live in the aquatic environment such as aquatic birds, lampreys, whales, seals, and perhaps swimmers and skin divers. Until the courts, aided by expert testimony, interpret the phrase "deleterious to fish and related forms," the regulation will have to be administratively interpreted.

Another defect of the regulation is the failure to define toxic conditions and the methods used to determine toxicity. Specific test procedures (toxicity bioassay procedures) are outlined by Hart, Doudoroff, and Greenbank(*74*), the Doudoroff Committee(*75*), ASTM(*76*), and the APHA(*77*).

V. Trends in Research

The pressure on our water resources is going to increase faster than our population. Thus, we will be hard pressed to find ways to maintain the growth our necessary industrial establishment requires, and yet allow us to preserve fishery resources of increasing importance for recreation and food supply. The answers sought should tell us how much of a waste load can be delivered to our waters and yet allow these waters to maintain large and useful fish and shellfish populations.

Research in the future must provide precise data on the limits of tolerance of sensitive species and of the most sensitive stages in the life cycle of species of greatest interest. On the other hand, such sensitive species and such life cycle stages (eggs, larvae, etc.) cannot be expected to be amenable to culture in the laboratory; thus, an attractive alternative to the direct use of sensitive species would be the use of species and/or life-cycle stages which are amenable to easy handling in the laboratory coupled with the imposition of a safety factor or margin of safety in the standards set. Ideally, the safety factor should be based on biological principles, e.g., specific toxicity mechanisms, or on a body of correlation data relating bioassay results and actual response of natural populations. In actual practice, the safety factor(s) may have to be somewhat arbitrary.

For a direct assay of the long-term effects of a pollutant, toxicity bioassays must be conducted using fertilized fish eggs, carried through maturity and spawned to produce the second generation. It is of little use to determine the concentration of waste in which tolerant adult fish can survive when it is well known that newly hatched young are killed at lower concentrations. Research by many workers strongly suggests that studies with freshly hatched larvae are perhaps most instructive in determining the acute toxicity of substances.

At the same time, studies should be carried on to determine the effects of wastes in the food chain. It may often be the case that the fish in all of its stages may be more tolerant than the phytoplankton of zooplankton organisms which are the fishes' food. It is thus important to evaluate the effects of the wastes under study on the microscopic food chain organisms.

It is anticipated, of course, that research in industrial waste treatment and its allied disciplines of water chemistry and waste product recovery will be carried on by the appropriate groups. Let no one delude themself, this will be a costly research effort. One cost of ignorance will either be paid in deteriorating water quality if standards are set too liberally, or they will be paid in direct costs of producing goods if standards are set too conservatively. Only good research can spare these costs.

REFERENCES

1. A. R. Gaufin and C. M. Tarzwell, *Public Health Repts.* U.S., **67**, 57 (1952).
2. M. Katz and A. R. Gaufin, *Trans. Am. Fisheries Soc.*, **82**, 156 (1953).
3. M. Katz and W. C. Howard, *Trans. Am. Fisheries Soc.*, **84**, 228 (1955).
4. H. B. N. Hynes, *The Biology of Polluted Waters*, Liverpool Univ. Press, Liverpool, England, 1963.
5. A. F. Bartsch and W. M. Ingram, *Public Works*, **90**, 104 (1959).
6. T. S. English, *J. Water Pollution Control Federation*, **39**, 1337 (1967).
7. W. Van Horn, J. B. Anderson, and M. Katz, *Trans. Am. Fisheries Soc.*, **79**, 55 (1949).
8. D. W. M. Herbert, D. H. M. Jordan, and R. Lloyd, *Inst. Sewage Purif. J. Proc.*, Part 6, 1 (1965).
9. R. Lloyd, *Ann. Appl. Biol.*, **48**, 84 (1960).
10. R. Lloyd, *Water Waste Treat. J.*, **8**, 278 (1961).
11. R. Lloyd and D. W. M. Herbert, *J. Inst. Public Health Engrs.*, July 1962, 132 (1962).
12. R. Lloyd, *J. Exptl. Biol.*, **38**, 447 (1961).
13. J. R. E. Jones, *Fish and River Pollution*, Butterworth, Washington, D.D., 1964.
14. M. M. Ellis, *Bull. U.S. Fish. Bur.*, **48**, No. 22, 365 (1937).
15. P. Doudoroff and M. Katz, *Sewage Ind. Wastes*, **22**, 1432 (1950).
16. P. Doudoroff and M. Katz, *Sewage Ind. Wastes*, **25**, 802 (1953).
17. J. E. McKee and H. W. Wolf, *Water Quality Criteria*, California State Water Quality Control Board, Publ. No. 3-A, 1963.
18. P. Doudoroff, in *The Physiology of Fishes, Vol. 2, Behavior* (M. E. Brown, ed.), Academic, New York, 1957, p. 403.
19. M. Katz and G. G. Chadwick, *Trans. Am. Fisheries Soc.*, **9**, 394 (1961).
20. K. Wuhrmann, F. Zehender, and H. Woker, *Viertelsjahresschr. Naturforsch. Ges. Zuerich*, **92**, 198 (1947).
21. F. J. Brinley, *Biol. Bull.*, **53**, 365 (1927).
22. K. Wuhrmann and H. Woker, *Schweiz. Z. Hydrol.*, **11**, 210 (1948).

23. P. Doudoroff, *Sewage Ind. Wastes,* **28**, 1020 (1956).
24. P. Doudoroff, G. LeDuc, and C. R. Schneider, *Trans. Am. Fisheries Soc.,* **95**, 6 (1966).
25. G. E. Burdick and M. Lipschuetz, *Trans. Am. Fisheries Soc.,* **78**, 192 (1948).
26. D. W. M. Herbert and A. C. Wakeford, *J. Air Water Pollution,* **8**, 251 (1964).
27. H. J. Bandt, *Beitr. Wasser-, Abwasser Fischerei Chem.,* No. 1, 15 (1946).
28. P. Doudoroff, *Proc. 4th Pacific Northwest Ind. Waste Conf.,* State College of Washington, Pullman, 1952, p. 21.
29. R. Lloyd, Biological Problems in Water Pollution, 3rd Seminar, 1962, *PHS Bull. No. 999-WP-25,* 1965, p. 181.
30. D. W. M. Herbert and D. S. Shurben, *Ann. Appl. Biol.,* **53**, 331 (1964).
31. D. W. M. Herbert and J. M. Van Dyke, *Ann. Appl. Biol.,* **53**, 415 (1964).
32. R. Lloyd and D. H. M. Jordan, *Inst. Sewage Purification, J. Proc.,* Part 2, 167 (1963).
33. R. Lloyd and D. H. M. Jordan, *Inst. Sewage Purific., J. Proc.,* Part 2, 183 (1964).
34. K. E. Carpenter, *J. Exptl. Biol.,* **4**, 378 (1927).
35. B. A. Westfall, *Ecology,* **26**, 283 (1945).
36. G. Schweiger, *Arch. Fischereiwiss.,* **8**, 54 (1957).
37. J. F. Skidmore. *Quart. Rev. Biol.,* **39**, 227 (1964).
38. D. I. Mount, *Trans. Am. Fisheries Soc.,* **93**, 174 (1964).
39. D. L. Mount and C. E. Stephan, *J. Wildlife Management,* **31**, 168 (1967).
40. H. H. Shephard and J. N. Mahan, *Chem. Eng. News,* **43**, 108 (1965).
41. R. Carson, *Silent Spring,* Houghton Mifflin, Boston, 1960.
42. D. E. Mount and G. J. Putnicki, *Trans. 31st North American Wildlife and Nat. Resources Conf., March, 1966,* Wildlife Management Inst., Washington, D.C., 1966, p. 177.
43. M. Katz, *Trans. Am. Fisheries Soc.,* **90**, 264 (1961).
43a. L. Weaver, C. C. Gunnerson, A. W. Breidenbach, and J. J. Lichtenberg, *Public Health Rept. (U.S.),* **80**, 481 (1965).
44. T. E. Bailey and J. R. Hannum, *J. Sanit. Eng. Div., Proc. Am. Soc. Civil Engrs.,* **93**(SA5), Paper 5510, 27 (1967).
45. H. E. Johnson, Ph.D. Thesis, Univ. of Washington, 1967.
46. D. E. Ferguson, J. L. Ludke, and G. G. Murphy, *Trans. Am. Fisheries Soc.,* **95**, 335 (1966).
46a. P. Rosato and D. E. Ferguson, *Bioscience,* **18**, 783 (1968).
46b. J. L. Ludke, D. E. Ferguson, and W. D. Burke, *Trans. Am. Fisheries Soc.,* **97**, 260 (1968).
47. C. M. Weiss, *Sewage Ind. Wastes,* **31**, 580 (1959).
48. J. R. Brett, *J. Fish. Res. Board Can.,* **9**, 265 (1952).
49. P. L. Altman and D. S. Dittmer, eds., *Environmental Biology,* Federation of American Society for Experimental Biology, Bethesda, Md., 1966, Sec. 20, Parts II, III; Sec. 23, Parts IV, V; Sec. 24, Part II.
50. V. S. Kennedy and J. A. Mihursky, *Natural Resources Institute Contr., Univ. of Maryland,* 1967, p. 326.
51. E. C. Raney and B. W. Menzel, *Philadelphia Elec. Co. and Ichthylo. Bull. No. 1,* 1967.
52. Committee on Thermal Pollution, *J. Sanit. Eng. Div., Proc. Am. Soc. Civil Engrs.,* **93**(SA3), 85 (1967).
53. E. Naylor, *Advan. Marine Biol.,* **3**, 63 (1965).
54. J. R. Brett, *Quart. Rev. Biol.,* **31**, 75 (1956).
55. W. M. Ingram and W. W. Towne, *Eng. Bull. Purdue Univ.,* **44**, 678 (1960).
56. J. Cairns, *Ind. Wastes,* **1**, 180 (1956).
57. *Ibid.,* **1**, 150 (1956).

58. F. J. Trembley, Biological Problems in Water Pollution, 3rd Seminar, 1962, *PHS Bull. No. 999-WP-25,* Public Health Service, 1965, p. 334.
59. C. C. Coutant, *Proc. Penna. Acad. Sci.,* **36**, 58 (1962).
60. D. L. Belding, *Trans. Am. Fisheries Soc.,* **58**, 98 (1928).
61. J. E. Shelbourne, *Advan. Marine Biol.,* **2**, 1 (1964).
62. C. E. Woelke, personal communication, 1968.
63. A. L. Hawkes, *Trans. 26th North American Wildlife and Nat. Resources Conf.,* Wildlife Management Inst., Washington, D.C., 1961, p. 343.
64. H. Reichenbach-Klinke, *Muench. Beitr. Abwasser-, Fischerei-, Flussbiol.,* **9**, 73 (1962).
65. H. Mann, *Symp. de Monaco, April, 1964, Comm. Intern. Exploration Sci. de la Mer Mediterrané,* 1965, p. 371.
66. T. K. Nitta, K. Arakawa, T. Okubo, and K. Tabata, *Bull. Tokai Reg. Fisheries Res. Lab., Tokyo,* **42**, 23 (1965).
67. J. Boetius, *Medd. Denmark Fiskeri-og Havunder Søgelse n.s.l,* 1 (1954).
68. H. J. Bandt, *Phenolabwässer und Abwasser phenole, ihre Enstehung Schadwirkung und Abwasser technische Behandlung-Eine Monographische Studie, Wiss. Abhandl.,* Akademie Verlag, Berlin, (1958).
69. E. Schulze, *Abwasser. Intern. Rev. Ges. Hydrobiol.,* **46**, 84 (1961).
70. D. Shumway, personal communication, 1968.
70a. M. M. Ellis, *U.S. Bur. Fisheries Bull. No 22,* **48**, 365 (1937).
71. F. E. J. Fry, *Publ. Ontario Fisheries Res. Lab.,* **55**, 62 (1947).
72. F. E. J. Fry, in *Physiology of Fishes* (M. E. Brown, ed.), Vol. I, Academic, New York, 1957, pp. 1–63.
73. E. S. Gibson and F. E. J. Fry, *Can. J. Zool.,* **32**, 252 (1954).
74. W. B. Hart, P. Doudoroff, and J. Greenbank, Waste Control Lab., Atlantic Refining Co., Philadelphia, 1945.
75. P. Doudoroff and Committee, *Sewage Ind. Wastes,* **23**, 1380 (1951).
76. R. F. Weston, *ASTM Designatim D-1345-57,* adopted 1959, copyrighted supplement to book of ASTM standard. American Society for Testing Materials, Philadelphia, 1959, Part 10, p. 348.
77. *Standard Methods for the Examination of Water and Waste Water,* Part VI, 11th ed., Am. Public Health Assoc., New York, 1960.

Chapter 7 Chemical, Physical, and Biological Characteristics of Wastes and Waste Effluents

H. A. Painter
WATER POLLUTION RESEARCH LABORATORY
STEVENAGE, ENGLAND

I. Introduction

Knowledge of the composition of sewage and of trade waste waters permits a better assessment of which method of treatment (see Chaps. 8 and 9) should be applied before discharge to a river or stream. In particular, such knowledge will help to determine whether or not a trade waste would attack the sewer, whether it should be treated alone or in admixture with sewage—and in what proportions—and whether a single-stage or multistage process should be adopted. Further, a thorough survey of the processes from which the various components of a trade waste arise, together with their composition, often leads to recovery of materials and saving of water. An analysis of treated waste waters is necessary to assess potential toxicity and disease hazard to man, toxicity to fish and other river life, and trophic properties for algae, "sewage fungus," etc. This knowledge is also desirable, especially in light of the growing need for reuse and conservation of water, to help decide how to treat the effluents further for use as low grade water for industrial purposes and as potable water.

More fundamentally, however, it is hoped that a detailed knowledge of sewage composition before and after treatment will help unravel the biochemical and microbiological mechanisms involved in the treatment processes and that this in turn will lead to greater efficiency of these processes and to new processes.

The main source of pollution in sewage is human excreta, with smaller contributions from food preparation, washing, laundry, surface drainage, etc.; trade wastes, in general, consist of one or more strong spent liquors from the main industrial process together with comparatively weak waters from rinsing, washing, condensing, floor-washing, etc. Analyses of the various gross components of sewage are not available in enough detail, except perhaps for urine, to enable predictions of composition to be made with certainty, and therefore analyses of the sewage itself has to be made. Most investigators have studied specific substances in sewage for particular purposes, e.g., indole for odor, detergents for foaming, or pesticides for health hazards, and thus some data have been accumulated, but very few comprehensive analyses have so far been attempted.

II. Methods

A. Sampling

The value of analytical work can be considerably reduced if the sampling techniques and programs adopted do not adequately take account of the nature of the waste water and the wide, often rapid, fluctuations in flow, strength, and composition which occur in waste waters and effluents (see Sec. III. B). In purely domestic systems the variations follow a fairly regular pattern, but where trade wastes are discharged to the sewers other fluctuations will be observed. Continuous industrial processes often give rise to less variation in the waste water produced than occurs in sewage, while factories operating batch processes, or in which much washing water is used, often give rise to violent fluctuations in the volume and composition of waste. Treatment of wastes by the biological processes reduces the fluctuations, and there is usually little systematic variation in the composition of the effluent (see Sec. V. A).

Apparatus used for sampling range from vessels hand-operated such as beakers and buckets to sophisticated automatic devices which take constant volume samples at known time intervals or which take volumes proportional to the flow of waste water (see Chap. 10). The type of sampler must be suitable for the waste being sampled; for example, any tubes through which the liquid has to pass should be wide enough to prevent choking by suspended solids. Sometimes it is necessary to analyze the subsamples; in other cases it is important to know the total pollutional load on a treatment plant or discharged to a river, and for this purpose a composite sample is made by mixing the subsamples in amounts proportional to the flow of liquid at the time of sampling. The time interval between subsamples is commonly 1 hr for sewage, but for trade wastes more frequent samples may be required; to help decide the frequency of sampling, a knowledge of the manufacturing processes involved is invaluable. In view of postsampling changes (see Sec. II.B), it may be necessary to make provision for preservation of subsamples by refrigeration or addition of a bacterial inhibitor.

The collection of samples for bacterial examination requires sterile conditions (*1*), especially where biological effluents are involved. The qualitative estimation of viruses is usually made by using the Moore "swab" technique (*2*), in which pads of sterile cotton gauze are immersed in the flowing liquid for a number of hours before being withdrawn for examination.

B. Preservation of Samples

Because of changes which take place on standing, it is desirable that analyses be made immediately after collection. This is not usually possible, but it is essential that for such determinations as dissolved gases, volatile substances, and bacterial numbers, the analyses be made within a short time of collection. Of the changes which can occur when bacteria are present the more important are the absorption of dissolved oxygen, growth and death of bacteria, hydrolysis of urea to ammonia, oxidation of ammonia to nitrite and nitrate, reduction of nitrate to gaseous nitrogen, disappearance of sugars, and formation of volatile fatty acids. Urea was converted to ammonia at about 3 mg N/liter per hour at 12°C (*3*), dissolved oxygen was absorbed at about 3 mg/liter per hour at 18°C(*3*), and sugars were removed at about 4 mg/liter per hour at 20°C(*4*).

Changes in the bacterial count on storage vary widely from sample to sample, probably due to unknown factors rather than to incompetent operators(*5*). At 20°C the "total count" decreased on the average to 86% in 7 hr but rose to 110% of the original in 24 hr; the number of coli-aerogenes, however, rose to 342% in 7 hr and to 420% of the original number in 24 hr. At 6°C the total count fell to 58% in 7 hr and rose to 130% in 24 hr, the coli-aerogenes falling to 69% and rising to 216% in 7 and 24 hr, respectively(*6*). From similar results at 1, 4, and 7 C it was concluded that no advantage is gained in accuracy or conservation by storage at less than 7°C(*5*); indeed, storage in the frozen state at −15°C removed 97% of the total count in one day(*7*). To avoid this uncertainty, a membrane filtration technique may be used in the field when samples have to be sent some distance to a laboratory(*8*).

The effect of storage on the five-day biochemical oxygen demand (BOD) is consistently to lower the value obtained, but there is a wide scatter in the proportional reduction. In completely full bottles the reduction in 6 hr was about 5% at 0–3°C, 16% at 7–10°C, 12% at 20°C (*9*), and about 10% at 10 and 20°C(*10*). Beyond 6 hr no further decrease occurred in full bottles, but in open beakers at 20°C the BOD was lowered by 33% in 48 hr(*9*) and in half-full bottles by about 40% in 48 hr at 37°C(*10*). In contrast, it has been reported that storage at 4°C for longer than one day gives significantly lower values(*11*). However, storage at −5 to −15°C in the frozen state is reported to give virtually unchanged values, even after six months, for BOD, chemical oxygen demand (COD), total solids (TS), ammonia, nitrite, nitrate, and organic nitrogen (*7, 11–14*). Since the bacterial population was seriously depleted, it was necessary to reseed the thawed samples when determining BOD(*12*). The value for suspended solids (SS) initially increased on freezing but did not change further when the sample was stored for some days at 4°C(*11*). The pH value of sewage increased by 0.5–1 unit almost immediately on freezing but thereafter remained unchanged(*7*). The BOD, COD, and SS of raw sewage stored in half-full bottles at 1°C were also unchanged, within experimental error, for at least 6 days(*10*).

Other methods used to preserve samples do not allow the BOD or bacterial count to be estimated but are useful for determination of chemical and physical parameters. The addition of sulfuric acid kept the SS constant for eight days and the COD constant for at least 17 days (*15*), while the addition of the antibiotic polymyxin B, after heating the sample to 80°C for 1 hr, gave constant COD values for at least six months on storage at 22–27°C(*16*). Mercuric chloride, added at 3–50 mg/liter depending on the concentration of organic matter in the sample(*17*), preserved samples for at least two weeks for the determination of pH value, COD, TS, volatile solids (VS), ammonia, nitrite, nitrate, and organic nitrogen(*18*) but interfered in the determination of phosphate and phenol(*19*). It is undesirable to use mercuric chloride in gas–liquid chromatography methods and in the methods for organic carbon involving catalytic gaseous oxidation. Agents such as chloroform, Formalin, thymol, and potassium cyanide were ineffective(*18*).

Thus, the method chosen will depend on the situation, and indeed, two or more methods of preservation may have to be used for a single series of samples. Whatever method is chosen, the samples should be stored in the dark in full bottles.

C. *Separation and Concentration*

Some determinations can be carried out directly on the untreated sample, but because of the very low concentration of many constituents, and also sometimes for convenience, many samples are processed to separate or concentrate the constituent or to remove interfering compounds, before analyses are made. Methods of separation and concentration are given in detail in Chap. 11.

The constituents in suspended solids are best estimated on the dried solids, which are conveniently prepared by lyophilization (freeze drying) of the various solids fractions obtained by successive settlement, differential centrifugation, and ultrafiltration by membrane or Pasteur candle. For most constituents in solution, lyophilization is, again, a useful general concentrating process which prevents loss of heat-labile substances. The method has been used successfully with domestic sewage to yield fine, buff- or tan-colored powders(*20, 21*) and is especially useful when collecting composite samples over a long period(*20*). Recovery by this method of the total solids in solution was 73%(*21*), the highest proportional loss being of volatile acids and ammonia; the recovery of the total solids in raw sewage was 87–91% and losses were considered to be largely manipulative and nonselective(*22*). Sewage and biological effluents have also been concentrated at about 54°C under reduced pressure in rotary and cyclone evaporators designed to minimize loss of heat-labile substances(*4, 22, 23*). Losses in the distillate from effluents were low, 3–10%(*4, 23*), but precipitation of solids occurred in the concentrate and about 30% of the organic carbon was removed with the precipitate(*4*).

Since volatile compounds were preferentially lost by the evaporative methods, special means must be used to determine such compounds in the original sample. Volatile acids and bases have been separated by appropriate steam distillation, while some of the neutral volatile compounds have been extracted by solvents (*21*). Solvent extraction is also extensively used to separate fats, greases, and detergents from solids using petroleum ether, chloroform, or alkaline methanol; to separate nonvolatile acids and pesticides from solution; and to separate carbohydrates of differing degrees of complexity by successive extraction with aqueous ethanol, perchloric acid, 0.25 *N* HCl, and 72% H_2SO_4 (*21*). Examples of methods for removing interfering substances prior to analysis are the use of ion exchange resins to remove salts and amino acids before separating individual sugars by paper chromatography and the use of electrodialysis to eliminate nitrite and nitrate from concentrated effluents prior to the determination of sugars by the anthrone method.

D. Analytical Procedures

The methods used to determine such parameters as BOD, COD, SS, etc., are given in Chaps. 15 and 19 and in standard texts [e.g., (*24,25*)]. Hitherto most analyses for organic constituents have been made by conventional titrimetric and colorimetric procedures and, after suitable concentration and separation, by long-established chromatographic methods. Examples are amino acids by reaction with ninhydrin, and higher fatty acids by reversed phase column chromatography (see Chap. 11). More use is now being made of gas-liquid chromatography of aqueous solutions (see Chap. 28), e.g., for volatile fatty acids, and also of infrared spectroscopy (see Chap. 30), e.g., for degradation products of ABS† and for poly-β-hydroxybutyric acid in sewage microorganisms. The "wet" combustion method for organic carbon is being replaced by high temperature catalytic oxidation and determination of the carbon dioxide formed by a nondespersive infrared method (*25a*). Bacterial numbers have been determined by most probable numbers (MPN) or plate counts (see Chap. 13), and protozoa have been counted by direct microscopic observation.

III. Characteristics of Sewage

A. Physical Properties

Fresh sewage is normally turbid and appears gray to yellow-brown according to the time of day collected; if trade wastes are discharged

†Alkyl benzene sulfonates.

into the sewer the sewage sometimes takes on the color of the waste. When viewed in ultraviolet light, a colored flourescence is often seen which is probably due to minor constituents of packaged detergents. Sewage when fresh has a musty, but not offensive, odor; on staling, however, putrefaction sets in and objectionable odors are produced. Occasionally the odor of a trade waste is evident.

The temperature of sewage is normally a degree or two above that of the water supply; in winter in moderate climates the temperature range is 8–12°C and in summer 17–20°C(*6*, *26*, *27*). When hot discharges are made to the sewer, higher temperatures are observed, for example, as high as 30°C caused by a carpet factory effluent(*28*), and similarly, infiltration of storm or surface waters can be expected to cause decreases in temperature. In other climates different average annual temperatures prevail, e.g., in India values as high as 28–30°C(*29*) and in Bulgaria as low as 12°C(*30*) have been recorded. The pH value of sewage is usually above 7, the actual value depending on the degree of hardness of the water supply. In a soft water area in the United States the modal pH value was 7.2, range 6.7–7.5(*20*), while in a hard water area in the United Kingdom the modal value was 7.8, range 7.6–8.2(*31*).

Solids suspended in sewage range from colloids up to recognizable gross matter. In electron microscopic studies the sizes of particles in fractions produced by successive sedimentation, centrifugation, and filtration through membranes have been determined(*32*, *33*). Settleable solids, comprising about 50% of the suspended solids, were greater than 100 μm in diameter; the supracolloidal fraction, 30–37% of total suspended solids, contained particles in the range 1–100 μm; and colloidal solids, 17–20% of total suspended solids, consisted of particles in the range 1 nm to 1 μm(*20*,*33*). At pH values of 7 and above, the colloidal particles were negatively charged and had electromobilities of -0.55 to -3.75 μm/sec per V/cm, with an average of -1.73(*34*). Raw sewage also contains "floatable" solids, important when sewage is discharged directly into a river or the sea. Concentrations of 5–60 mg/liter have been recorded; only 1–2 mg of "floatables" per liter were present in settled sewage(*35*).

B. Variations in Composition

Changes occur in the flow, strength, and composition of sewage hourly, daily, and seasonally; of these, hourly changes are usually greatest. Variations in flow are normally larger, the smaller the community served; the hourly variation is usually 50–200% of the average(*36*) and can be as wide as 20–300%(*3*). The strength and composition also vary considerably during a day, and a fairly regular pattern is followed similar to that shown in Fig. 1 for a compact residential area in the United Kingdom(*3*). In a comparable, but larger, area in Los Angeles the flow pattern was similar and the BOD reached a peak of 300 mg/liter at about the same time and a minimum of 50 mg/liter at 7 a.m., but the strength was relatively higher than that of the U.K. sewage between midnight and 6 a.m.(*37*). The time at which peak values occur depends on the length of the sewer and the nature of the area served; for example, samples collected from the very long sewer of a large city in the United Kingdom(*38*) between 10 p.m. and 6 a.m. were consistently stronger than those collected during the other two-thirds of the day and there was less difference between highest and lowest values. The concentration of ammonia plus urea in the compact British area(*3*) showed two

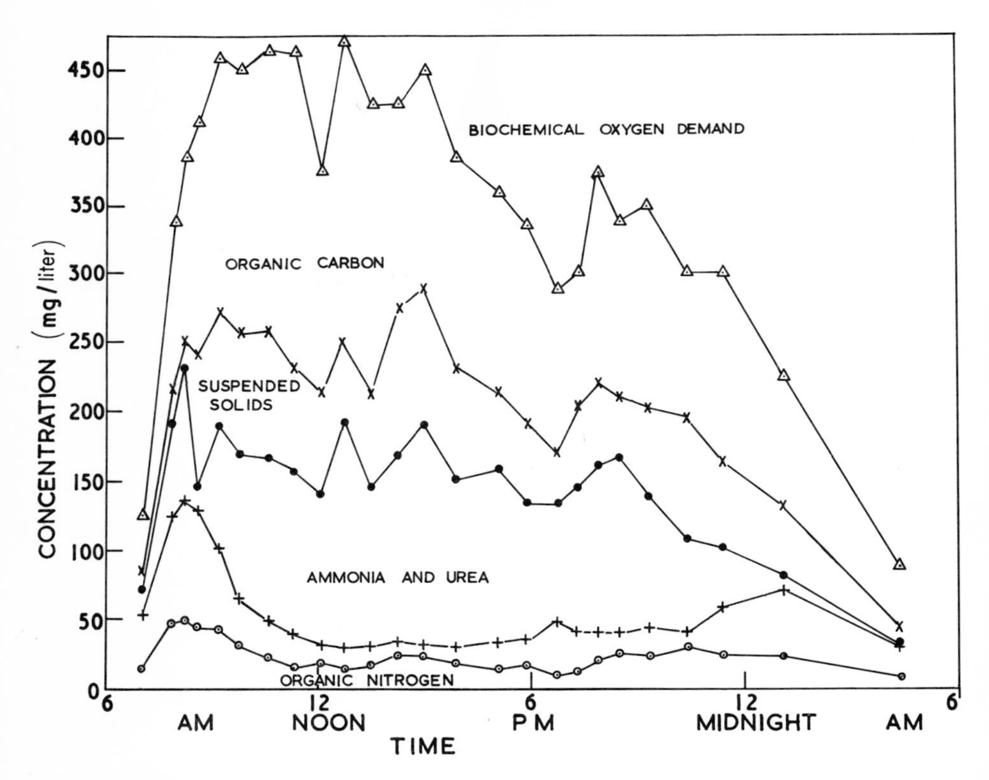

Fig. 1. Variations in BOD, and concentrations of organic carbon, suspended solids, ammonia plus urea, and organic nitrogen in a British domestic sewage during 24 hr. [Reprinted from Ref. 3, pp. 497 and 498, Crown Copyright, reproduced by courtesy of Controller, H. M. Stationery Office, London.]

distinct peaks (Fig. 1)—at 8–9 a.m. and around midnight—clearly reflecting the habits of the local population, but only the morning peak was observed in the Los Angeles area (*37*).

The concentration of individual constituents has also been found to vary considerably during a 24-hr period. For three British sewer systems the ranges of concentration of aldose sugar (expressed as glucose) were 6–110, 5–92, and 5–160 mg/liter, of ketoses (expressed as fructose) the ranges were < 5–32, < 5–27, and 5–30 mg/liter, while the ranges for volatile fatty acids for two systems were 3–48 and 3–56 mg/liter, expressed as acetic acid (*39*). The concentration versus time curves roughly paralleled the BOD–time curves. The grease content of Los Angeles sewage varied during a day from 15 to 150 mg/liter, averaging 60 mg/liter (*37*).

Variations in strength occur from day to day but are not as great in separate sewer systems as in systems into which significant proportions of surface or storm water enter. There is some evidence (*40*) that the flow of sewage in the United States is higher on Mondays than on other days of the week and that the concentration of detergents in U.K. sewage is higher on Mondays than during the rest of the week (*41*).

There is little evidence (*20*) of consistent differences in strength and composition between the seasons of the year, though there appeared to be more grease and amino acids in suspension in summer than in winter (*40*). The bacterial content, however, is higher in the warmer than in the colder months (*42, 43*), and viruses are present in highest concentration in late summer and fall (*2*).

C. Gross Composition

The polluting strength of sewage is assessed by such parameters as five-day BOD, COD, SS, and ammonia content. The strength varies widely and depends on such factors as the quantity of water used per head of population, the amount of groundwater and surface water entering the sewer, and local habits. In the United States 450–900 liters of water per person are used daily (*43a*), whereas in the United Kingdom the average is about 180 liters (*43a*), so that American sewage is usually weaker than British sewage. The average BOD of American raw sewage is usually between 100 and 300 mg/liter (*36*), while in Britain it is between 250 and 700 mg/liter and, for settled sewage, between 170 and 400 mg/liter (*44*). In general, the settlement of sewage removes about 50% of the suspended solids, 20–25% of the BOD, and about 33% of the COD (*4, 20*). The gross analyses of carefully composited samples of domestic sewage in the United States and the United Kingdom are given in Table

1, and the detailed analyses are given in Sec. III. E. The cited British settled sewage was stronger than most, since it came from an area where the daily use of water was only about 120 liters/head. The American values were typical of those cited in the literature, for example, the average BOD of 21 Connecticut raw sewages was 194 mg/liter, range 43–325 mg/liter (*45*). The total solids content depends to a large extent on the hardness of the water; the British sewage (Table 1) came from a hard water area (hardness about 400 mg/liter), while the American sewage presumably came from a soft water area. A range of concentration of total solids of 304–2740 mg/liter was reported for the Connecticut sewages (*45*).

TABLE 1
General Properties of Domestic Sewage[a, b]

	Settled samples from one U.K. town[c]		Unsettled samples from one U.S. town[d]			
			Winter–spring		Fall–winter	
Characteristic	Mean of 7	Range	Mean of 28	Range	Mean of 15	Range
Flow	0.15	0.12–0.19	5.6	3.4–8.8	7.5	3.8–13.8
pH	7.8[e]	7.6–8.2	7.2[e]	6.8–7.5	7.2[e]	6.7–7.5
BOD	370	340–390	147	75–276	136	46–216
COD	670[f]	—	288	159–436	282	97–443
Permanganate value	69	65–81	—	—	—	—
Total solids	1309[g]	—	453	322–640	481	294–676
Suspended solids	146	145–146	145	83–258	146	58–236
Volatile SS	—	—	120	62–208	125	54–174
Ammonia (as N)	46	41–53	21[h]	4–35[h]	—	—
Organic N	22	16–23	12	—	9	—
Organic C	219	186–247	—	—	—	—

[a]24-hr composite samples.
[b]Values expressed as milligrams per liter, except flow (million liters/day) and pH.
[c]Taken from Ref. *4* by courtesy of Controller, H.M. Stationery Office, London.
[d]Taken from Ref. *20* by courtesy of Water Pollution Control Federation, Washington, D.C.
[e]Modal value.
[f]Calculated from BOD = 0.55 × COD, found to hold for this sewage.
[g]Single determination.
[h]Taken from Ref. *45*.

The average weight of oxygen taken up by the sewage discharged per person in one day was calculated to be 0.12 lb(*44*) for raw sewage and 0.10 lb(*3*) for settled sewage in the United Kingdom, while U.S. values for raw sewage of 0.113, 0.126(*44*), and 0.15 lb [calculated from Ref. *20*] have been reported. Similarly, the per capita daily ammonia load in the United Kingdom was 0.013 lb N(*3*), which agrees well with the average value of 0.012 lb N for several American cities(*46*).

D. Inorganic Constituents

The inorganic content of sewage depends on the nature of the water supply from which it is derived as well as on the nature of the polluting material. The analysis of two domestic sewages—one from a soft water area, the other from a hard water area—is given in Table 2; over 95%

TABLE 2
Concentration of Some Inorganic Constituents of Domestic Sewage[a]

Constituent	Concentration, mg/liter: Whole sewage, U.S.A.[b] (soft water area)	Concentration, mg/liter: Settled sewage, U.K.[c] (hard water area)
Cl	20.1	68
Si	3.9	—
Fe	0.8	0.8
Al	0.13	—
Ca	9.8	109
Mg	10.3	6.5
K	5.9	20
Na	23	100
Mn	0.47	0.05
Cu	1.56	0.2
Zn	0.36	0.65
Pb	0.48	0.08
S (all forms)	10.3	22
Phosphate (as P)	6.6	22

[a]24-hr composite samples.
[b]Taken from Ref. *40*, p. 415, by courtesy of Water Pollution Control Federation, Washington, D.C.
[c]Taken from Ref. *3*, p. 498, by courtesy of Controller, H.M. Stationery Office, London.

of the American composite was identified and over 85% of the inorganic portion was in solution(*40*). Other reports give values about the same as those in Table 2, e.g., 0.1–0.8 mg Cu/liter, 0.2–0.6 mg Zn/liter(*47*), 9–17 mg P/liter(*48*), 6–30 mg P/liter(*49*); about half the phosphate was in the condensed form derived from packaged detergents(*49*). Minor elements, found in a Texas sewage, were, in milligrams per liter: Sr, 0.5; Cr, 0.4; B, 0.4; Ba, 0.1; As, < 0.1; and Ag, Cd, Mo, Ni, Co, V, and Be all at < 0.02 or lower(*48*). Concentrations of metals can be much higher when certain trade wastes are present, e.g., Ni, 0.15 mg/liter(*50*); Cu, 5 mg/liter; Zn, 4 mg/liter(*47*).

Flowing sewage usually contains 1–2 mg/liter dissolved oxygen(*51*), and presumably carbon dioxide and nitrogen are also present. Hydrogen sulfide is not present if sewers are normally vented, if there are no blockages, and if the velocity is 1.75 fps or more(*52*).

E. Organic Constituents

1. General

Two attempts—one in the United States and one in the United Kingdom—have been made to assign to chemical groups all the organic matter in domestic sewage. Difficulties were encountered in that some compounds inevitably showed up in two or more groups and because there were unknown losses especially during concentration of the soluble portions. Over 80% of the volatile matter was accounted for by the American group(*20*) and about 75% of the organic matter by the British group(*4, 21*); more of the soluble carbon (80–88%) was identified than the insoluble (70%). Somewhat less (60–70%) of the total organic carbon was accounted for in two mixed domestic-industrial sewages(*53*), and only 40–70% of the organic nitrogen present in the various sewages was identified(*4, 21, 54*). Of the total organic carbon in raw sewage, most (31–41%) was in solution, while 30% was in settleable solids, 19–24% in supracolloidal solids, and 10–14% in colloidal solids; the distribution of organic nitrogen was variable. In settled sewage about half of the organic carbon and one-third to one-half of the organic nitrogen were in solution.

The major groups present in solution (Table 3) were sugars, free and bound amino acids, volatile and nonvolatile acids, anionic detergents, and unspecified, ether-soluble neutral compounds, while minor groups included bases, amphoterics, phenols, sterols, and various nitrogen-containing substances. Sugars were consistently the largest group in solution in British sewage, but in the United States they ranked only second, acids taking top place. This may have been due to postsampling

TABLE 3
Organic Constituents in Solution in Domestic Sewage[a]

Constituent	Mean of three samples in U.K.[b] Concentration, mg/liter	Mean of three samples in U.K.[b] Proportion as carbon of total soluble carbon, %	Mean of winter-spring and fall-winter samples in U.S.A.[c] Concentration, mg/liter	Mean of winter-spring and fall-winter samples in U.S.A.[c] Proportion of soluble volatile solids, %
Sugars	70	31.3	10	12.5
Nonvolatile acids	34	15.2	28.5	35.6
Volatile acids	25	11.3	0.3	0.4
Amino acids: free	5	3.1	9.0 (free and bound)	11.2 (free and bound)
bound	13	7.6		
Detergents (ABS)	17	11.2	4.0	5
Uric acid	1	0.5	0.33	0.4
Phenols	0.2	0.2	0.11	0.14
Cholesterol	Present	—	0.04	0.05
Creatine–creatinine	6	3.9	0.18	0.22
Organic carbon	90	100	—	—
Volatile solids	—	—	80	100

[a]24-hr composites analyzed.
[b]Calculated from Ref. *21*, p. 153.
[c]Calculated from Ref. *20*, p. 1159.

changes in the American study resulting in an increased acid content at the expense of the sugars (see Sec. II. B) and possibly some loss of volatile acids during lyophilization (see Sec. II.C).

In all three fractions of suspended solids, fats, carbohydrates, and proteins were the main identified constituents and together accounted for 60–80% of the organic matter present (Table 4). The fat (acids plus esters) content of U.K. sewage was about five times the fat plus grease content of American sewage; fats comprised about half of the organic carbon in each solids fraction of British sewage but only about a quarter in the case of American sewage. This is in keeping with the known greater use of soap per person in the United Kingdom(*55*). The solids in American sewage contained relatively more carbohydrate and protein than did the solids in British sewage. In general, the content of organic nitrogen increased and the C:N ratio decreased with decreasing particle size. Various minor constituents were present in suspension including adsorbed anionic detergents, amino sugars, amide-nitrogen, and insoluble salts or derivatives of volatile acids.

TABLE 4
Organic Constituents in Suspension in Whole Domestic Sewage[a]

Constituent	Average of two samples in U.K.[b] Concentration, mg/liter	Average of two samples in U.K.[b] Proportion as carbon of total carbon in suspension, %	Mean of winter-spring and fall-winter samples in U.S.A.[c] Concentration, mg/liter	Mean of winter-spring and fall-winter samples in U.S.A.[c] Proportion of volatile matter in suspension, %
Fats	140	50	26	22.4
Proteins	42	10	27.5	23.7
Carbohydrates	34	6.4	34	29.3
Lignin	Present	—	4	3.4
Alcohol-soluble (unspecified)	—	—	8.5	7.3
Detergents (ABS)	5.9	1.8	0.3	0.3
Amino sugars	1.7	0.3	—	—
Amides	2.7	0.6	—	—
Soluble acids	12.5	2.3	—	—
Organic carbon:				
by analysis	211	100	—	—
by addition	151	71.4	—	—
Volatile matter:				
by analysis	—	—	116	100
by addition	—	—	101	87

[a]24-hr composites analyzed.
[b]Calculated from Ref. *21*, p. 156.
[c]Calculated from Ref. *20*, p. 1157.

2. Carbohydrates

Sugars identified in solution in sewage were glucose, sucrose, and lactose, with smaller proportions of galactose, fructose, xylose, and arabinose; these sugars accounted for 89–96% of the total sugars present, 50–120 mg/liter (*4, 21*). Glucose predominated in composite samples, accounting for over 50% of the sugar content, and in afternoon samples sucrose predominated; the hexose : pentose ratio was between 10 and 12 (*4*).

Carbohydrates in suspension (30–38 mg/liter) were mainly high molecular weight polymers, cellulose, and starches, and on hydrolysis yielded glucose, galactose, arabinose, xylose, and rhamnose, with smaller proportions of mannose, maltose, ribose, fucose, and six or more unidentified sugars; the hexose:pentose ratio was 2–2.6(*21*). There were no marked qualitative differences between the three solids fractions.

3. Fats and Grease

There were wide variations in the quantitative results obtained by various authors for the detailed composition of fats and greases even in composite samples, although there was good agreement qualitatively. For example, total fat and grease was usually 40–100 mg/liter, but values as low as 16 and as high as 1480 mg/liter were reported for sewages containing industrial wastes(*56*). Free fatty acids reported include all the saturated ones from C_8 (caprylic) to C_{14} (myristic), including those with odd numbers of carbon atoms(*57*), the saturated acids C_{16}, C_{18}, and C_{20}, and the unsaturated acids $C_{16}-2H$, $C_{18}-2H$ and $C_{18}-4H$(*56–58*). A number of these acids bound as esters have also been reported(*56*); free acids as a proportion of total free plus ester acid varied from 5%(*40*) to 70%(*21*). The major acids were palmitic, stearic, and oleic, which together formed the majority—over two-thirds(*21*) or even as much as 90%(*58*); myristic, lauric, and linoleic acids were present in relatively low proportions. However, the methods used were not capable of identifying unequivocally the named acids; for example, it has been reported(*59*) that feces contain several different C_{18} hydroxy acids at low concentrations, all of which are indistinguishable from stearic acid in the chromatographic method used, so that the results must be treated with some reserve.

4. Proteins and Amino Acids

Proteins and amino acids comprise the largest single nitrogen-containing group in sewage, the proportion of total organic nitrogen present in this form being reported variously as 65–80%(*40*), 48–70%(*20*), 40–70%(*4*), and 44%(*60*). It is suspected that some of these values may be low due to formation of humin-like material during hydrolysis prior to colorimetric or chromatographic determination(*20, 21*). Most reports give the concentration of free amino acids as under 5 mg N/liter, though values as high as 9(*61*) and 18 mg N/liter(*62*) have been reported for grab samples. The content of bound amino acids (peptide and protein) is higher, at 4–15 mg N/liter for composite samples, with a range of 1–26 mg N/liter for grab samples. Some 18 individual amino acids have been identified in hydrolyzed suspended solids(*40, 60, 62, 63*), each at concentrations in whole sewage of 1–4 mg/liter or 0.1–0.8 mg N/liter.

5. Soluble Acids

The first six straight-chain members of the volatile fatty acid series, and isobutyric and isovaleric acids, have been detected in sewage (*4, 64–66*). Acetic acid at 6–37 mg/liter is always the main constituent, followed by propionic acid (1–8 mg/liter), butyric(s) (0.4–2.4 mg/liter), valeric(s) (0.4–1.7 mg/liter), formic acid (0.2–1.0 mg/liter), and caproic acid (0.3 mg/liter). Over 90% of the total volatile acidity was accounted for by these acids. It is important to note that the acidity of sewage samples stored anaerobically soon increases at the expense of sugars, and this could give rise to abnormally high concentrations of volatile acids, especially of acetic acid (see Sec. II. B).

The nonvolatile acid fraction in solution can be as much as one-half the total acidity(*4*). About 20 acids were indicated by chromatographic methods; those identified with some certainty were lactic, citric, gallic (*20, 21*), glutaric, glycollic, oxalic, pyruvic, succinic, benzoic, phenyllactic, phenylacetic, and *p*-hydroxybenzoic acids(*4*). Some of these acids were present at the 1- to 2-mg/liter level, but most were present at about one-tenth of this concentration.

6. Miscellaneous N Compounds

A number of N-containing compounds other than amino acids have been found in sewage. Of those derived from urine, urea was the most abundant in fresh material; concentrations as high as 55 mg N/liter were reported(*3*) but more usual values were in the range 2–16 mg N/liter (*61*). Changes in urea content are fairly rapid, aerobically and anaerobically, leading to higher ammonia content and lower organic carbon content (see Sec. II. B). Uric acid was present at a fairly constant level of 0.2–1 mg/liter(*4, 67, 68*); hippuric acid was also detected(*21*), while a fourth urinary constituent, creatine–creatinine, was found at 5–7 mg/liter(*21*) in U.K. sewage but at only 0.2 mg/liter in U.S. sewage(*20*). Indole and skatole, derived from feces, were present at very low concentrations, 0.002 mg/liter(*69*), while aliphatic amines were found at the 0.1-mg/liter level(*4*).

Amino sugars, like glucosamine, were detected in hydrolyzed solids at 1.2–2.2 mg/liter, and muramic acid (3-O-α-carboxyethylglucosamine), present only in the cell walls of some bacterial species, was detected in sewage at about 0.5 mg/liter(*4*). Seven members of the B group of vitamins have been detected by microbiological methods in extremely low concentrations (micrograms per liter): thiamine, 29; riboflavin, 25; niacin, 135; cobalamin, 0.8(*70*); biotin 0.3(*71*); pantothenic acid and folic acid have also been detected(*70*). A review of enzymes in sewage

(*72*) led to the conclusion that lipase, diastase, catalase, pepsin, and trypsin are probably present in the free state; many others, e.g., urease, are present, but most are contained within bacterial and plant cells. No doubt the introduction of enzyme-containing washing powders will cause the concentration of enzymes in sewage to increase. Enzymes will, of course, be detected by the same methods as are used for proteins.

7. Miscellaneous Constituents

Sterols as a group were thought to be normally present in sewage at 0.1–0.2 mg/liter and seldom in excess of 1–2 mg/liter(*73*). Cholesterol was reported at 0.04–0.26(*74*) and 0.03–0.05 mg/liter(*20*), while the concentration of coprostanol ranged from 0.096 to 0.75 mg/liter(*74*). The group vaguely called lignins was present at only 1.5 mg/liter in American sewage(*20*), but in British sewage the value was about 5 mg/liter(*21*).

Various pigments derived from plants, e.g., chlorophyll and lycopene (*75*), from urine, e.g., urochrome B(*76*), and from feces, e.g., stercobilin, have been detected in very low concentrations in sewage.

The increasing use of synthetic detergents has been reflected in their increasing concentration in sewage; by 1967 the average concentration of alkyl benzene sulfonate (ABS) in U.K. sewage reached an estimated 11.3 mg/liter (as ABS)(*77*), and values as high as 40 mg/liter(*3*) have been obtained in grab samples. It is of interest to note that since biologically "softer" types of ABS have replaced the original "harder" types, the concentration in sewage reaching disposal plants in the United Kingdom has decreased by about 15%(*78*). In the United States lower values for ABS are usually recorded, e.g., 4(*20*), 1.0–5.0(*79*), 7.4 mg/liter (*80*), and no decrease was observed when the "softer" types replaced the "harder"(*79*).

The concentration of non-ionic detergents has been estimated at 1–2 mg/liter (as alkyl phenol with 8 to 11-ethylene oxide units) in sewage of purely domestic origin(*81*), but in areas where industrial use is high, e.g., for wool processing, concentrations as much as 10 times higher can be inferred(*82*). The concentration of "polyglycols" would also be high in these areas(*82*).

Low concentrations of phenols have been found in domestic sewage, e.g., 0.2–1.0(*4*) and 0.1 mg/liter(*20*), possibly derived from the use of antiseptics; sewage containing waste waters from gas manufacture had concentrations as high as 3 mg/liter(*53*).

Various organochlorine pesticides have been detected in domestic sewage at exceedingly low concentrations, e.g. (micrograms per liter),

BHC, 0.01–0.05; dieldrin, 0.01; DDT, 0.1; TDE and DDE up to 0.4(*83*). In areas where the pesticides are synthesized the concentrations reached 20 times these values(*84*), and in areas of extensive use, e.g., for carpet making, values of the order of 100 times those in domestic sewage can be inferred(*85*).

F. Microorganisms

The organisms present in sewage originate from feces, soil, and water and range from viruses through bacteria and fungi to protozoa and worms. The identity and concentration of these organisms are imprecisely known; for obvious reasons effort has been concentrated on organisms pathogenic to man and the degree to which they are removed by treatment. Many species of the various types of organisms have been described and in some cases the numbers of individuals of the species have been reported, but nothing like a comprehensive analysis of the total number of organisms present, even of bacteria, has been achieved.

Bacteriophages of the enterobacteria have been regularly detected(*86*); the number of coliphages has been put at around 400/ml(*87*). Phages died off much less rapidly than the host, some surviving in sewage for as long as four months(*88*), and this has led to the proposal that they be used as an indicator of survival of human viruses and of the presence of pathogenic bacteria in water and seawater. There was no evidence that coliphages reduced the number of *Escherichia coli* in sewage(*87,89*).

More than 70 serologically distinct human enteric viruses have been detected in sewage(*90*), the most commonly encountered being polio and Coxsackie viruses. In one study, up to 60% of healthy children had viruses in their feces(*91*), so that viruses must be considered to be normal constituents of sewage; negative results were considered to be due to insensitivity of the methods used. Even in areas having no experience of the paralytic disease, polio virus was isolated from the sewage(*92*). Although low proportions of samples containing virus have occasionally been reported, e.g., 8%(*93*), usually 80–90% of the samples were positive(*94*). The concentrations of enteroviruses have not been properly established because of lack of suitable methods. The best estimate so far is 500 virus units/100 ml in summer and about 20% of this in winter(*2*). Assuming that each virus unit weighs 0.1% of a bacterial cell, which weighs about 10^{-12} g, then 500 units/100 ml is equivalent to only 0.5×10^{-5} μg/liter. It is still undecided whether the concentrations in sewage are infectious; only infectious hepatitis has been proved to have been transmitted via sewage-contaminated water(*95*), but there was no significant relationship between incidence of the disease and employment at a waste treatment plant(*96*).

Because of deficiencies in the methods so far evolved it cannot yet be said whether the spectrum of species of bacteria reported to be in sewage is a true reflection of the bacterial population. A very large number of species have been reported, with fecal types predominating. The genera most frequently reported [e.g., (*97,98*)] are: *Aerobacter*, *Achromobacter*, *Alcaligenes*, *Azotobacter*, *Bacillus*, *Bacteriodes*, *Brevibacterium*, *Clostridium*, *Chromobacter*, *Corynebacterium*, *Escherichia*, *Flavobacterium*, *Micrococcus*, *Nitrobacter*, *Nitrosomonas*, *Paracolobactum*, *Proteus*, *Pseudomonad*, *Salmonella*, *Serratia*, *Shigella*, *Sphaerophorus*, *Streptococcus*, and *Xanthomonas*.

Values for the so-called "total count" of bacteria ranged from 1 to 38 $\times$ 10^6/ml(*99*); the average for 1 year for a completely domestic sewage was 11.7 $\times$ 10^6/ml, range 3–18 $\times$ 10^6/ml(*6*). If one bacterial cell weighs 10^{-12} g, then the count of 11.7 $\times$ 10^6/ml represents 11.7 mg dry weight/liter. It is of interest that only 2.7% of the total number of suspended particles were found, on direct microscopic observation, to be viable bacteria(*6*); in an average sewage this represents about 4 mg of suspended solids per liter. Increases of up to 100% in total count occurred on aeration for one day (*99*), though anaerobic storage sometimes gave a decreased count(*6*).

As many as 3 $\times$ 10^6/ml total coliforms have been reported, the usual values being 0.5–1 $\times$ 10^6/ml(*6,43,100*); fecal streptococci were present in lower numbers, usually 5000–20,000/ml(*42,100*). The average numbers of *Pseudomonas aeruginosa* and *Clostridium perfringens* in South African sewage were 102 and 507/ml, respectively(*101*). Members of the *Salmonella* group occurred in over 60% of samples and were found in every month of the year(*102*). The numbers present varied considerably, namely, 8(*101*), 357(*103*), to 400–1200/100 ml(*104*). The occurrence of *Shigella* appears to be rarer, with only two positives out of 96 communities in the United Kingdom(*105*).

The commonly observed "sewage fungus" is really a collection of organisms including filamentous bacteria such as *Sphaerotilus natans*, *Zoogloea ramigera*, and *Beggiatoa alba*, fungi such as *Leptomitus lacteus* and *Fusarium aqueductum*, and protozoa such as *Carchesium spectabile*(*106*). Other frequently found fungi were white yeasts, *Geotrichum candidum*, *Margarinomyces heteromorphum*, and *Penicillium lilacinum*. Less frequently observed were red yeasts, *P. ochrochloron*, *Aspergillus versicolor*, *Pullularia pullulans*, and *Phoma* spp. and many others including the rotifer parasite, *Zoophagus insidians*; in all, a total of 112 species were reported(*106,107*). Of 30 strains of yeast the more important were *Trichosporon cutaneum*, *Rhodotorula glutinis*, *R.*

mucilaginosa, *Candida parapsilosis*, and *C. tropicalis*; none was pathogenic to man, though some have been associated with tropical diseases (*108*). The only estimate of the concentration of yeasts was 20–70 cells of 18 species/ml(*109*).

The protozoan fauna of sewage has been studied only infrequently. Over a score of species were identified in sewage(*110*) and in sedimentation tanks(*111*); the more abundant were flagellates (species of *Bodo*, *Trepomonas*, and *Polytoma*) and the ciliates (*Carchesium*, *Colpidium*, *Cyclidium Glaucoma*, *Paramecium*, and *Vorticella*). The "sewage amoeba," *Endamoeba moshkovskii*, practically indistinguishable from *E. hystolytica* which causes human dysentery, has been frequently found(*112*). The total number of protozoa in sewage is low, usually fewer than 1/ml.

The eggs of many parasitic worms are present in sewage; for example, it was estimated that 5 liters of sewage in the United Kingdom contain one to six tapeworm eggs, *Taenia saginata*(*113*). Other worms identified include *Ascaris lumbricoides*, *Trichurus trichiura*, *Echinococcus granulosus*(*114*), *Diphyllobothrium latum*, and *Hymenolepsis nana*(*115*). Nematodes (Rhabditoidea) were present in settled sewage at 40–200/liter (*116*).

IV. Trade Waste Waters

A. General

The total polluting load of industrial waste waters in the United States has been estimated to be at least as great as that of domestic sewage. The volume and strength of trade wastes vary considerably from industry to industry and even within each industry there are wide variations, though the general properties of wastes from a given industry are usually similar. For many installations there are diurnal variations associated with batch production, a weekly pattern with little flow at weekends, or seasonal variations associated with availability of raw materials. Many trade wastes, especially from the food industry, are similar to domestic sewage and can be purified alone or in admixture with sewage by the usual biological processes. Other wastes are distinctly different, for example, explosive wastes and metal-pickling wastes, and while some of these do not interfere if added to sewage in sufficiently low proportions, others cannot be purified by the usual methods either alone or in admixture with sewage. Thus, some wastes are discharged, with or without treatment, to sewers and others are discharged to rivers or the sea, again with or without prior treatment.

It must be emphasized that a given trade discharge must be examined in detail, not only by analysis of suitable samples but also by examining in detail the processes that produce each of the waste waters that together form the discharge. Such an examination often leads to savings of salable produce and of water and to reduction in the strength of the waste. Indeed, in planning a new factory, the production and treatment of waste waters should be one of the factors considered. The nature of wastes from unit processes which together form the discharge is not dealt with here; for this and for other details the reader is referred to standard texts on the subject(*44*,*117*,*118*).

B. Chemical Composition

For convenience, trade wastes are here divided according to the classification introduced by Nemerow(*118*). The values for the strengths of a selection of trade wastes, given in Table 5, are for combined waste liquors and washings and are to be taken only as guides, since there will be variations between installations. There is relatively little information on the identity of compounds present in trade wastes, but for many wastes a knowledge of origin leads to an approximate composition.

1. Apparel Industries

The main constituents of wastes from cotton kiering, dyeing, and finishing are starch, gum, wax, dextrin, and detergents together with glucose, pectin, alcohols, acetic acid, methylcellulose, carboxymethylcellulose, polystyrene, dyestuffs, sodium hydroxide, chloride, sulfate, and silicate. Wastes from the scouring, finishing, and dyeing of wool are usually colloidal, containing high concentrations of grease, dried perspiration, and soap. Retting of flax yields wastes containing formic, acetic, propionic, and butyric acids, as well as some unidentified acids(*119*), while the most important constituents of waste waters from the viscose industry are sulfuric acid, sodium sulfate, hemicelluloses, and hydrogen sulfide. Laundry wastes contain high concentrations of soap, detergent, and alakali, with lower concentrations of grease, carboxymethylcellulose, and blue and optical "whiteners"(*119a*). Tannery waste waters are usually highly colored and alkaline and contain hair, particles of flesh, and sometimes chromium and arsenic.

TABLE 5
Comparative Strengths of Waste Waters from Industry[a]

Type of waste	5-day BOD, mg/liter	COD, mg/liter	Suspended solids, mg/liter	pH value
Apparel				
Cotton	200–1000	400–1800	200	8–12
Wool scouring	2000–5000	2000–5000[b]	3000–30,000	9–11
Wool composite	1000	–	100	9–10
Tannery	1000–2000	2000–4000	2000–3000	11–12
Laundry	1600	2700	250–500	8–9
Food				
Brewery	850	1700	90	4–6
Distillery	7000	10,000	Low	–
Dairy	600–1000	150–250[b]	200–400	Acid
Cannery: citrus	2000	–	7000	Acid
pea	570	–	130	Acid
Slaughterhouse	1500–2500	200–400[b]	800	7
Potato processing	2000	3500	2500	11–13
Sugar beet	450–2000	600–3000	800–1500	7–8
Grass silo[c]	50,000	12,500[b]	Low	Acid
Farm	1000–2000	500–1000[b]	1500–3000	7.5–8.5
Poultry	500–800	600–1050	450–800	6.5–9
Materials				
Pulp: sulfite	1400–1700	84–10,000	Variable	
Kraft	100–350	170–600	75–300	7–9.5
Paperboard	100–450	300–1400	40–100	
Strawboard	950	850[b]	1350	
Coke oven	780	1650[b]	70	7–11
Oil refinery	100–500	150–800	130–600	2–6

[a]The values given herein are from a wide number of sources too numerous to quote.
[b]Permanganate value.
[c]Undiluted with washings.

2. Food Industries

With few exceptions these wastes contain adequate nitrogen and phosphate for self-purification. Brewery wastes contain, in addition to unchanged carbohydrate substrate, low concentrations of the B group vitamins: thiamine, biotin, pyridoxine, riboflavin, pantothenic acid, nicotinamide, *p*-aminobenzoic acid, inositol, and choline; ergosterol and lipids have also been noted (*120*). In distillery wastes there are few suspended solids, and soluble sugar comprises over half the organic matter.

Dairy wastes are usually neutral but readily become acidic by conversion of lactose, the main constituent, to lactic and other acids; casein and fats are also present. Wastes from canneries are essentially similar to domestic kitchen wastes, with fruit sugars and acids such as citric acid as the main components; for example, it was found that 80–90% of the polluting load of wastes from a pineapple cannery was due to sugars, mainly sucrose(*121*). Slaughterhouse and meat-packing waste waters are dark reddish brown and contain a high proportion of blood, as well as excreta, undigested food, grease, and hair. About 95% of the organic matter in sugar beet wastes is sucrose and most of the remainder is raffinose; only small amounts of nitrogen are present and the waste is practically devoid of phosphate(*122*); the color of the waste is due to betanin (*123*). Sugar refinery wastes contain, besides sucrose, free amino acids, about 4 mg N/liter; bound amino acids, 12 mg N/liter; amines, 15 mg N/liter; and fatty acids, 40 meq. Individual amines identified were ethanolamine, methylamine, allylamine, and amylamine; the acids were acetic, propionic, butyric, pyruvic, adipic, malonic, and oxalic(*124*). The main organic constituents of effluent from grass silos were the sugars—glucose, galactose, fructose, xylose, and arabinose; the acids—acetic and lactic; and ammonia and many other volatile nitrogen compounds(*125*).

3. Materials Industries

These wastes are in general highly colored and much more difficult to treat biologically than domestic sewage. Wastes from industries using wood as a raw material have as the main pollutants carbohydrates, lignins, and their decomposition products. In a waste of BOD 1500 mg/liter, the hexose concentration was 500 mg/liter, pentose 90 mg/liter, and lignin 2000 mg/liter(*126*). Sugars identified include glucose, galactose, mannose, fructose, arabinose, xylose, rhamnose, cellobiose, and other oligosaccharides; aldonic acids such as galacturonic acid were also present(*127*). Many decomposition products of lignin were present, including vanillin, vanillic acid, coniferaldehyde, ferulic acid, *p*-hydroxybenzoic acid, and furfural and its derivatives. Simpler acids found were formic, 30 mg/liter; acetic, 160 mg/liter(*126*); butyric, oxalic, and laevulinic. Low concentrations of methyl and ethyl alcohols, acetone, acetaldehyde, methylglyoxal, formaldehyde, limonene, terpene, *p*-cymene, anisole, and guaiacol have also been detected in concentrated spent-sulfite liquors. Kraft pulp waste waters contain resin acids, mercaptans, and sulfides, all toxic to fish. In paper mill wastes starch is the major source of BOD.

Important wastes in steel mill operation come from the by-product coke plant; the main components are, in the undiluted ammoniacal liquor: total ammonia, 300–10,000 mg N/liter; phenols, 150–3000 mg/liter; tar acids (as phenol), 25–1400 mg/liter; thiocyanate, 80–1300 mg CNS/liter; thiosulfate, 20–1300 mg S_2O_3/liter; cyanide, trace–60 mg CN/liter; N bases (as pyridine), about 100 mg/liter.

The composition of the ammoniacal liquor produced in making coal gas is similar to that from steel mill operation but stronger. Some 30 or more individual phenols and about 20 pyridine and quinoline bases have been identified in both sorts of ammoniacal liquors(*128, 129*). The pickling of steel yields a waste containing 1–5% H_2SO_4 and 4–30% $FeSO_4$ but little organic matter(*129a*). Metal-plating wash wastes are not large in volume but are usually fairly acid or alkaline and the concentration of metals can be quite high. For example, concentrations of copper as high as 300 mg/liter, of chromium up to 600 mg/liter, and of nickel 32 mg/liter have been reached; the concentration of cyanide was 15 mg/liter(*130*), but if the spent plating liquors are discharged much higher concentrations are found. Oil refinery wastes contain phenols (up to 400 mg/liter) (*130a*), hydrocarbons (benzene, cyclohexane, etc.), alcohols, ethers, ketones(*130b*), and, from some refineries, carbon disulfide, acrylonitrile, ethylene oxide, propylene oxide, and naphthenic and sulfonic acids.

4. Chemical and Pharmaceutical Industries

An extremely wide variety of substances are present in waste waters from the chemical industry; many of the wastes, usually low in volume, are appreciably acidic or alkaline, highly colored, low in suspended solids, and often contain toxic compounds. Antibiotic waste waters (spent liquor plus washings) are fairly strong, with BOD values of 5000–20,000 mg/liter, and contain sugars in high concentration and vitamins, steroids, and antibiotics in low concentration. Other pharmaceutical wastes, COD 40,000–60,000 mg/liter, contained acetic acid, methanol, xylene, tars, chlorides, and bromides(*131*). The manufacture of synthetic resins produced wastes of BOD 500–1400 mg/liter, the main constituents being formaldehyde, 100–600 mg/liter; methanol, 400–1000 mg/liter; and phenol, 10–50 mg/liter(*132*). In a few synthetic resin wastes, thiourea, a powerful inhibitor of nitrification, is present, and similar inhibitors, containing nitrogen and sulfur, are present in wastes from rubber manufacture. In a waste (BOD 2500 mg/liter) from a plant making fertilizers and pesticides the concentration of phenols was 800 mg/liter; dinitro-*o*-cresol, 200 mg/liter; and glycollic acid, 1000 mg/liter(*133*). In the manufacture of the pesticides 2,4-D and DDT, the wastes contained highly toxic chlorophenols and chlorohydrocarbons. Wastes from explosives

(TNT) factories are extremely acid, pH 1–3, very dark red, low in suspended solids, and fairly high in chemical oxygen demand (500–800 mg/liter) but low in BOD. An example of an organometallic compound in a waste is methyl mercury chloride present at 15–20 mg/liter in waters from the production of vinyl chloride and acetaldehyde (*134*).

C. Microorganisms

Microorganisms occur less frequently in trade wastes than in domestic sewage. The important industries discharging wastes containing bacteria are food, antibiotic, cellulose, paper, and leather industries.

Besides the usual animal intestinal bacteria, many species of *Salmonella* have been identified in slaughterhouse wastes (*135*); in cattle runoff water the total coliform count was $80\text{–}130 \times 10^4$/ml (*136*). In papermill wastes the "total" bacterial count was 10^4 to 2×10^8/ml (*137*), and 14 species of *Clostridia* were identified at 100 to nearly 10^6/ml; all species produced slimes and bad odors (*138*). Other bacteria identified were species of *Flavobacterium*, *Pseudomonas*, *Aerobacter*, *Bacillus*, and sulfate reducers; fungi observed were species of *Rhodotorula*, *Monilia*, *Penicillium*, and *Aspergillus* (*138*). *Sphaerotilus natans* is also frequently found in cellulose waste waters, and among the many other bacteria positively identified are *Bacterium thiogenes*, *Bacterium album*, *Bacterium bovista*, *Cellfalcicula viridis*, *Cellvibrio ochraceus*, *Bacillus mesentericus*, *Bacillus subtilus*, and *Thiobacillus thioparus* (*139*). Anthrax is sometimes associated with tannery wastes. In penicillin production wastes, gram negative bacteria predominated and about one-third of the total bacteria was penicillin resistant, a larger proportion than is resistant in domestic sewage (*140*).

The acidic waters which drain from ore deposits and coal mines contain sulfur- and iron-oxidizing bacteria, such as *Thiobacillus thio-oxidans* and *T. ferro-oxidans*. Fungi and yeasts are also sometimes present, e.g., *Pullularia pullulans*, *Spicaria divaricata* (*141*), *Rhodotorula* spp., and *Trichosporon* spp. (*142*). From drainage water from a copper mine a flagellate, *Eutrepia*, was isolated, having a high tolerance to copper (*142*).

V. Effluents from Biological Treatment Processes

A. Chemical Composition

Biological methods of treating sewage and waste waters (see Chap. 9) are capable of producing effluents with BOD 15–20 mg/liter, COD 50–70 mg/liter, and SS 15–30 mg/liter or less and in which the ammonia has been converted to nitrate; tertiary treatment can further reduce the strength to BOD < 10 mg/liter, COD < 40 mg/liter, and SS < 10 mg/liter. The inorganic constituents of effluents are much the same as those in sewage, unless special procedures have been adopted to remove nitrate and phosphorus. It appears that the total solids content increases by about 290–340 mg/liter between the water supply and the sewage effluent derived from it (*143*) and the salinity increment is about 7 g of chloride per capita per day (*144*). The average values for the inorganic increments for five U.S. cities (*80*) were, in milligrams per liter: total solids, 291; inorganic N, 20; Ca, 23; Mg, 7; K, 9.3; Na, 57; phosphate (as P) 24.3; sulfate (as S), 11; and Cl, 56. Most (60–95%) of the phosphate present was in the form of orthophosphate (*49*). There is an abundant literature on the general analysis of biological effluents, typical values being given in Table 6, but there is an extreme dearth of information on the nature of the organic compounds present.

The results of two comprehensive analyses of effluents from trickling

TABLE 6[a]

Typical Analyses of Sewage Effluents after Conventional Primary and Secondary Treatment in Three Towns in Southeast England

Constituent	Concentration, mg/liter (except pH)		
	1	2	3
Total solids	728	640	931
Suspended solids	15	—	51
Permanganate value	13	8.6	16
BOD	9	2	21
COD	63	31	78
Organic carbon	20	13	—
Detergent (ABS)	2.1	0.6	1.2
Ammonia (as N)	4.1	1.9	7.1
Nitrate (as N)	38	21	26
Nitrite (as N)	1.8	0.2	0.4
Chloride	69	69	98
Sulfate	85	61	212
Total phosphorus (as P)	9.6	6.2	8.2
Sodium	144	124	—
Potassium	26	21	—
pH	7.6	7.2	7.4

[a]Reprinted in abridged form from Ref. *143*, p. 3, Crown Copyright, reproduced by courtesy of Controller, H.M. Stationery Office, London.

TABLE 7
Organic Constituents in Sewage Effluents after Conventional Primary and Secondary Treatment[a]

Constituent	Settled effluent from two trickling filters in U.K.[b]		Filtered effluents from activated sludge plants and trickling filters in five cities in U.S.A.[c]
	In solution	Total	
BOD	—	12–29	—
COD	—	40–90[f]	23–85
Suspended solids	—	8–15	4–76[g]
Volatile matter	35–90	—	53–131
Organic carbon	14–36	16.5–40.7	—
Carbohydrate	0.6–1.2	1.4–3.0	0.8–2.4
Protein and amino acid	0.6–0.7	2.0–3.6	1.6–7.4
Ether extractables[d]	3.9–6.8	4.0–7.0	0.8–5.2
Detergents (ABS)	2.1–6.5	2.2–6.6	1.5–12.5
Fats	0	0.08–0.16	—
Amino sugars	0	0.2–0.4	—
Tannins and lignins	—	—	0.5–1.7
Identified, %[h]	24–26	28–30	~ 35

[a]Results expressed as milligrams per liter, except where otherwise stated.
[b]Calculated from Ref. *4*.
[c]Taken from Ref. *23*.
[d]Contains at least 65% strong acids.
[e]Calculated assuming volatile matter contains 40% carbon.
[f]Estimated.
[g]Before paper filtration.
[h]Sum total of carbohydrates down to tannins.

filters and activated sludge plants are given in Table 7, which shows that only 28–35% of the organic matter was identified. A larger proportion of organic carbon was present in solution than was the case with sewage, and the solids contained far less fatty material than did sewage solids. Carbohydrates, free and bound amino acids, acids, and ABS comprised the bulk of the identified soluble material, while carbohydrates, proteins, amino sugars, and fats made up the identified material in suspension.

The same amino acids have been identified in hydrolyzed suspended solids from effluents as in sewage solids(*145*), but no analyses appear to have been made of individual amino acids or sugars in solution, or of fats and carbohydrates in suspension. Only 2–20% of the total organic acids was volatile, and this portion included all the acids up to caproic acid at concentrations, in badly purified effluents, up to the following values (milligrams per liter): isovaleric, 0.5; acetic, 0.2; formic, isobutyric, butyric, and caproic, 0.1; propionic, 0.07; and valeric, 0.04(*65*); in well purified effluents the concentrations were usually between 1/10th and 1/100th of these. Three or four unidentified acids not present in a sewage were detected in the derived effluent(*66*). Non-ionic detergents were present at 0.5–1.0 mg/liter in effluents derived from domestic sewage, while in effluents from sewage containing wool-processing wastes as much as 6 mg/liter were present, together with polyglycols up to 8.8 mg/liter(*82*). The sterols, coprostanol and cholesterol, were detected in filter effluents at 102 and 57 μg/liter, respectively and in activated sludge effluents at 8 and 15 μg/liter, respectively(*74*). Uric acid was also found at the low concentration of 5–12 μg/liter(*68*), and the condensed hydrocarbon, pyrene, was detected at 0.4–1.0 μg/liter(*75*).

Dieldrin, DDT, and BHC were present in sewage effluents from 16 works at about 0.1 μg/liter, and TDE and DDE were also detected(*84*). Effluents from a wool-processing area contained as much as 1.9 μg/liter of dieldrin and 0.8 μg/liter of DDT, and concentrations of BHC up to 14 μg/liter were found in effluents from an area serving a pesticide factory (*84*). Up to 10 μg/liter of dieldrin were reported in effluents from a works treating sewage containing a high proportion of wool mill or carpet factory effluents(*85*).

The strength of effluents varies throughout 24 hr but usually not so markedly as does the strength of sewage. For example, the concentration of ABS varied in an activated sludge effluent from 1 to 3 mg/liter(*49*) when that in the influent ranged from 4 to 17 mg/liter, and the concentration in a filter effluent ranged from 0.8 to 2.5 mg/liter compared with a range in the influent of 2–16 mg/liter(*41*). On the other hand, changes in the concentration of total phosphate – a constituent not normally removed – were almost identical with those in the influent (*49*).

Such seasonal variations as occur are due largely to the effect of temperature on the efficiency of the purification processes. Considering influents having the same BOD, the suspended solids and BOD of effluents from activated sludge plants were about 25% higher at 12°C than at 18°C (*27*), and the BOD values of effluents from filters were similarly about 25% higher at the lower temperature. Whether any qualitative differences occur does not seem to have been investigated. Also, at the lower temperatures there is much more likelihood that the concentration of ammonia would be higher and the concentration of nitrate lower.

B. Microorganisms

All the organisms found in sewage (see Sec. III.F) can be expected to be present in treated sewage, together with organisms which enter from the air, though they will usually be present in considerably lower concentrations.

Trickling filters were inefficient in removing phages and viruses; only about 50% coliphages (*87*) and 12–53% enteroviruses (*146*) were removed, and on some occasions viruses were isolated from effluent when no virus was seen in the raw sewage. Activated sludge was much more effective in removing viruses; only 5% of effluent samples contained viruses compared with 31% of sewage samples (*147*), only 20% of the plaque-forming count in sewage was detected in effluents (*146*), and as much as 98% of added Coxsackie virus was removed by laboratory-scale plants (*95*).

Both types of biological treatment were more effective in removing bacteria than in removing viruses; efficient plants eliminated from 90% to over 99% of the "total count." The range of counts for a number of effluents over a year was $93–560 \times 10^5$/ml; there were higher numbers in winter than in summer (*6*). The "total viable count" was only 0.4–1% of the total particle count compared with 2.7% for sewage. Similar reductions in numbers were observed with coliforms (*42*), *Shigella* (*148*), and *Clostridium perfringens* (*101*), smaller reductions resulted with *Salmonella* (70%) and *Mycobacterium tuberculosis* (66–88%) (*148*), while with *Pseudomonas aeruginosa* increases in numbers were reported (*101*). The variety of fungal species observed in final filter effluents was as great as in the sewage treated (*149*), and it can be inferred that other species not identified in sewage but present in filters, e.g., *Sepedonium* and *Ascoidea*, would be present in effluents. The protozoan fauna of purification plants was more diverse than in sewage, with over 80 genera described (*111*), and these would no doubt appear in effluents. The more important additional protozoa were the flagellates *Mastigamoeba*, *Monas*, *Oikomonas*, *Pleuromonas*, *Cercobodo*, *Astasia*, and *Anisonema* and the ciliates *Chilodon*, *Cinetochilum*, *Aspidisca*, and *Opercularia*. The total number present in effluents is much larger than in sewage, with values of 100/ml not uncommon; this concentration is equivalent to about 1 mg dry weight/liter.

The eggs of parasitic worms are removed from sewage by sedimentation for 2 hr but not by subsequent biological treatment, so that effluents from well-operated plants should be free from such eggs (*113,150*). A larger number—up to 500/liter—of nematodes, as well as a larger variety of genera, are found in effluents than in sewage (*116*), and although about 100 viable bacteria were associated with each worm, no *Salmonella*, *Shigella*, or enteroviruses were recovered from the worms.

Electron microscopic examination of the colloidal fraction of activated sludge effluents revealed fragments of bacterial cell wall as the dominant material; also present was cellular debris of phages, viruses, flagella, and other organisms (*33,151*).

VI. Conclusions

It would be complacent to think that the problem of the composition of sewage and waste waters had been solved. Undoubtedly, some advances have been made in that 70–90% of the organic matter in domestic sewage has been identified and much of the inorganic matter, too, has been accounted for. However, some of the methods employed in the investigations do not give unequivocal values, some are nonspecific, while some compounds react in more than one test, so that the high proportions identified and the reported concentrations of some groups of compounds have to be treated with some reserve. Even so, it is unlikely that any major group in solution has been overlooked, though the complexity of the individual substances under the generic headings is probably greater than has so far been demonstrated; for example, as many as 150 ninhydrin-reacting substances, presumably fairly low molecular weight peptides, were found recently in urine (*152*). In suspension, however, it is likely that important groups still remain to be estimated; e.g., no reference could be found to nucleic acids present both in bacteria and in plant debris. (From the probable number of bacteria present it can be calculated that the concentration of nucleic acids in sewage should be 3–6 mg/liter.) Other compounds not apparently reported quantitatively but likely to be present are alkaloids, purines, pyrimidines, teichoic acids (in bacterial cell walls), poly-β-hydroxybutyric acid, pigments from urine, feces, and plants, and the so-called humic acids; this list could easily be extended. The importance of teichoic acids, muramic acid (shown to be present) (*4*), and nucleic acids is their possible use as indicators of the bacterial content of wastes.

At least as important as the identity of the organic matter is its state of organization, but on this topic little information is available. Something is known of the complexity of polymerization of the carbohydrates in suspension, but nothing is known about how much is bound to protein, bound to lignin, and present in bacteria; a similar situation exists with proteins and fats. It is important that this should be investigated since the degree of complexity is likely to determine the ease and rate of breakdown of the organic matter. It is undoubtedly a difficult task and will be solved only by a many-sided attack, possibly using combinations of a wide variety of techniques such as differential centrifugation, electrophoresis, histology, electron microscopy, enzymic hydrolysis, and spectrophotometry.

Similarly, little has been discovered about the distribution between the various ionic and chelated forms in which such constituents as trace metals and cyanide are present in waste waters. This again is a difficult problem because of the low concentrations and equilibria involved;

physicochemical solutions have been attempted but it seems that a microbiological approach might also yield results, since the toxicity of the metals to bacteria and fungi is moderated by chelation.

There has been some success in eliminating the inhibitory components in trade wastes discharged to sewers, e.g., thiourea, metals, and pentachlorophenol, but much more could be done. More could also be done in ascertaining possible deficiencies of trace metals and vitamins, as well as of nitrogen and phosphorus, for those trade wastes treated biologically before discharge.

Far less has been attempted, and gleaned, about the more challenging question of the composition of treated effluents than about sewage; it is strange, too, that the technically superior method of concentration—lyophilization—does not appear to have been applied to effluents. Since only about 30% of the organic matter has been identified, it is impossible to say whether the bulk of the pollution in effluents is derived from sewage passing through the treatment process unchanged or whether it consists of metabolic and degradation products of organisms in the filter or activated sludge plant. If the latter explanation is true this would seem to set the lower limit for the strength of effluents readily obtained from biological treatment processes. There is very little indication as to what the other 70% unidentified matter is, but there is a strong likelihood that it consists of a large number of compounds at very low concentrations. About 40% of the soluble volatile matter was nondialyzable, suggesting that it was of high molecular weight (*23*). Since organic compounds in effluents might survive subsequent treatment, for the production of potable water it is imperative that more effort be devoted to establishing their identity.

The nature of microbial populations of waste waters has by no means been satisfactorily established, and advances in techniques are necessary before the problem can be successfully tackled. Possible avenues for advance are the use of immunological reactions and the simultaneous use of a far wider range of isolation media, following the recent finding that 40% of isolates from activated sludge made on a medium containing a sludge extract failed to grow on the isolation media normally employed (*153*). A more sensitive method for detecting viruses is also needed.

Much of the work surveyed has been done using conventional methods, and the application of more sophisticated and fruitful methods has been rather slow. Such methods as gas-liquid chromatography of aqueous solutions, infrared and fluorescence spectroscopy, electrophoresis, and automated analysis [especially to study diurnal variations (see Chaps. 20, 27, 28, 30, and 31)] may well be applied with advantage in the future. Workers in the field of waste water treatment should be alert to the possible application of these and other methods as they are developed.

REFERENCES

1. Ministry of Health and Ministry of Housing and Local Government, *Reports on Public Health and Medical Subjects, No. 71, The Bacteriological Examination of Water Supplies*, 3rd ed., H.M. Stationery Office, London, 1956, p. 20.

2. S. A. Kollins, *Advan. Appl. Microbiol.*, **8**, 145 (1966).

3. H. A. Painter, *Water Waste Treat. J.*, **6**, 496 (1958).

4. H. A. Painter, M. Viney, and A. Bywaters, *Inst. Sewage Purif., J. Proc.*, **1961**, 302.

5. R. M. Cody, R. G. Tischer, and H. K. Williford, *J. Water Pollution Control Federation*, **33**, 164 (1961).

6. T. G. Tomlinson, J. E. Loveless, and L. G. Sear, *J. Hyg.*, **60**, 365 (1962).

7. A. E. Zanoni, *Public Works*, **96**, 72 (1965).

8. E. W. Taylor, N. P. Burman, and C. W. Oliver, *J. Inst. Water Engrs.*, **9**, 248 (1955).

9. H. A. C. Montgomery, *Water Res.*, **1**, 631 (1967).

10. R. C. Loehr and B. Bergeron, *Water Res.*, **1**, 577 (1967).

11. F. J. Agardy and M. L. Kiado, *Purdue Univ. Eng. Bull., Ext. Ser. No. 121*, 1966, p. 226.

12. W. J. Fogarty and M. E. Reeder, *Public Works*, **95**, 88 (1964).

13. O. Sturz, *Dt. Gewasswerk. Mitt.*, **8**, 57 (1964); through *Water Pollution Abstr.*, **38**, Abstr. No. 1601 (1965).

14. P. E. Morgan and E. F. Clarke, *Public Works*, **95**, 73 (1964).

15. J. H. Winneberger, W. I. Saad, and P. H. McGauhey, *A Study of Methods of Preventing Failure of Septic-Tank Percolation Fields, 1st Ann. Rept. Sanit. Eng. Res. Lab.*, University of California, Berkeley, Calif., 1961, p. 5.

16. G. Berg, G. Stern, D. Berman, and N. A. Clarke, *J. Water Pollution Control Federation*, **38**, 1472 (1966).

17. Department of Scientific and Industrial Research, *Notes on Water Pollution, No. 8*, H.M. Stationery Office, London, March, 1960; also *Inst. Sewage Purif. J. Proc.*, **1960**, 483.

18. D. H. R. Hellwig, *Intern. J. Air Water Pollution*, **8**, 215 (1964).

19. Water Pollution Research Laboratory, Stevenage, England, unpublished work.

20. J. V. Hunter and H. Heukelekian, *J. Water Pollution Control Federation*, **37**, 1142 (1965).

21. H. A. Painter and M. Viney, *J. Biochem. Microbiol. Technol. Eng.*, **1**, 143 (1959).

22. J. V. Hunter and H. Heukelekian, *Purdue Univ. Eng. Bull., Ext. Ser. No. 106*, 1961, p. 150.

23. R. L. Bunch, E. F. Barth, and M. B. Ettinger, *J. Water Pollution Control Federation*, **33**, 122 (1961).

24. *Standard Methods for the Examination of Water, Sewage, and Industrial Wastes*, 12th ed., American Public Health Association, New York, 1965.

25. Ministry of Housing and Local Government, *Methods of Chemical Analysis as applied to Sewage and Sewage Effluents*, H.M. Stationery Office, London, 1956.

25a. H. A. C. Montgomery and N. Thom, *Analyst*, **87**, 689 (1962).

26. T. G. Tomlinson and H. Hall, *Rept. Public Works Munic. Sewage Congr.*, London, 1950, p. 600.

27. C. E. Keefer, *J. Water Pollution Control Federation*, **34**, 1186 (1962).

28. D. Evers, *Textile Inst. Ind.* **3**, 237 (1965).

29. V. Kothandaraman, V. P. Thergaonkar, T. Koshy, and S. V. Kanapati, *Environ. Health (India)*, **5**, 356 (1963).

30. K. Dimorski and G. Khitov, *Tr. Nauchn.-Issled. Inst. Vodosnabzh. Kahaliz. Sanit., Tekhn. Sofia*, **2**, 75 (1965); through *Water Pollution Abstr.*, **39**, Abstr. No. 727 (1966).

31. H. A. Painter, unpublished.
32. W. Rudolfs and J. L. Balmat, *Sewage Ind. Wastes*, **24**, 247 (1952).
33. D. A. Rickert and J. V. Hunter, *J. Water Pollution Control Federation*, **39**, 1475 (1967).
34. S. D. Faust and M. C. Manger, *Purdue Univ. Eng. Bull., Ext. Ser. No. 115*, 1963, p. 684.
35. J. Scherfig and H. F. Ludwig, *Intern. Conf. Water Pollution Res., 3rd, Munich, 1966*, **3**, 217 (1967).
36. Water Pollution Control Federation, *Manual of Practice No. 11, Operation of Waste Water Treatment Plants*, Washington, D.C., 1961, p. 4.
37. F. R. Bowerman and F. D. Dryden, *J. Water Pollution Control Federation*, **34**, 475 (1962).
38. T. G. Tomlinson and H. Hall, *Inst. Sewage Purif. J. Proc.*, **1955**, 40.
39. H. A. Painter, R. S. Denton, and C. Quarmby, *Water Res.*, **2**, 427 (1968).
40. H. Heukelekian and J. L. Balmat, *J. Water Pollution Control Federation*, **31**, 413 (1959).
41. G. E. Eden and G. A. Truesdale, *Inst. Sewage Purif. J. Proc.*, **1961**, 30.
42. L. A. Allen, E. Brooks, and I. L. Williams, *J. Hyg.*, **47**, 303 (1949).
43. R. J. Burm and R. D. Vaughan, *J. Water Pollution Control Federation*, **38**, 400 (1966).
43a. L. Klein, *River Pollution, Part 3, Control*, Butterworth, London, 1966, p. 82.
44. B. A. Southgate, *Treatment and Disposal of Industrial Wastes*, Department of Scientific and Industrial Research, H.M. Stationery Office, London, 1948.
45. J. W. Masselli, N. W. Masselli, and M. G. Burford, *The Effect of Industrial Wastes on Sewage Treatment*, New England Interstate Water Pollution Control Commission, 1965, pp. 33–35.
46. H. E. Babbitt, *Sewerage and Sewage Treatment*, 6th ed., Wiley, New York, 1947, p. 305.
47. J. W. Masselli, N. W. Masselli, and M. G. Burford, *The Occurrence of Copper in Water, Sewage and Sludge and Its Effect on Digestion*, New England Interstate Water Pollution Control Commission, 1961, pp. 27, 28.
48. D. Vacker, C. H. Connell, and W. N. Wells, *J. Water Pollution Control Federation*, **39**, 750 (1967).
49. M. S. Finstein and J. V. Hunter, *Water Res.*, **1**, 247 (1967).
50. T. Stones, *Inst. Sewage Purif. J. Proc.*, **1959**, 252.
51. B. Gustafsson and N. Westberg, *Intern. Conf. Water Pollution Res., 2nd, Tokyo, 1964*, **1**, 221 (1965).
52. F. H. Miller, W. T. Barron, and E. K. Goffigon, *Water Works Wastes Eng.*, **2**, 45 (1965).
53. Department of Scientific and Industrial Research, *Water Pollution Research, 1961*, H.M. Stationery Office, London, 1962, p. 48.
54. P. V. R. Subrahmanyan, C. A. Sastry, A. V. S. P. Rao, and S. C. Pillai, *J. Water Pollution Control Federation*, **32**, 344 (1960).
55. J. Prat and A. Giraud, *The Pollution of Water by Detergents*, Organization for Economic Co-operation and Development, Paris, 1964, p. 9.
56. R. C. Loehr and T. J. Kukar, *Intern. J. Air Water Pollution*, **9**, 479 (1965).
57. O. J. Sproul, J. W. Caskey, and D. W. Ryckman, *Ind. Water Wastes*, **7**, 139 (1962).
58. C. S. Viswanathan, B. M. Bai, and S. C. Pillai, *J. Water Pollution Control Federation*, **34**, 189 (1962).
59. A. J. James, J. R. W. Webb, and T. D. Kellock, *Biochem. J.*, **74**, 21P (1960).
60. B. M. Bai, C. V. Wiswanathan, and S. C. Pillai, *J. Sci. Ind. Res. (India)*, **21C**, 72 (1962).

61. A. M. Hanson and T. F. Flynn, *Purdue Univ. Eng. Bull., Ext. Ser. No. 117*, 1964, p. 32.
62. L. Kahn and C. Wayman, *J. Water Pollution Control Federation*, **36**, 1368 (1964).
63. C. S. Sastry, *Environ. Health* (*India*), **7**, 111 (1965).
64. E. Hindin, D. S. May, R. McDonald, and G. H. Dunstan, *Water Sewage Works*, **111**, 92 (1964).
65. J. J. Murtaugh and R. L. Bunch, *J. Water Pollution Control Federation*, **37**, 410 (1965).
66. H. F. Mueller, T. E. Larson, and W. J. Lennarz, *Anal. Chem.*, **30**, 41 (1958).
67. G. J. Kupchik and G. P. Edwards, *J. Water Pollution Control Federation*, **34**, 376 (1962).
68. J. O'Shea and R. L. Bunch, *J. Water Pollution Control Federation*, **37**, 1444 (1965).
69. W. Rudolfs and B. Heinemann, *Sewage Works J.*, **11**, 587 (1939).
70. E. G. Srinath and S. C. Pillai, *Current Sci.* (*India*), **35**, 247 (1966).
71. H. Y. Neujahr and J. Hartwig, *Acta Chem. Scand.*, **15**, 954 (1961).
72. M. C. Sridhar and S. C. Pillai, *J. Sci. Ind. Res.* (*India*), **25**, 167 (1966).
73. E. Stumm-Zollinger and G. M. J. Fair, *J. Water Pollution Control Federation*, **37**, 1506 (1965).
74. J. J. Murtaugh and R. L. Bunch, *J. Water Pollution Control Federation*, **39**, 404 (1967).
75. P. Wedgwood, *Inst. Sewage Purif. J. Proc.*, **1952**, 20.
76. P. Sattelwacher and E. Furstenau, *Gesundh. Ingr.*, **82**, 16 (1961).
77. G. E. Eden, G. A. Truesdale, and G. V. Stennett, *Water Pollution Control*, **67**, 107 (1968).
78. Ministry of Housing and Local Government, *9th Progress Report of Standing Technical Committee on Synthetic Detergents*, H.M. Stationery Office, London, 1967.
79. E. F. Barth and M. B. Ettinger, *J. Water Pollution Control Federation*, **39**, 815 (1967).
80. R. L. Bunch and M. B. Ettinger, *J. Water Pollution Control Federation*, **36**, 1411 (1964).
81. Ministry of Technology, *Notes on Water Pollution*, No. 34, H.M. Stationery Office, London, September, 1966; also *Water Pollution Control*, **66**, 294 (1967).
82. S. J. Patterson, K. B. E. Tucker, and C. C. Scott, *Intern. Conf. Water Pollution Res., 3rd, Munich, 1966*, **2**, 103 (1967).
83. Water Pollution Research Laboratory, Stevenage, England, unpublished work.
84. Ministry of Technology, *Notes on Water Pollution, No. 36*, H.M. Stationery Office, London, March, 1967; also, *Water Pollution Control*, **66**, 633 (1967).
85. A. V. Holden and K. Marsden, *Inst. Sewage Purif. J. Proc.*, **1966**, 295.
86. S. Damilano and Z. Bernardi, *Igiene Mod.*, **57**, 63 (1964); through *Water Pollution Abstr.*, **39**, Abstr. No. 192 (1966).
87. G. C. Ware and M. A. Mellon, *J. Hyg.*, **54**, 99 (1956).
88. E. M. J. Carstens, *Inst. Sewage Purif. J. Proc.*, **1963**, 467.
89. W. A. Pretorius, *J. Hyg.*, **60**, 279 (1962).
90. H. H. Malherbe and M. Strickland-Cholmley, in *Symposium on Transmission of Viruses by the Water Route* (G. Berg, ed.), Robert A. Taft Sanitary Engineering Center, Cincinnati, Ohio, Wiley (Interscience), New York, 1966, p. 379.
91. C. Berg, *Health Lab. Sci.*, **3**, 86 (1966).
92. R. Ozere, R. Faulkner, and C. E. V. Rooyen, *Can. Med. Assoc. J.*, **85**, 1419 (1961).
93. W. N. Mack, W. L. Mallman, H. H. Bloom, and B. J. Krueger, *Sewage Ind. Wastes*, **30**, 957 (1958).
94. S. M. Kelly and W. W. Sanderson, *Sewage Ind. Wastes*, **31**, 683 (1959).
95. N. A. Clarke and P. W. Kabler, *Health Lab. Sci.*, **1**, 44 (1964).
96. Safety Committee, California Water Pollution Control Assoc., *J. Water Pollution Control Federation*, **37**, 1629 (1965).

97. F. F. Dias, *J. Indian Inst. Sci.*, **45**, 36 (1963).
98. E. M. Russell, E. P. Munro, and J. C. Peacock, *Water Waste Treat. J.*, **8**, 575 (1962).
99. L. A. Allen, *Proc. Soc. Agr. Bacteriol.*, **1940**, 8.
100. W. J. Benzie and R. J. Courchaine, *J. Water Pollution Control Federation*, **38**, 410 (1966).
101. O. J. Coetzee and N. Fourie, in *Proc. Resolutions and Papers of the Conference on the Problems Associated with the Purification, Discharge and Re-use of Municipal and Industrial Effluents*, National Institute for Water Research, C.S.I.R., Conf. No. S5, Pretoria, South Africa, 1965, p. 93.
102. J. H. McCoy, *Intern. J. Air Water Pollution*, **7**, 597 (1963).
103. O. J. Coetzee and T. Pretorius, *Public Health (Johannesburg)*, **65**, 415 (1965).
104. J. H. McCoy, *Proc. Soc. Water Treat. Exam.*, **6**, 81 (1957).
105. H. D. Holt, *Monthly Bull. Ministry Health Lab. Serv.*, **19**, 29 (1960).
106. W. B. Cooke, *Sewage Ind. Wastes*, **26**, 539 (1954).
107. W. B. Cooke and F. J. Ludzack, *Sewage Ind. Wastes*, **30**, 1490 (1958).
108. W. B. Cooke, H. J. Phaff, M. W. Miller, M. Shifrine, and E. P. Knapp, *Mycologia*, **52**, 210 (1960).
109. W. B. Cooke, *Mycologia*, **57**, 696 (1965).
110. V. Stadecek, *Sci. Papers Inst. Chem. Technol., Prague, Fac. Technol. Fuel Water*, **3**, 229 (1959).
111. A. N. Barker, *The Naturalist*, July-September, 1943, p. 65.
112. E. C. Bovee and D. E. Wilson, *Develop. Ind. Microbiol.*, **4**, 350 (1963).
113. P. H. Silverman, paper presented to the Oxford Meeting of the British Association for the Advancement of Science, 1954.
114. H. Sinnecker, *Z. Ges. Hyg.*, **4**, 98 (1958); through *Public Health Eng. Abstr* , **40**, 33 (S. 33) (1960).
115. E. S. Didenk, G. P. Kondrat'eva, Z. V. Gorbunova, and V. I. Gorskaya, *Coll. Rept. 3rd Sci. Conf. Tallinn Res. Inst., Tallinn, USSR, 1964*, p. 314; through *Biol. Abstr.*, **46**, 1349 (1965).
116. S. L. Chang and P. W. Kabler, *J. Water Pollution Control Federation*, **34**, 1256 (1962).
117. W. Rudolfs, ed., *Industrial Wastes, Their Disposal and Treatment*, Reinhold, New York, 1953.
118. N. L. Nemerow, *Theories and Practice of Industrial Waste Treatment*, Addison-Wesley, Reading, Mass., 1963.
119. K.-C. Menzel and I. Thomas, *Faserforsch. Textiltech.*, **8**, 138 (1957); through *Chem. Zentr.*, **128**, 13846 (1957).
119a. R. E. Wagg, *J. Inst. Public Health Eng.*, **58**, 155 (1959).
120. F. Knorr, *Brauwissenschaft*, **18**, 191 (1965).
121. N. C. Burbank and J. S. Kumagai, *Purdue Univ. Eng. Bull., Ext. Ser. No. 118*, 1965, p. 365.
122. J. E. Laughlin, *Preprint, Ind. Water Waste Conf., 4th*, Texas Water Pollution Control Assoc., 1964, D.2.
123. R. E. Pailthorp, *J. Water Pollution Control Federation*, **32**, 1201 (1960).
124. E. Leclerc and F. Edeline, *Centre Belge Etude Doc. Eaux Bull. Mensel*, **114–115**, 201 (1960).
125. W. Moore, H. F. Walker, E. G. Gray, and E. M. Weir, *Water Waste Treat. J.*, **8**, 226 (1961).
126. E. F. Eldridge, *Trans. 2nd Seminar Biological Problems on Water Pollution, Tech. Rept. W60–3*, Robert A. Taft Sanitary Engineering Center, Cincinnati, Ohio, 1959, p 255.
127. T. E. Maloney and E. L. Robinson, *Tappi*, **44**, 137 (1961).

128. J. R. Catchpole, *Spec. Rept. No. 6*, Iron and Steel Institute, London, 1958, p. 219.
129. M. A. Hughes, *J. Appl. Chem. (London)*, **12**, 450 (1962).
129a. R. D. Hoak in *Industrial Wastes, their Disposal and Treatment* (W. Rudolfs, eds.), Reinhold, New York, 1953, pp. 255–282.
130. M. G. Burford and J. W. Masselli, in Ref. (*117*), p. 289.
130a. D. W. Ryckman, E. D. Edgerly, and N. C. Burbank, *Ind. Water Wastes*, **7**, 89 (1962).
130b. H. C. Schutt and J. Loftus, *Oil Gas J.*, **64**, 70 (1966).
131. R. R. Melcher, *Biotechnol. Bioeng.*, **4**, 147 (1962).
132. K. G. Singleton, *Inst. Sewage Purif. J. Proc.*, **1965**, 498.
133. S. H. Jenkins and H. A. Hawkes, *Intern. J. Air Water Pollution*, **5**, 407 (1961).
134. K. Irukayama, *Intern. Conf. Water Pollution Res., 3rd, Munich, 1966*, **3**, 153 (1967).
135. W. Pollach, *Wien Tierärztl. Monatsschr.*, **51**, 161 (1964); through *Zentr. Bakteriol. Parasitenk., I, Ref.*, **197**, 556 (1965).
136. J. R. Miner, R. I. Lipper, L. R. Fina, and J. W. Funk, *J. Water Pollution Control Federation*, **38**, 1582 (1966).
137. J. Paluch, *Acta Microbiol. Polon.*, **14**, 327 (1965); through *Chem. Abstr.*, **64**, 19180 (1966).
138. L. L. Wolfson and R. J. Michalski, *Tappi*, **47**, 197 (1964).
139. B. Smyk and E. Rozycki, *Gaz, Woda Tech. Sanit.*, **32**, 21 (1958).
140. D. A. Cornelson, I. Sechter, E. Balteanu, and V. Avrem, *Arch. Roumaines Pathol. Exptl. Microbiol.*, **17**, 271 (1958); through *Water Pollution Res. Abstr.*, **33**, Abstr. No. 380 (1960).
141. B. Marchlewitz, *Zentr. Bakteriol. Parasitenk., I, Orig.*, **184**, 293 (1962).
142. H. L. Ehrlich, *J. Bacteriol.*, **86**, 350 (1963).
143. G. E. Eden and G. A. Truesdale, in *Symposium on Conservation and Reclamation of Water*, Institute of Water Pollution Control, London, November, 1967.
144. M. Rebhun, *Intern. J. Air Water Pollution*, **9**, 253 (1965).
145. F. Jursik, *Sci. Papers Inst. Chem. Technol., Prague, Fac. Technol, Fuel Water*, **4**, 221 (1960).
146. S. M. Kelly and W. W. Sanderson, *J. Water Pollution Control Federation*, **32**, 1269 (1960).
147. H. H. Bloom, W. N. Mack, B. J. Krueger, and W. L. Mallman, *J. Infect. Diseases*, **105**, 61 (1959).
148. P. W. Kabler, *Environ. Health (India)*, **4**, 258 (1962).
149. W. B. Cooke, *Purdue Univ. Eng. Bull., Ext. Ser. No. 96*, 1958, p. 26.
150. H. Leibmann, *Intern. Conf. Water Pollution Res., 2nd, Tokyo, 1964*, **2**, 269 (1965).
151. R. B. Dean, S. Claesson, N. Gellerstadt, and N. Bowman, *Environ. Sci. Technol.*, **1**, 147 (1967).
152. P. B. Hamilton, paper presented at Technicon European Symposium on Automation in Analytical Chemistry, Brighton, England, November 14, 1967.
153. T. B. S. Prakasam and N. C. Dondero, *Appl. Microbiol.*, **15**, 461 (1967).

Chapter 8 Chemical and Physical Purification of Water and Waste Water

J. E. Singley

COLLEGE OF ENGINEERING
UNIVERSITY OF FLORIDA
GAINESVILLE, FLORIDA

I. Processes Used for the Production of Drinking and Process Water

A. General

In the not too distant past it was possible for most of the smaller and a few of the larger municipalities in this country to have readily available an adequate supply of pure water that required no treatment. As the cities grew, as unpolluted supplies became more scarce, and as waters of higher quality were required to satisfy the public, water treatment plants became commonplace. As the pollutants have become more varied and as the public's insistance upon better quality has increased, the plants have become more sophisticated. With most water supplies it is no longer possible to provide quality water with a simple filtration or aeration plant. The possible solutions to a given water treatment problem may involve a complex combination of chemical and physical factors that require a specialized understanding of treatment units and chemicals.

1. U.S. Plants

The first treatment plant in the United States, at Poughkeepsie, New York in 1871, was based on the slow sand filtration plants in use in Europe. These proliferated, particularly in the New England states, where many are still in use.

The number of public water supplies increased until today there are over 22,000 serving a population in excess of 165,000,000 or over 80% of the population of the United States.

The properties of the raw water supply dictate the type of treatment plants required. Surface waters are usually turbid and may be colored. In addition, they are frequently saturated with oxygen and carbon dioxide and contain various microscopic organisms. The minimum treatment for these waters is disinfection, usually chlorination. Groundwaters, on the other hand, are usually highly mineralized and may contain hydrogen sulfide and excess carbon dioxide.

Some waters may only require filtration, but the most common treatment for the removal of turbidity, color, or microorganisms is coagulation followed by flocculation, sedimentation, filtration, and disinfection. The majority of the treatment plants in the United States utilize this series of operations.

Plants in those parts of the country that have hard water supplies reduce this hardness to an acceptable level by softening. This is accomplished by the lime, lime–soda, or ion exchange process.

Of the almost 21 billion gal of water supplied by public water supply systems, about 14.2 billion gal are drawn from surface supplies and about 6.4 billion gal from groundwaters (*1*).

Although it is generally accepted that surface waters may be polluted, an opinion that is, unfortunately, most frequently correct, it is also generally held that groundwaters are pure and wholesome, an opinion that may or may not be correct. As a matter of fact, more and more groundwaters are becoming contaminated.

One other source of raw water is the ocean itself or brackish waters adjacent to the ocean. This inexhaustible supply has been the focus in the last few years of many research projects which have resulted in a large number of plants in fresh water-poor areas of the world; desalination plants are utilized in only a very few locations in the United States, however.

The municipal water treatment industry is much larger than might be thought. The original investment in facilities is estimated to have been approximately $17 billion at the time of installation, with a current replacement value in excess of $40 billion.

2. Water Consumption

Man's basic physiological requirement for water is about 3 quarts of water/day, but his primitive requirements are about 4 gpd (gallons per day). Rural use was about 50 gpd per capita in 1960, and urban use was about 155 gpd per capita. The urban use includes water produced by municipal facilities but used by industry and for other requirements such as fire-fighting, watering parks, etc. (*1*).

The remarkable aspect of water use is the very large quantity required to support man. If we take the weight of the gross national product, except water, i.e., fuel, lumber, food, cars, chemicals, bricks, cement, steel, etc., the amount required in the United States is about 20 tons per capita/year. The amount of water required is over 2000 tons per capita/year. This includes agricultural and industrial use for a total of about 1500 gpd per capita.

The requirements by various industries are also rather large. The paper industry, for example, uses from 10,000 to 100,000 gal of water/ton of paper; the steel industry may use up to 100,000 gal/ton of steel; textiles up to 30,000 gal/ton of cloth; and cement up to 750 gal/ton of cement (*2*). These industries may use relatively untreated waters for many purposes, but other industrial uses, such as in beverages, foods, boiler feed waters, and chemical plants, may have very rigid quality requirements that need a high degree of treatment.

Water quality, then may be quite variable, depending upon the use to which the water will be put. Potable water has no exact standards but most state boards of health have adopted standards very similar to the United States Public Health Service drinking water standards, the latest of which were issued in 1962(*3*). These standards are required of all drinking water on interstate carriers and have been accepted by the American Water Works Association as standards for all public water supplies. A summary of these standards is shown in Table 1. The ability to distribute waters that meet these standards is a function of the raw water quality and the type of treatment facilities available. Table 2 shows representative examples of the water quality of some of the largest U.S. cities(*4*). It should be noted that, regardless of the chemical and physical characteristics of the water distributed, the water is always bacteriologically safe.

Industrial water quality may vary from essentially untreated cooling water to demineralized water for high pressure boiler feed. In the case of boiler feed waters there are three major chemical criteria that are dictated by the boiler water temperature. The first of these is the total dissolved solids (TDS), an excess of which can form scale on the interior of the boiler and thereby result in localized overheating and tube failure. The higher the temperature and pressure of the boiler the lower must be the TDS. The level of calcium and magnesium salts is another important factor, as the solubility of several of their salts decreases with increasing temperature. The scales they form are hard and dense. The other chemical factor of concern is the dissolved oxygen content, as

TABLE 1
Representative Examples from *U.S. Public Health Service Drinking Water Standards, 1962*[a]

Substance	Maximum concentration, mg/liter
Alkylbenzene sulfonate (ABS)	0.5
Arsenic	0.01
Chloride	250.0
Copper	1.0
Cyanide	0.01
Iron	0.05
Manganese	0.05
Phenols	0.001
Sulfate	250.0
Total dissolved solids	500.0
Zinc	5.0

[a]Abstracted from Ref. *3*.

TABLE 2
Representative Examples of Water Quality of Some Major U.S. Cities[a]

City and state	Population[b] in 1000's	Source[c]	Treatment[d]	Range of hardness, ppm	Total dissolved solids, ppm
Birmingham, Ala.	441	S	C, Cl	61–120	101–250
Los Angeles, Calif.	2458	M	S, Cl	61–120	251–500
San Diego, Calif.	600	S	C, S, Cl	> 180	> 500
San Francisco, Calif.	1600	S	Cl, F	< 61	< 100
Washington, D.C.	1100	S	C, Cl, F	61–120	101–250
Jacksonville, Fla.	247	G	Cl	> 180	251–500
Miami, Fla.	550	G	C, S, Cl, F	61–120	101–250
Atlanta, Ga.	600	S	C, Cl	< 61	< 100
Chicago, Ill.	4425	S	C, Cl, F	121–180	101–250
Boston, Mass.	2000	S	Cl	< 61	< 100
Detroit, Mich.	3078	S	C, Cl	61–120	101–250
New York, N.Y.	8350	M	C, Cl	< 61	< 100

[a]Extracted from data of Dufor and Becker, Ref. *4*.
[b]1960 census.
[c]S, surface; G, ground; M, mixed.
[d]C, clarification; Cl, chlorination; S, softening; F, fluoridation.

oxygen greatly accelerates the rate of corrosion of the boiler steel. It is obvious, then, that these are the factors that determine the required quality of boiler feed waters.

Process waters for cooling vary so widely in quality requirements that it is impossible to suggest any but the most general requirements. These are the obvious points that the water should not produce deposits, either inorganic or organic, nor should it be corrosive. The treatment may vary from softening and coagulation to none at all.

Process waters for various industries may likewise vary widely in the treatment required.

B. Coagulation and Flocculation

In order to achieve the water quality criteria established for a given water it is frequently necessary to provide for treatment to remove undesirable constituents and to provide a means of disinfection, particularly for drinking water. There are some sources of supply that are satisfactory without further treatment; in general, these are groundwaters or impounded supplies in sparsely populated areas. The most common type of treatment is the removal of turbidity by coagulation, flocculation, sedimentation, and filtration.

1. Definitions

Coagulation may be defined operationally as the reactions that take place upon the addition of a coagulant to water and result in the formation of insoluble products of reaction between the coagulant and the impurity to be removed. Flocculation is the process of building the coagulated particles into a floc that is large enough and dense enough to settle.

These two processes also may be defined in terms of the reaction mechanisms involved. In this context coagulation is the destabilization of the suspended colloids by compression of the double layer, thereby allowing the colloidal particles to approach each other closely enough so that they may be agglomerated by the Van der Waal's forces of attraction. Flocculation is the further agglomeration of the particles by interparticle bridging(*5*). Many practitioners use the two terms synonymously.

2. History

Coagulants were originally added to waters ahead of rapid sand filters in order to duplicate the slimy, gelatinous "schmutzdecke" that was characteristically present on slow sand filters. The first such use is credited to Isiah Hyatt, who was issued U.S. Patent 293,740 in 1884 for the use of ferric chloride(*6*). The use of metal salts, principally salts of aluminum and iron, as coagulants was first studied systematically by Fuller at Louisville, Kentucky in the early 1900's(*7*) and later in the 1920's by Theriault, Clark, and Miller at the U.S. Public Health Service laboratories in Cincinnati, Ohio(*8–12*). These studies led to an appreciation that the results were not due to the coating on the filters but rather to chemical or physical interactions between the metal ions or their hydrolysis products and the impurities to be removed. Later studies have improved our understanding of the details of the reactions but have not changed our ideas about the basic reactions involved to any significant extent. Among the principal contributors to improved coagulation mechanism theories have been A. P. Black, V. LaMer, E. Matejevic, W. Stumm, and R. F. Packham. They and their co-workers have studied many phases of the coagulation and flocculation process. Many others have made contributions over the years.

One of the areas of research interest has been the development of various coagulant aids. These are substances that can be added to the water to improve formation of the floc. They may provide additional condensation nuclei, as in the case of addition of clays, limestone, or fly ash, or they may improve the cross-linking between the small agglomerates in the flocculation step. Such aids are frequently organic polymers, either natural polymers such as starches, gums, or algal derivatives or long-chained synthetic polymers. They are referred to as coagulant aids since they are used along with a primary coagulant such as one of the metal salts or a charged polymer.

3. Coagulation and Flocculation Theories

Since the impurities removed by coagulation and flocculation are colloidal, they are destabilized by overcoming or reducing the repulsive forces that keep the particles in suspension. These forces are electrostatic in nature and result from factors such as ionization or specific adsorption of charged species. Under natural water conditions turbidity, color, bacteria, algae, protozoa, and colloidal organic debris all are charged negatively and therefore are destabilized by positively charged coagulant species. These species result from the metal cations themselves or from their hydrolysis products. They may react specifically with negative sites at the solid–liquid interface or may be absorbed at this interface. In either case they reduce the particle charge. This has been explained in terms of double-layer compression as based on the Stern fixed double-layer model of colloids. The presence of positively charged species in the vicinity of the colloidal particles reduces the overall charge and the distance through which the charge is effective. The higher this counter-ion concentration, the weaker the repulsive forces between particles and the smaller the charged envelope. Agglomeration then occurs as a consequence of Van der Waal's forces of attraction when the particles approach each other closely enough for these forces to be effective. Particle transport results from Brownian movement or induced agitation(*13*). The chemical reaction or adsorption step is thought to be very rapid, with the overall rate controlled by the transport step(*14*).

Particle transport by Brownian movement alone is far too slow to be useful in water treatment, so mechanical agitation is required. This is recognized in practice; a rapid mix is usually provided at the point where the coagulant is added to the raw water, and further agitation is provided by mechanical flocculators prior to sedimentation and filtration. The flocculators must provide a more gentle agitation than the rapid mix, as the floc that is being built into large enough particles to settle is relatively fragile.

Long chain polyelectrolytes are adsorbed at the solid-liquid interface and may build a floc simply by adsorbing onto several colloids and other polymer chains, thus forming a complex cross-linked mass containing the colloidal impurities. Much the same mechanism may result from coagulation with metal salts at high pH values where the solubility product of the insoluble hydroxide has been exceeded. The gelatinous hydroxide polymers may entrap the colloids by a "sweep" action as they settle through the colloidal suspension.

4. Coagulants

There are two major types of coagulants used in water treatment processes, inorganic metal salts and organic polymers.

a. Inorganic

The metal salts most commonly used are aluminum sulfate, sodium aluminate, ferrous [iron(II)] sulfate or copperas, ferric [iron(III)] sulfate, ferric [iron(III)] chloride, and chlorinated copperas [iron(III) chlorosulfate]. Other polyvalent metal salts could be used but economics favors those listed. The most common of those listed is aluminum sulfate or "filter alum," as it is generally called.

The metal salts hydrolyze when added to water according to the following typical reactions, as exemplified by Al^{+3}:

$$Al^{3+} + H_2O \rightleftharpoons Al(OH)^{2+} + H^+ \tag{1}$$

$$Al(OH)^{2+} + H_2O \rightleftharpoons Al(OH)_2^+ + H^+ \tag{2}$$

$$Al(OH)_2^+ + H_2O \rightleftharpoons Al(OH)_{3(s)} + H^+ \tag{3}$$

$$Al(OH)_{3(s)} + H_2O \rightleftharpoons Al(OH)_4^- + H^+ \tag{4}$$

There is some evidence to suggest that some of the charged intermediates may polymerize to form highly charged species such as $Al_8(OH)_{20}^{4+}$. It has been shown that the anion has some effect on the coagulation reaction. In the interest of simplification, though, we can consider the hydrolysis of the cation as the only reaction of interest.

The predominant species for the hydrolysis of aluminum and iron(III) ions are shown in Figs. 1 and 2, respectively, as functions of pH. Thus, the predominate species for aluminum is $Al(OH)_3$ in the pH range of concern in water treatment processes, i.e., pH 5–9(*15*). For iron(III) the predominant species over this range is either $Fe(OH)_4^-$, $Fe(OH)_5^{2-}$, or $Fe(OH)_6^{3-}$(*16, 17*). These were determined for systems prior to the attainment of equilibrium. Equilibrium considerations indicate that the predominant species are $Al(OH)_3$ and $Fe(OH)_3$ under the conditions of concern in water treatment processes.

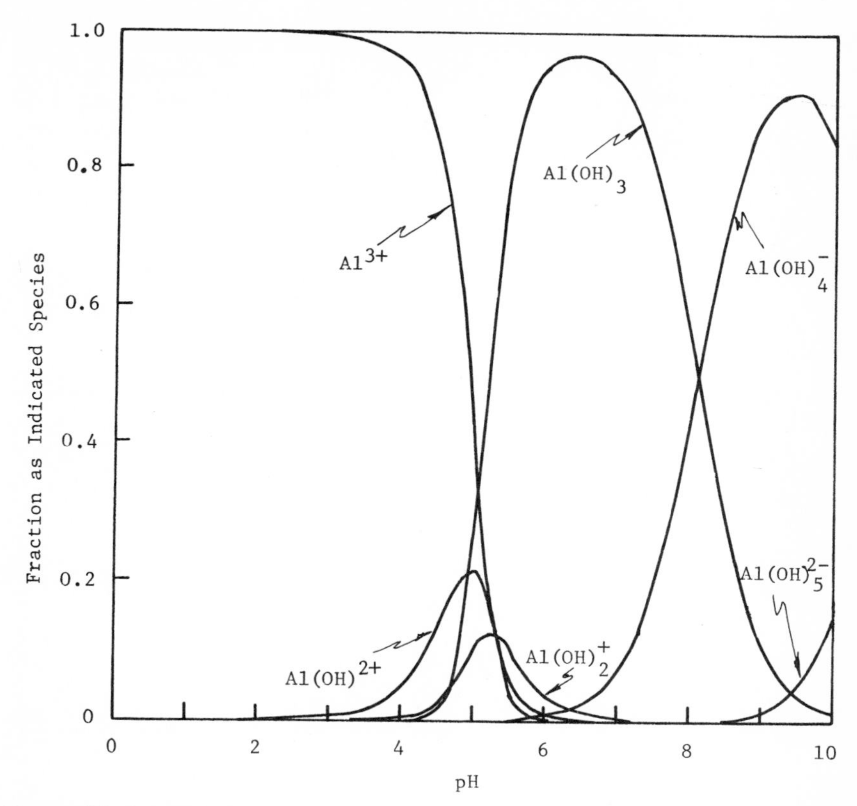

Fig. 1. Distribution of hydrolysis species of Al^{3+} as a function of pH. Total Al^{3+} concentration of 1×10^{-4} *M*. The fraction of the total Al^{3+} concentration as the indicated species is shown for nonequilibrium conditions. [Reprinted from Ref. *15*, by courtesy of the *Journal of the American Water Works Association*.]

Alum reacts with the alkalinity in water as follows:

$$Al_2(SO_4)_3 + 3Ca(HCO_3)_2 \rightleftharpoons 2Al(OH)_3 + 3CaSO_4 + 6CO_2 \quad (5)$$

Actually, the formation of $Al(OH)_3$ might better be shown as:

$$Al^{3+} + 3HCO_3^- \rightleftharpoons Al(OH)_3 + 3CO_2 \quad (6)$$

This shows that the alkalinity is reduced when alum is added as a coagulant. One part of alum, as $Al_2(SO_4)_3 \cdot 14H_2O$, reduces the alkalinity, as $CaCO_3$, by 0.55 parts.

Similar reactions with alkalinity can be written for iron(III) salts. One part of $Fe_2(SO_4)_3$ (90% pure) will reduce the alkalinity by 0.68

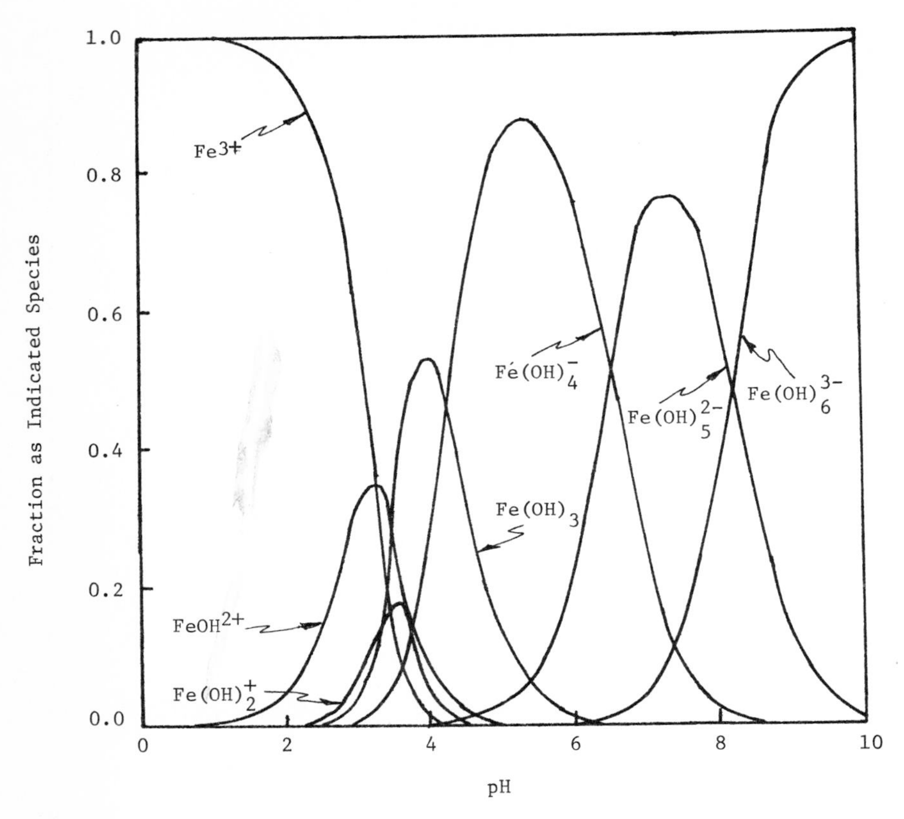

Fig. 2. Distribution of hydrolysis species of Fe(III) as a function of pH. Total Fe(III) concentration of 1×10^{-4} *M*. The fraction of the total Fe(III) concentration as the indicated species is shown for nonequilibrium conditions. [Reprinted from Ref. *17*, by courtesy of the *Journal of the American Water Works Association*.]

parts as $CaCO_3$. This reduction in alkalinity sometimes requires that additional alkalinity be provided in coagulation plants in order to maintain the pH within the optimum range.

Under equilibrium conditions, which may obtain in the course of transit through the flocculation and sedimentation steps in treatment plants, the predominant species of iron(III) and aluminum are the neutral hydroxides, as shown in Figs. 3 and 4, which show the equilibrium solubility domains for the hydroxides. The barred areas indicate the pH concentration range of interest in water treatment and, as such, show that the solubilities of both $Al(OH)_3$ and $Fe(OH)_3$ usually are exceeded.

The choice between the various metal salts is dictated by local availa-

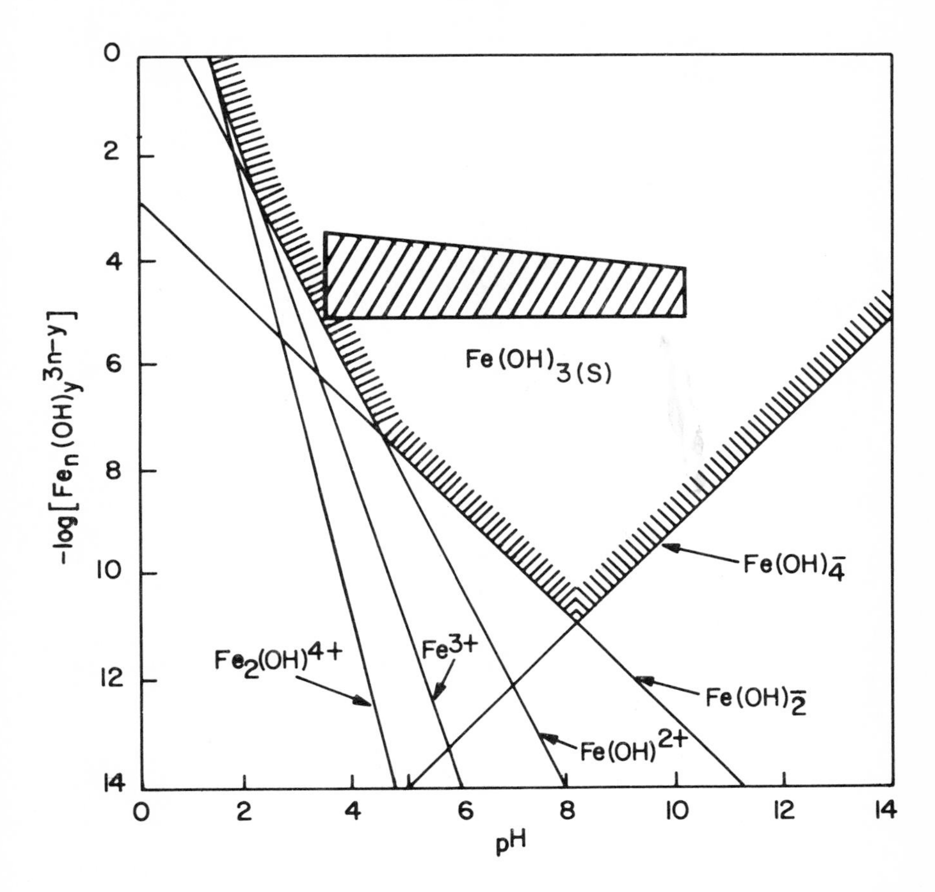

Fig. 3. Solubility domain of $Fe(OH)_3$. The barred area is occupied by solid $Fe(OH)_3$. The limiting lines are determined from the appropriate equilibrium constants.

bility, economics, convenience, and application to the specific treatment problem. Those commercially available are:

Aluminum sulfate: available commercially as $Al_2(SO_4)_3 \cdot 14H_2O$, although the pure crystal would have 18 moles of water. The forms available are granular and liquid. The granular form contains 17% aluminum as Al_2O_3 and the liquid about one-half of that. Liquid alum has the advantage of ease of handling, although it is somewhat more expensive on a solids basis due to the cost of transporting 50% water. This differential in cost is reduced or eliminated when one takes into account the cost of getting the solid alum into solution before its use. An addit-

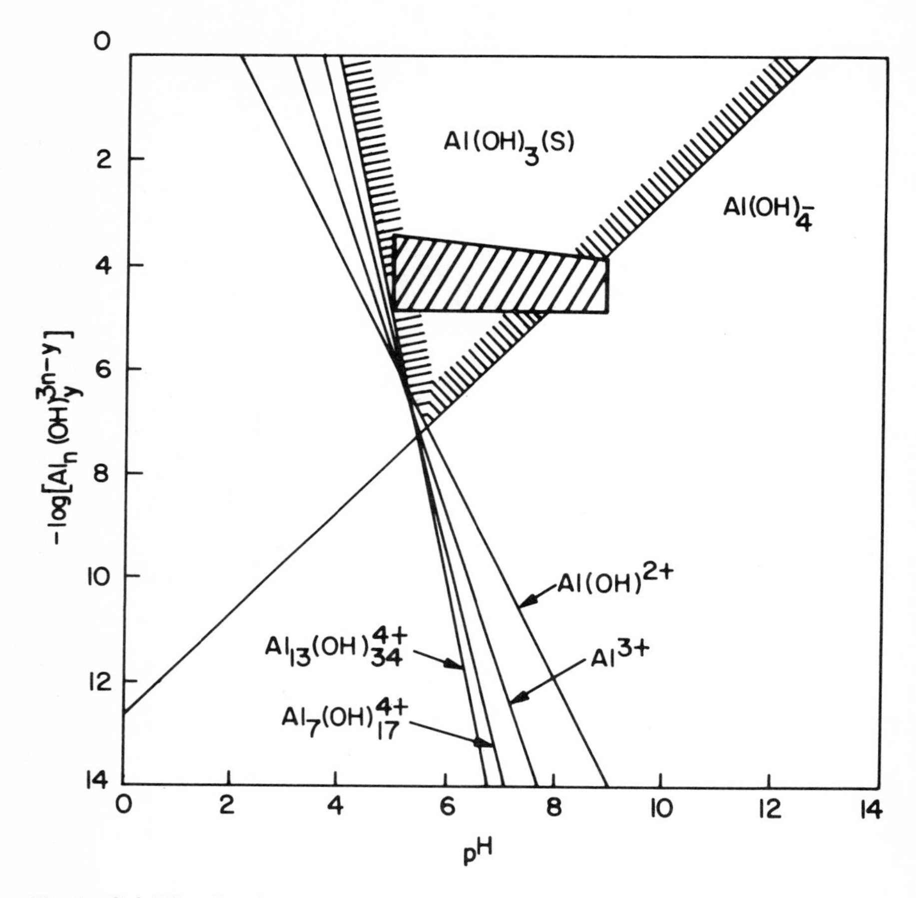

Fig. 4. Solubility domain of $Al(OH)_3$. The barred area is occupied by solid $Al(OH)_3$. The limiting lines were calculated from the appropriate equilibrium constants. Comparison of Figs. 3 and 4 shows the much lower solubility of $Fe(OH)_3$ at all pH values.

ional advantage is the increased accuracy and ease of controlling the dosage, since it is easier to meter a solution than to maintain accurate dry feeding into a solution pot.

Ferric sulfate: Available commercially as $Fe_2(SO_4)_3 \cdot H_2O$, contains a minimum of 20% Fe^{3+} or 28.5% Fe_2O_3. It is sold as a powder either in bags or in bulk and must be dissolved before use. The feed solution must be reasonably concentrated or the ferric ion will hydrolyze to $Fe(OH)_3$ prior to addition to the water to be treated and the effectiveness will be decreased significantly. Approximately 50% solutions can be made up conveniently and will remain stable for several days since the pH is low enough to prevent hydrolysis. Because of this low pH, solutions of ferric sulfate are very corrosive and must be handled in corrosion-proof equipment or equipment lined with plastic or rubber.

Some advantages of ferric sulfate are: (a) it is frequently less expensive for specific cases, particularly in color removal, (b) it is effective over a much broader pH range than aluminum salts, and (c) it is more effective in coagulating oxidized manganese. Some disadvantages are: (a) it is not available in liquid form, (b) solutions are very corrosive, (c) solutions cause red stains when spilled, and (d) more careful control of effluent quality must be practiced as any ferric ion passing into the distribution system may manifest itself as "red water" which will cause consumer complaints.

Ferrous sulfate: available commercially as $FeSO_4 \cdot 7H_2O$, copperas, or as a solution. A by-product of the pickling of steel with sulfuric acid, it is an inexpensive coagulant in steel-producing areas but must be oxidized to the plus three, ferric state before use since $Fe(OH)_2$ is fairly soluble—about 4 mg/liter as Fe. It can be oxidized readily at pH values above 7 by oxygen or chlorine. Theoretically, each milligram of $FeSO_4 \cdot 7H_2O$ requires 0.03 mg of oxygen or 0.126 mg of chlorine to oxidize it to the ferric state.

Ferric chlorosulfate: commonly called chlorinated copperas, $FeClSO_4$, produced as discussed above by adding chlorine to ferrous sulfate to oxidize it to the ferric ion.

Sodium aluminate: available as $NaAlO_2$ plus an excess of sodium hydroxide or soda ash to increase the pH when it is dissolved in water. This is necessary to prevent precipitation of $Al(OH)_3$. Sodium aluminate is used in the same way as alum, the difference being its alkaline nature as contrasted to the acidic character of alum.

The various metal salts currently offer several advantages over other possible coagulants, but these may be overcome within the next few years, for many applications, by the synthetic coagulants which can be tailored for specific functions. The current advantages are: (a) lower cost per unit of water produced, (b) acceptability by public health authorities, (c) techniques and art of use established, and (d) plants and equipment designed for their use. The principal disadvantages are: (a) rather large residuals may be carried into the distribution system, (b) large volumes must be handled, frequently entailing hand labor, (c) reactions are pH sensitive, which requires careful control and attention to changes in raw water quality, and (d) sludge disposal.

b. Organic

The use of organic substances for water clarification dates back to at least 2000 B.C. when the Egyptians used various ground nuts and seeds to coagulate or adsorb impurities in drinking water. Current practice is based upon two classes of organic coagulants, the natural and the synthetic polymers. The natural products include starches and cellulose derivatives. They are used principally as coagulant aids rather than as primary coagulants. The hydroxyl groups on both the starches and the cellulose derivatives may act as binding sites for attachment of foreign particles but are relatively ineffective in many cases because the negative charges on the impurities to be removed are not attracted to the negative sites on the polymers. Some of the starches have been modified to produce positively charged sites on the polymer chain and are consequently much more effective as coagulants.

The synthetic organic polymers, frequently referred to as "polyelectrolytes," are becoming increasingly important in municipal and industrial coagulation, as they have been designed specifically for use as coagulants. These polymers have long carbon skeletons with recurring active sites that are either positively or negatively charged or that may contain electronegative atoms in the chain itself, such as the polyethylene oxide polymers. Thus, there are three types of polymers as defined in terms of their charges: cationic, anionic, and non-ionic.

It has been shown that the polyelectrolytes destabilize colloids by adsorption onto the surface of the colloid and by interparticle bridging. Adsorption conforms to the Langmurian isotherm indicating monolayer geometry. The kinetics of adsorption of a cationic polymer by an essentially monodispersed latex suspension was shown to be dependent upon the rate of particle transport. The dosage required was a function of surface area (*18*).

There are many polyelectrolytes available commercially, but their exact structure is not easily determined as they are usually proprietary compounds. The type that theoretically shows the greatest promise as a primary coagulant, the cationics, is the least available – at least in numbers – since the U.S. Public Health Service Advisory Committee on Coagulant Aids in Water Treatment has certified a very limited number of synthetic cationic polyelectrolytes as safe for use in potable waters. As a matter of fact, at the end of 1969 only one such product had been so certified. The principal problem is due to the toxicity of the monomers that have been used in the production of these compounds (*19*, *20*).

There are several advantages in the use of polyelectrolytes for coagulation and flocculation. These include: (a) smaller volumes required, which reduces the shipping, handling, and storage expenses; (b) simplified dosage control, as metering of the solutions is relatively simple; (c) less critical pH control, as the coagulation and flocculation properties of the polyelectrolytes are relatively unaffected by changes in the pH of the system as contrasted to the great sensitivity of the metal salts system to the pH range; (d) reduced overall chemical cost in many cases; and (e) better results, as heavier flocs are formed particularly when used in conjunction with the metal salts where the dose of the metal salt may be reduced by substituting a small quantity of a polyelectrolyte for a large part of the metal salt to produce a larger, heavier, and more adsorbtive floc.

The disadvantages include: (a) dosage control is more critical, as underdosing produces a weak, ineffective floc and overdosing restabilizes the colloids and may produce a poor quality water; (b) technology is largely underdeveloped and many areas need considerable research attention before some basic problems can be overcome; and (c) current treatment plants are not designed to take advantage of the use of the polyelectrolytes as they were designed for metal salt coagulation. In particular, the polyelectrolytes require a greater energy input in the rapid mix step for maximum efficiency and this is provided in few present-day plants.

5. Applications

a. Color Removal

Color in water is derived from the degradation of organic materials and from the extraction of organic materials from the soil. These compounds are complex and have defied exact analysis, but it is known that at least some of them are aromatic polyhydroxy methoxy carboxylic acids, with structures similar to tannic acids (*21*). They are negatively charged in natural waters and frequently occur along with chelated iron or manganese. There are some waters, though, that are high in color but very low in iron and manganese.

Colored waters are obtained from surface supplies that drain swampy areas. The intensity of the color is quite variable from such sources. For example, the Braden River, which supplies the water treatment plant at Bradenton, Florida, varied over a period of several days from a value of 50 platinum–cobalt color units to a value of over 250. In dry periods the color is low, but it increases rapidly when rains wash the highly colored water from the swamps.

Coagulation or precipitation of color with metal salts optimally occurs at lower pH values than coagulation of turbidity. For aluminum salts the optimum pH range is from about 5.5 to 7.0, while for ferric salts the optimum range is from about 3.5 to 4.5. Coagulation at a higher pH seems to "fix" the color, i.e., make it even more difficult to remove. This may be an indicator effect since natural organic color is a pH indicator or, at least, is pH sensitive. The color for a given water is higher as the pH is increased. For this reason it has been recommended that the color be measured at a standard pH of 8.3 for comparison purposes (*22*).

The removal of natural organic color by either aluminum or iron can be explained on the basis of a direct precipitation reaction in which the color molecules react by replacing one or more of the hydroxide ions from the insoluble aluminum or iron hydroxide. Packham and his co-workers at the Water Research Association in England (*22a*) have proposed that the humic and fulvic acid fractions of color are removed by the formation of insoluble basic aluminum or iron humatates and fulvates. This mechanism is supported by the fact that coagulation of color occurs at low pH values where the iron(III) and aluminum hydroxides are reasonably soluble and do not precipitate (*23*).

The exact pH for the coagulation of a given water must be determined experimentally, generally by the use of the "jar test," whereby a series of small-scale coagulation tests are run in laboratory "jars." A good approximation of the pH and of the required dosage of ferric sulfate can be predicted from the color of the raw water by use of relationships developed by Black et al. in 1963 (*24*). In order to prevent passage of high levels of aluminum and iron ions through the filters, it is necessary to increase the pH to a value above 6.0 prior to filtration. This reduces the solubility of both iron and aluminum, as can be seen from Figs. 3 and 4, which show the solubilities as a function of pH. Since the finished waters must have the pH raised anyway before distribution for stabilization, this does not present any problem except that it must be done before filtration rather than after, as is practiced in some plants.

b. Turbidity Removal

Hydrolyzable metal salts are used for the removal of turbidity. The reaction between the metal ions and the tubidity may be a cation exchange reaction of the simple metal cation at the surface or it may be absorption at the interface. Most recent studies have indicated an adsorption mechanism. The adsorbed metal species is certainly one of the hydroxylated metal ions, as the pH range for the reaction of the metal salts and turbidity is such that the concentration of the unhydrolyzed metal ion is very small—far too small to be effective in destabilization of the colloidal turbidity particles.

Turbidity occurs in almost all surface supplies but rarely in groundwaters because of the filtering action of the soil through which the water percolates on its way to the underground aquifer. The composition of turbidity is quite varied and may range from inorganic substances such as clays, silts, sulfur, calcium carbonate, silica, manganese dioxide, and ferric hydroxide to organic substances such as bacteria, algae, dispersed oils and greases, and other debris. Larger particles may be kept in suspension by turbulence but settle rapidly when the turbulence is reduced.

Smaller particles, particularly those in the colloidal size range, will remain suspended for longer times even in quiescent waters. It is this fraction of the turbidity that must be removed by treatment. Turbidities vary widely even for the same water supply. A heavy rain usually increases turbidity due to increased erosion, so that water obtained from a river may vary from relatively low values, 0–10 Jackson candle units (JCU), for periods of little or no rainfall to high values, up to several thousand Jackson candle units, for periods after heavy rains. Pretreatment by passage through a reservoir, lake, or sedimentation basin will reduce the turbidity and level off the values. The effect of this factor may be illustrated by comparing the influent raw water at the water treatment plants of Minneapolis and St. Paul, Minnesota, which both use water from the Mississippi River. Prior to the St. Paul intake, the water passes through a series of lakes. The average turbidity of the plant influent for Minneapolis in 1967 was 6 and for St. Paul, 1. Whereas both cities soften the water before distribution, Minneapolis must also coagulate for the removal of turbidity.

U.S. Public Health Service Drinking Water Standards, 1962 (*3*) and most state standards recommend potable water turbidities of less than 5 JCU, although the more stringent Quality Water Goals of the American Water Works Association, published in December, 1968, recommend turbidities of only 0.1 JCU (*25*). In order to achieve these goals it is necessary for most plants that treat surface waters to employ coagulation, settling, and filtration for the removal of turbidity. Filtration alone is not satisfactory, as the very small colloidal particles that commonly occur will pass through the filters.

The coagulation of such waters with aluminum salts is usually accomplished in the pH range from 5.5 to 7.0 and with iron(III) salts from 3.5 to 9.5. The dosages vary from less than 1 ppm to almost 100 ppm, or from about 5 lb/1 million gal to 800 lb/1 million gal. The exact dose varies not only from plant to plant but from day to day or even hour to hour within the same plant. The determination of the dosage as a function of the raw water quality is still somewhat of an art, although a given plant may be able to make the required changes based on experience as the raw water quality changes. The jar test is still used by most plants as the principal tool in predicting dosages, although more sophisticated techniques such as microelectrophoresis and pilot filters are slowly gaining favor.

Plants designed for turbidity removal are frequently referred to simply as "filtration" plants but involve much more than simple filtration. There are two basic types, the conventional settling basins and up-flow, solids contact, or sludge blanket units.

The conventional plants normally have separate rapid mix, flocculation, and settling basins, with the settling basins designed so as to allow the sludge to settle to the bottom as the water passes through the basin (see Fig. 5). The up-flow units are designed so that the treated water passes up through a blanket of sludge. Detention time in the conventional units varies from 2 to 4 hr, whereas it is about 1 hr in the up-flow units.

Fig. 5. Montgomery, Alabama municipal water treatment plant. A conventional coagulation plant for removal of turbidity from Tallapoosa River water; capacity, 20 mgd. One of the settling basins is empty to show construction details.

This is an advantage in many cases in that much less space is required for the up-flow units. The increased contact of the precipitate with the sludge increases the degree of reaction and makes this type of unit more effective. The unit is used in both coagulation and softening plants.

The major application for the polyelectrolytes in municipal water treatment has been in the removal of turbidity. They have been used in conventional plants as primary coagulants and as filter conditioners to prevent uncoagulated particles from passing through the filters. In plants employing high rate filters they have been used as the sole coagulant with no settling prior to filtration. This can be done with the polyelectrolytes but not with the metal salts, as the floc formed using the polyelectrolytes is much tougher and resists breakup.

C. Softening

The presence of polyvalent cations in water causes "hard" water, so called because it is hard to form a lather with soap. Polyvalent salts of the long chain fatty acids present in soap are insoluble. In the days when all washing was done with soaps, the problem of hard water was more of a nuisance than it is today when the soaps have been replaced largely with synthetic detergents whose polyvalent metal salts are relatively soluble. The predominant polyvalent metal ions in waters used for potable supplies are calcium and magnesium, with the hardness expressed in terms of calcium carbonate. There are several ways of differentiating between types of hardness: (a) there are calcium hardness and magnesium hardness, (b) the synonymous terms carbonate, bicarbonate, or temporary hardness, with carbonate preferred, and (c) the analogous, synonymous terms noncarbonate, sulfate, or permanent hardness, with noncarbonate preferred.

The disadvantage of hard water, in addition to the cost of soap wasted, is the reduction in the life of hot water heaters, irons, washing machines, dish washers, and all other appliances that come in contact with hot water. In contrast to most other salts, calcium carbonate, magnesium carbonate, calcium hydroxide, magnesium hydroxide, and calcium sulfate decrease in solubility as the temperature is increased. This leads to precipitation of these salts in any environment in which the temperature is increased and the consequent localized overheating where scales are formed from the precipitated salts.

A water may be classified as hard if it contains more than 120 ppm of divalent ions, usually calcium and magnesium, as $CaCO_3$. The American Water Works Association Quality Water Goals recommend a maximum total hardness of 80 ppm for municipal purposes, but the values for softened waters range from 70 to 120 ppm, the final quality being determined in each municipality by such factors as public acceptance and economics. The hardness of waters in the 100 largest cities in the United States in 1962, treated or untreated, varied from 0 to 738 ppm with a median value of 90 ppm. Only 13 cities had waters with a hardness exceeding 180 ppm, and 28 soften their water supply (*4*). The distribution of hard waters in the United States is shown in Fig. 6.

There are two principal methods of softening water for municipal purposes: lime or lime-soda and ion exchange. The choice between the two is based on such factors as the raw water quality and the local cost of the softening chemicals.

1. History

The first softening plant in the United States was built in Oberlin, Ohio in 1903 and was followed by other plants in Ohio and Florida. The first plant in North America was built in Winnipeg, Canada in 1901, and England had over 50 plants by 1900. The early plants used the lime softening process with batchwise treatment. Later, settling basins were used in the continuous treatment method, which greatly increased the amount

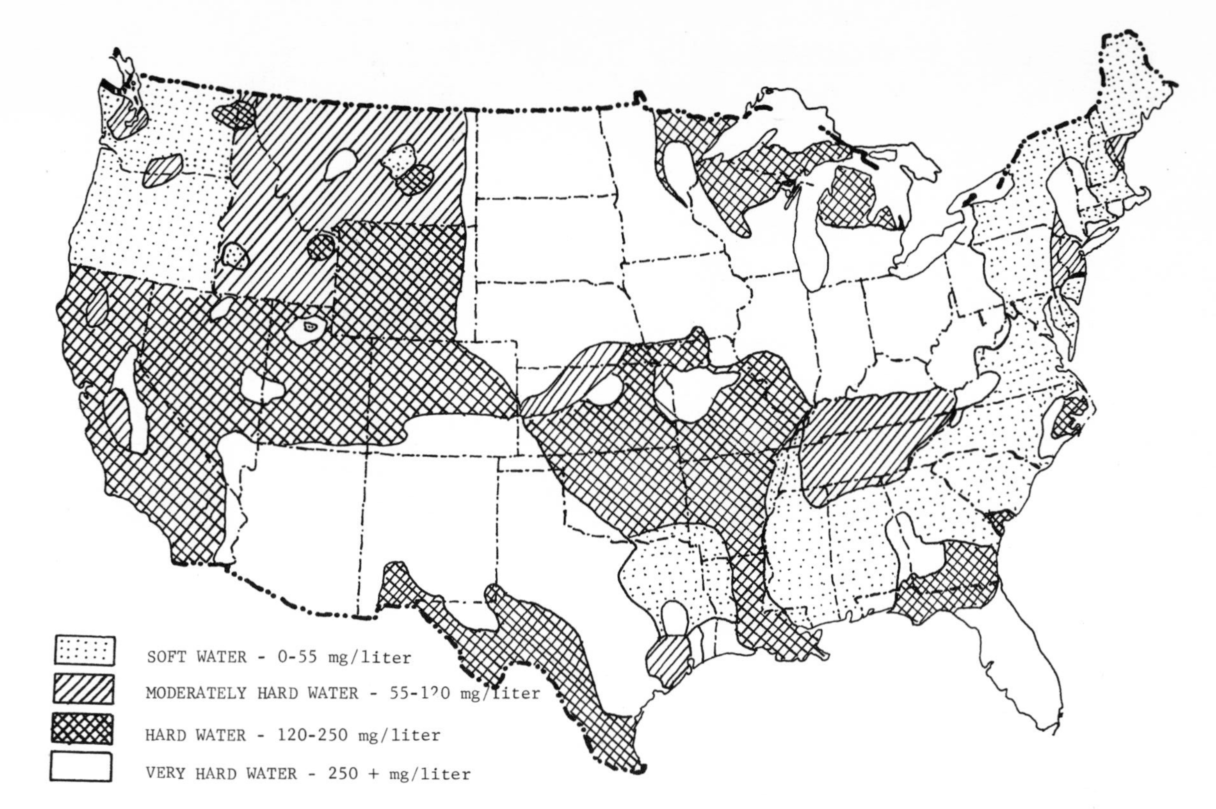

Fig. 6. Distribution of hard water in the United States. The areas shown define approximate hardness values for municipal water supplies.

Fig. 7. New softening plant, city of Gainesville, Florida. This plant will be in operation in 1971 with a lime-soda softening capacity of 30 mgd.

of water that could be treated in a given size facility. The growth in the number of softening plants was rapid thereafter, until today more than 1000 U.S. municipalities soften water. The majority of these plants are in the Midwest and in Florida (see Fig. 7).

2. Lime–Soda Process

The lime–soda process is based upon precipitation of calcium as calcium carbonate and magnesium as magnesium hydroxide. The lower limits of softening by this process, then, are based on the solubility of these precipitates. Table 3 shows the solubilities of calcium carbonate and magnesium hydroxide as functions of pH and temperature. When lime alone is used, only the carbonate hardness is reduced. The additional use of soda ash can reduce the noncarbonate hardness. This is shown by the following reactions, which cover the steps in the process:

Step 1. The reaction of any free carbon dioxide with the added lime:

$$CO_2 + Ca(OH)_2 \rightleftharpoons CaCO_3 \downarrow + H_2O \qquad (7)$$

This reaction provides no softening but takes place preferentially since carbon dioxide is the strongest acid present in the system.

Step 2. The reaction of calcium carbonate hardness with lime, when the

TABLE 3
Solubilities of Calcium Carbonate and Magnesium Hydroxide as Functions of Temperature and pH

	$CaCO_3$ solubility, mg/liter		$Mg(OH)_2$ solubility, mg/liter	
pH	77°F	140°F	77°F	140°F
7	970	240	—	—
8	79	20	—	—
9	8.0	2.1	13,000	5500
10	1.0	0.3	130	55
11	0.4	0.13	1.3	0.6
12	0.3	0.11	0.01	—

calcium may be represented as the bicarbonate at the pH values normally occurring:

$$Ca(HCO_3)_2 + Ca(OH)_2 \rightarrow 2CaCO_3 \downarrow + 2H_2O \tag{8}$$

Step 3. The reaction of magnesium carbonate hardness with lime, which may be represented as taking place in two steps to emphasize the stoichiometry:

$$Mg(HCO_3)_2 + Ca(OH)_2 \rightarrow CaCO_3 \downarrow + MgCO_3 + 2H_2O \tag{9}$$

$$MgCO_3 + Ca(OH)_2 \rightarrow CaCO_3 \downarrow + Mg(OH)_2 \downarrow \tag{10}$$

Combining the above two equations yields:

$$Mg(HCO_3)_2 + 2Ca(OH)_2 \rightarrow 2CaCO_3 \downarrow + Mg(OH)_2 \downarrow + 2H_2O \tag{10a}$$

These equations point out clearly that 2 moles of lime are required to remove 1 g atom of magnesium, or twice as much as is required for calcium removal.

Step 4. In the reaction of calcium noncarbonate hardness with soda ash, it is necessary to use soda ash to provide the CO_3^{2-}, as it is assumed that the available CO_3^{2-} would be used up in the reactions represented by Eqs. (8) and (9). The noncarbonate hardness may be represented as sulfate, although any anion except carbonate or bicarbonate could be present.

$$CaSO_4 + Na_2CO_3 \rightleftharpoons CaCO_3 \downarrow + Na_2SO_4 \tag{11}$$

Step 5. The reaction of magnesium noncarbonate hardness with lime and with soda ash. This is a two-step reaction since the reaction represented by Eq. (12) produces a soluble calcium salt that can be precipitated by reaction with carbonate ion.

$$MgSO_4 + Ca(OH)_2 \rightleftharpoons Mg(OH)_2 \downarrow + CaSO_4 \tag{12}$$

$$CaSO_4 + Na_2CO_3 \rightleftharpoons CaCO_3 \downarrow + Na_2SO_4 \tag{13}$$

It is to be noted that Eqs. (11) and (13) are the same but represent two different sources of $CaSO_4$.

From these reactions it can be seen that the addition of lime always serves three purposes and may serve a fourth. It removes, in order, CO_2, calcium carbonate hardness, and magnesium carbonate hardness, as represented by Eqs. (7), (8), and (10a). When some magnesium noncarbonate hardness must be removed, lime converts the magnesium noncarbonate hardness to calcium noncarbonate hardness, as seen in Eq. (12). Soda ash, then, removes noncarbonate hardness according to the reaction in Eqs. (11) or (13).

The effluent quality can be predetermined, within limits, since the amounts of lime and soda ash required can be calculated from the stoichiometry. The minimum hardness is determined by the solubilities of calcium carbonate and magnesium hydroxide, which are functions of pH, temperature, and ionic strength. Under average conditions this value is around 50–70 ppm, but it may be as low as 30 ppm when no magnesium is present. The carbonate alkalinity is usually 35–50 ppm, so the noncarbonate hardness can be determined from the final total hardness that is desired. Since soda ash is two to three times as expensive as lime, on an equivalent weight basis, it is desirable to maximize the noncarbonate hardness in the finished water. A further factor in the economics of lime–soda softening is that the amount of magnesium to be retained is maximized also, since it requires twice as much lime to remove magnesium as it does to remove calcium. Unfortunately, all of the hardness left in cannot be present as magnesium since magnesium hardness above 40 ppm causes scaling problems in hot water heaters. For this reason the magnesium concentration is usually reduced to this level (*26*).

Since softening is usually accomplished at high pH values and the reactions do not go to completion, the effluent from the treating unit is usually supersaturated with calcium carbonate. This would cement the filter medium and coat the distribution system. In order to avoid these problems, the pH must be reduced, thereby converting the insoluble calcium carbonate to the soluble bicarbonate. This is accomplished, in practice, by recarbonation, i.e., the addition of carbon dioxide. The reaction may be represented as follows:

$$CaCO_3 + CO_2 + H_2O \rightleftharpoons Ca(HCO_3)_2 \tag{14}$$

or

$$CO_3^{2-} + CO_2 + H_2O \rightleftharpoons 2HCO_3^- \tag{14a}$$

The lime–soda process may be modified for industrial applications that require relatively low hardness by increasing the temperature to near the boiling point. This reduces the total hardness to less than 25 ppm with no excess chemicals added and to less than 10 ppm with an excess of lime and soda ash. This is known as the hot lime–soda process. For a hardness of 0 ppm the lime–soda process may be followed by sodium cation exchange, and for hardness of less than 1 ppm phosphates may be added after lime–soda softening to precipitate $Ca_3(PO_4)_2$.

Among other modifications of the basic process are undersoftening, split recarbonation, and split treatment. In undersoftening, the pH is raised to between 8.5 and 8.7, where only calcium is removed and no recarbonation is required.

In split recarbonation the treatment is split into two units in series. In the first, or primary, unit the lime and soda ash are added and the water is settled and recarbonated just to the minimum pH of maximum carbonate ion concentration, about 10.3. The effluent then enters the second, or secondary, unit, where it contacts recycled secondary sludge to precipitate almost pure calcium carbonate. This pure calcium carbonate sludge is recycled in the secondary unit. The effluent is settled, recarbonated to the pH of stability, and filtered. The advantages over conventional treatment are reductions in lime, soda ash, and carbon dioxide requirements, very low alkalinities, and reduced maintenance costs because of the stability of the effluent. The major disadvantage is the requirement for twice the normal plant capacity.

Split treatment is used where the magnesium content of the water is high. This process also requires two units in series, which doubles the size of the plant. The raw water stream is split so that the ratio of the influent to the primary unit to the influent to the secondary unit is such that when all of the magnesium is removed in the primary and none in the secondary, the mixed effluents will contain 10 ppm magnesium as magnesium or 41 ppm as $CaCO_3$. For example, a raw water containing 35 ppm magnesium would require treatment of 25/35 of the water in the primary unit and 10/35 in the secondary unit. Sufficient lime is added in the primary unit to remove all of the magnesium, or at least to less than 1 ppm. This requires a pH of 11.1–11.3 with an excess of about 70 ppm of caustic alkalinity. The very clear effluent from the primary unit is mixed with the bypassed raw water in the secondary unit with no further chemical addition. The excess lime in the primary effluent is used to remove the calcium hardness from the bypassed raw water. This process has the advantage of using the excess lime required for magnesium removal, whereas in conventional treatment this excess must be wasted by recarbonation. Another advantage is the very low alkalinity of the finished water (*27*).

The large amount of sludge, containing calcium carbonate and magnesium hydroxide, produced by softening plants presents a disposal problem since it can no longer be discharged into the nearest stream or sewer. Each ton of lime used for softening produces 2.5–3.5 tons of sludge, depending on the relative amounts of magnesium and calcium and total hardness removed. Some of this sludge can be recycled to improve the completeness of reaction, but a large quantity must be disposed of. The methods used are lagooning, drying for land fill, agricultural liming, and lime recovery by recalcination. In recalcination the magnesium hydroxide is separated from the calcium carbonate by selectively dissolving it with recarbonation:

$$Mg(OH)_2 + 2CO_2 \rightleftharpoons Mg(HCO_3)_2 \quad (15)$$

The calcium carbonate is then concentrated and heated in a kiln, according to the following reaction:

$$CaCO_3 \overset{\Delta}{\rightleftharpoons} CaO + CO_2 \uparrow \quad (16)$$

The carbon dioxide can be used for recarbonation and the lime for further softening. The excess lime produced can be disposed of profitably because there is usually a market for lime. Miami, Florida and Dayton, Ohio, among others, have operated profitable recalcination plants for many years.

3. Ion Exchange Softening

In ion exchange softening, the calcium and magnesium ions are exchanged for monovalent ions, usually sodium or hydrogen. This is termed cation exchange softening, and current practice utilizes ion exchange resins based on highly cross-linked synthetic polymers with a high capacity for exchangeable cations. In cases where demineralization is required or the anion concentration must be reduced, anion exchange resins are employed. In general, these resins exchange monovalent anions such as OH^-, Cl^-, or HCO_3^- for polyvalent anions, or OH^- for all other anions, as in demineralization (*28*).

The exchange reactions are equilibria that favor the formation of a polyvalent ion-resin complex over a monovalent ion-resin complex or that may favor a given ion-resin complex over resin complexation with other ions of similar charge. The typical exhaustion steps may be represented as follows:

$$Ca^{2+} + 2NaR \rightleftharpoons CaR_2 + 2Na^+ \quad (17)$$

for a sodium cycle resin, or

$$Ca^{2+} + 2HR \rightleftharpoons CaR_2 + 2H^+ \quad (18)$$

for a hydrogen cycle resin. The regeneration steps are simply the reverse of the exhaustion steps, where the concentration of the regenerant ion is increased to a level that will reverse the reactions. Normally, about 15–25% solutions of the regenerants are required. For regeneration of a sodium-saturated resin the usual salt is sodium chloride and for hydrogen cycle either hydrochloric or sulfuric acid.

Any iron or manganese that is present in the soluble, reduced form as a divalent ion will be exchanged along with the calcium or magnesium.

The order of exchange of ions increases with atomic weight for a given valency. For example, $Ba^{2+} > Sr^{2+} > Ca^{2+} > Mg^{2+}$ and $Cs^+ > Rb^+ > K^+ > NH_4^+ > Na^+ > Li^+$.

The theoretical amount of 98% NaCl, the commercially available grade, required to regenerate a resin is 0.17 lb/kgrain of hardness removed. A kilograin of hardness, as $CaCO_3$, amounts to 1/7 lb. The actual salt consumption varies from 0.275 to 0.40 lb/kgrain, depending upon the specific resin and the regeneration conditions. The chemical cost of softening a given water can be calculated from the amount of hardness to be removed and thus from the amount of salt required in the regeneration. It should be kept in mind that the water passing through the ion exchanger is softened to zero hardness and is mixed with the amount of untreated water required to produce the quality desired.

In cases where the total dissolved solids may be too high, a combination of sodium and hydrogen cation exchange may be used. The exchange of sodium for either calcium or magnesium increases the total dissolved solids, but the exchange of hydrogen reduces it.

Another modification that is used is the removal of carbonate hardness by lime softening followed by removal of the necessary amount of noncarbonate hardness with sodium ion exchange. This takes advantage of the economics of each process. The lime–soda process is expensive when soda ash must be used for noncarbonate hardness removal, but the use of lime for carbonate softening is the least expensive method. On the other hand, ion exchange is just as efficient for noncarbonate as for carbonate softening to that it can be used economically for this part of the overall system.

One modification of the regeneration step is the use of seawater, which contains about 2.7% NaCl, as the regenerant. This requires coagulation and filtration for clarification, followed by chlorination of the seawater before use as a regenerant. However, the cost is much less than that of purchasing salt. The reduced concentration of the regenerant solution reduces the efficiency by 40–50% of that obtained in the normal process. Natural brines may also be used.

The quality of the effluent, then, can be predetermined from essentially zero hardness to the hardness of the raw water, and the total dissolved

TABLE 4
Comparison of Various Softening Processes

	Lime–soda		Ion exchange	
Properties	Cold	Hot	Na	H
Minimum hardness attainable	30 ppm	10 ppm	0 ppm	0 ppm
Total dissolved solids	Decreased appreciably	Decreased appreciably	Increased slightly	Decreased
Na Content	Depends on soda ash used	Depends on soda ash used	Increased significantly	Decreased significantly
Cost:				
Carbonate hardness[a]	Low	Low	Same for both CH and NCH	
Noncarbonate hardness[a]	High	High	Higher or lower than lime–soda—depends on CH/NCH ratio	
Capital	High	High	Lower than lime–soda	

[a]NCH, noncarbonate hardness; CH, carbonate hardness.

solids can be controlled also. Even though the exchange equilibria are shifted in favor of the softening reactions, very high concentrations of calcium and magnesium in the influent will allow some hardness to exist in equilibrium with the sodium form of the resin.

The two principal types of equipment used are fixed bed and continuous regeneration units. The continuous regeneration units have two tanks. In the softening tank the freshly regenerated resin is periodically fed into the tank countercurrent to the water flow. Part of the resin is periodically pulsed to a regeneration tank, where it is regenerated, washed, and then returned to the softening tank. This gives continuous softening by eliminating the interruptions for regeneration of the fixed bed exchangers.

4. Comparison of Lime–Soda and Ion Exchange Softening

The quality of the water produced by the various softening processes as compared to the quality of the raw water is shown in Table 4. The comparative costs for removal of both carbonate and noncarbonate hardness are also shown.

D. Filtration

1. General

The most common type of water treatment, except disinfection, is filtration. This was almost the only form of treatment practiced prior to the advent of chlorination and can be traced back to the Egyptians, who filtered water through earthenware vessels before the birth of Christ. The use of filtration was widely practiced in Europe in the 1800's and spread to the United States in the later part of that century with the first plants being based on simple filtration. Today, almost all supplies are filtered as an integral part of the purification process. It is important to note that most efficient filtration can be accomplished only after adequate preparation by coagulation, as most of the filters will not remove colloidal impurities by themselves. If an improperly coagulated water is filtered the effluent may be as turbid, as colored,or as bacteriologically unsafe as the influent, although many water authorities consider the filter as the primary barrier to passage of unsafe water. It is an effective barrier only when preceded by adequate pretreatment.

2. Types of Filters

There are several types of filters that are in use and they may be classified in several ways, first in terms of variation of filter media, second in terms of filter rate, and third in terms of operating conditions. The principal types of filters used today are discussed briefly.

The media used are sand, anthracite coal, and diatomaceous earth. The first two are graded carefully according to size and uniformity; the filters using these media are constructed so that the bottom layer is coarse gravel, with layers of smaller gravel above that and then graded layers of decreasing particle size of sand or coal as the filter bed is filled. The latest developments include filters having both sand and anthracite layers. Such a filter is shown in cross section in Fig. 8. Other modifications include mixtures of garnet, sand, and anthracite.

Sand filters are of two classes, "slow" and "rapid." The older of these is the slow sand filter, which depends on a layer of microscopic organisms, called a "schmutzdecke," composed of bacteria, fungi, and algae, to provide the filtering action. This necessitates a very slow passage of water through the filter and consequently larger filter areas. For example, a 1-mgd (million gallons per day) plant would require about 11,000 ft^2 of filter surface area with the rate between 3 and 8 million gal $acre^{-1}$ day^{-1} or 0.048–0.128 gpm (gallons per minute) ft^{-2}. No pretreatment is required, as the slow passage through the filter permits colloidal particles to attach to the "schmutzdecke." The very large filter areas required for the amounts of water needed in larger cities have favored the use of improved filters, though many small towns, particularly in the New England states, still use slow sand filters.

Rapid sand filters have filter rates about 40 times those of slow sand

Fig. 8. Cross-sectional view of a dual media filter. This view shows the use of gravel covered by sand and then by anthracite coal. These filters can be used at relatively high rates and frequently are exemplified by long filter runs. [Courtesy of the Taulman Company.]

filters. A 1-mgd plant would require only about 350 ft^2 of filter area with filtration rates customarily designed for and operated at 2 gpm ft^{-2}. This is the most common type of filter, but it must be preceded by pretreatment to agglomerate any colloidal particles. Studies have shown that impurities that have been well flocculated penetrate only a few inches into the filter bed before resistance builds up to the point that the filter must be cleaned. This is done by reversing the flow of the water through the filter and washing out the residue. About 2–5% of the water filtered is required for this backwashing.

Mixed media filters, called "high rate filters," can be used at up to 6–8 gpm ft^{-2} and achieve much deeper penetration into the bed as the particles in the surface layer are much larger.

Diatomaceous earth filters may be of either the vacuum or pressure type and are used principally on small supplies. The design and operation rate is about 1 gpm ft^{-2}, which would require about 700 ft^2 for a 1-mgd plant. The diatomaceous earth layer is usually about $\frac{1}{16}$ to $\frac{1}{8}$ in. in depth and is supported by septums of wire mesh. The filter particles bridge openings in the septum and provide a porous filter medium. Pretreatment is not needed, as even bacterial particles are removed in passage through the filter.

3. Theories of Filtration

The theories of water filtration can be classified as chemical or physical. The physical theories involve such factors as media size, filtration rate, and water temperature. Theories which involve such factors as interactions between the dispersed phase and the media based on the chemical and surface properties of the two phases are considered chemical theories (*29–32*).

There are two steps in the removal of a particle from the aqueous phase by a filter. These are transport of the suspended particle to the surface of the medium particle and interaction of the particles to form a bond that is strong enough to resist hydraulic stresses. The physical theories have attempted to combine the measurable parameters into mathematical equations for predicting filter performance. The parameters normally included are concentration of suspension, particle size of the medium, temperature, flow rate, flow time, head loss as a function of time, and many others. The problem has been that the theories will satisfy only a limited set of conditions or describe only a limited number of systems.

Chemical theories have concentrated on the colloidal chemical aspects of interaction; studies have attempted to relate filter performance to charge neutralization and optimization of the electrostatic attractive forces. Recent studies have focused attention on chemical bridging or specific adsorption models. These consider specific chemical interactions and may, therefore, be invoked in explaining many data previously thought to be contradictory. There are many aspects of the coagulation-flocculation-sedimentation-filtration sequence that need study before optimum results can be obtained, since too little attention has been paid to the overall process and the obvious interrelationships.

4. Effluent Quality

Slow sand filters can handle only relatively clear waters having initial turbidities of less than 10 JCU, but rapid sand filters can handle much higher turbidities and still produce effluents with turbidities less than 1 JCU. The quality of the effluent is controlled by the pretreatment efficiency. The Quality Water Goals of the American Water Works Association suggest a maximum turbidity of 0.1 JCU and point out that "today's consumer expects a sparkling, clear water. The goal of less than 0.1 unit of turbidity ensures satisfaction in this respect. There is evidence that freedom from disease organisms is associated with freedom from turbidity, and that complete freedom from taste and odor requires no less than such clarity. Improved technology in the modern treatment processes make this a completely practical goal" (*25*).

E. Stabilization

1. Definition

In the context of water treatment, the term stabilization means the control of the chemical characteristics of a water so that the water reaches the consumer with no change in composition and without causing any changes in the distribution system. Another way of expressing this is to define stabilization as the production of a water that is neither corrosive nor encrusting.

Water that has been softened by the lime or lime–soda process has a high pH and is saturated or supersaturated with calcium carbonate. This water is encrusting, i.e., it will precipitate calcium carbonate in the filter beds or in the distribution system. On the other hand, water that has been coagulated with aluminum or iron(III) salts has a low pH and is corrosive.

There are several ways of preventing either corrosion or encrustation. Encrustation may be controlled by reducing the pH of the water so that the excess $CaCO_3$ is converted to $Ca(HCO_3)_2$ or, more precisely, so that the CO^{2-} is converted to HCO_3^-. This is done after softening by recarbonation, as discussed in Sec. I.C. If the pH is reduced to a level below that at which the water is just saturated with calcium carbonate, corrosion may result (*33*).

Corrosion may be prevented by cathodic protection, whereby a potential is imposed on the system so that the metal to be protected becomes the cathode and therefore is unable to lose electrons. Corrosion of iron, for example, requires that the following reaction occur:

$$Fe^0 \rightleftharpoons Fe(II) + 2e \tag{19}$$

Thus, electrons must be lost by the iron for it to become oxidized. In the presence of oxygen in the water it may be further oxidized to Fe(III), which usually remains at the site as a tubercle. In the absence of oxygen the Fe(II) is carried on through the system and becomes obvious upon exposure to the atmosphere – in a household appliance or in an industrial process – where the characteristic reddish brown stain appears. Cathodic protection may be accomplished by imposing a direct current on the system or by using a more active metal than that to be protected, as a sacrificial anode. Magnesium is commonly used to protect iron. The magnesium electrode is buried near the iron to be protected and is connected to the iron by a conducting wire. Zinc may be used, such as in galvanizing, where physical contact between the two electrodes is desirable. The metal may also be protected by providing and maintaining a physical barrier such as paint or cement. This is the most effective method available, but it is impossible or prohibitively expensive to use this method for many systems.

Protection of the system and the consumer from such problems has been accomplished by chemical control for many years by the waterworks industry, as a matter of course.

2. Chemical Controls

The development of chemical control of corrosion and encrustation was based on the recognition that a thin "eggshell" film of $CaCO_3$ would provide a physical barrier to corrosion. Maintenance of such a film would be evidence of the stability of the system toward $CaCO_3$ precipitation or solution. The bases for determination of the conditions for stability are those that define the solubility of $CaCO_3$. These are: the K_{sp}^0 for $CaCO_3$, the ionic strength, the CO_3^{2-} concentration, the concentration of Ca^{2+}, and the temperature.

The solubility product constant, K_{sp}, is influenced by both the temperature and the ionic strength. The concentration of CO_3^{2-} can be determined from the alkalinity and the pH, if the value of K_2 for H_2CO_3 is corrected for ionic strength and temperature.

The equation for calculation of the pH of saturation, pH_s, was developed by Langelier (*34*):

$$pH_s = pK_2 - pK_{sp} + pCa + pAlkalinity \tag{20}$$

Since the values of pK_2 and pK_{sp} are functions of temperature and ionic strength, appropriate values for a specific water can be determined from a knowledge of these two factors. The correction of the thermodynamic constants, pK_2^0 and pK_{sp}^0, for changes in ionic strength may be satisfactorily approximated by using total dissolved solids. Values for these corrected constants were tabulated by Larson and Buswell(*35*).

The saturation index, SI, is usually determined as a measure of the extent of over- or undersaturation as,

$$SI = pH - pH_s \tag{21}$$

where the pH is the actual pH of the water and pH_s is obtained from Eq. (20). If the SI is (+) the water is encrusting, and if it is (−) the water is corrosive.

3. Practice

The protection of the system, the water quality, and the consumer is of such importance that water utilities use more than one protective device. Commonly, the system is protected by cement or other coatings and chemical control is practiced. Softened waters are routinely recarbonated to reduce the pH to the pH_s. Coagulated waters are treated with lime, soda ash, or caustic soda to increase the pH to the pH_s, and waters low in calcium carbonate may even be passed through limestone beds in order to saturate them.

The attainment of $pH = pH_s$ may not be adequate to protect the system, however. Visual inspection on a routine basis is necessary to assure safety. In general, it is common practice to maintain the saturation index slightly on the positive side, but observation of the thickness of the calcium carbonate layer may dictate changes in this practice. If the coating becomes too thick the carrying capacity is reduced and the SI must be lowered to allow for re-solution of the layer. On the other hand, it may be necessary to increase the pH of the water to produce a thicker layer. The pH_s, though, is a useful approximation of the conditions required for protection.

F. Disinfection

The most important single process in water treatment is the practice of disinfection, which can be defined as the reduction of the bacterial population to a safe level. This is in contrast to sterilization, which means the elimination of all bacterial life.

1. History

The first disinfection of a public water supply in the United States was achieved in Jersey City, New Jersey in 1903 using a hypochlorite solution. Many other cities followed this example, and in 1912 the use of gaseous chlorine was started. The incidence of waterborne disease has decreased with the increase in the disinfection of public water supplies. The few outbreaks of disease that have been traced to water supplies since the early years of this century have been related to supplies that were not disinfected or to contamination entering the distribution system after disinfection but prior to consumption.

2. Reactions

When chlorine is added to water it reacts according to the following equation:

$$Cl_2 + H_2O \rightleftharpoons H^+ + Cl^- + HOCl \qquad (22)$$

Hypochlorous acid has a K_a^0 of 2.5×10^{-8} and is therefore affected by the pH of the water:

$$HOCl \rightleftharpoons H^+ + OCl^- \qquad (23)$$

Figure 9 shows the distribution of two species for chlorine in water. There is essentially no chlorine present as Cl_2 at the pH values of concern in water supplies.

As far as disinfection with chlorine is concerned, the active species are Cl_2 and HOCl. The ionic form, OCl^-, is relatively ineffective.

When Cl_2 is added to a natural water there are many reactions that may take place in addition to the hydrolysis of the Cl_2. For example, most organic compounds are oxidized by Cl_2 or HOCl due to their high oxidation potential. In addition, ammonia or ammonia derivatives react to form chloramines in the following sequence:

$$NH_3 + Cl_2 \rightleftharpoons NH_2Cl + H^+ + Cl^- \qquad (24)$$

$$NH_2Cl + Cl_2 \rightleftharpoons NHCl_2 + H^+ + Cl^- \qquad (25)$$

$$NHCl_2 + Cl_2 \rightleftharpoons NCl_3 + H^+ + Cl^- \qquad (26)$$

Other reducing agents, such as H_2S, Mn^{2+}, and Fe^{2+}, will be oxidized by chlorine also. This represents a loss of chlorine for disinfection and is referred to as the "chlorine demand" of a water, i.e., the difference between the amount of Cl_2 added and the amount that can be determined analytically as residual chlorine. Residual chlorine includes chlorine in all reducible forms, such as Cl_2, HOCl, and OCl^-. A typical chlorine demand curve is shown in Fig. 10. The figure also illustrates "break point" chlorination, which is the addition of enough chlorine to satisfy the demand of all of the reducing agents present, including organic compounds

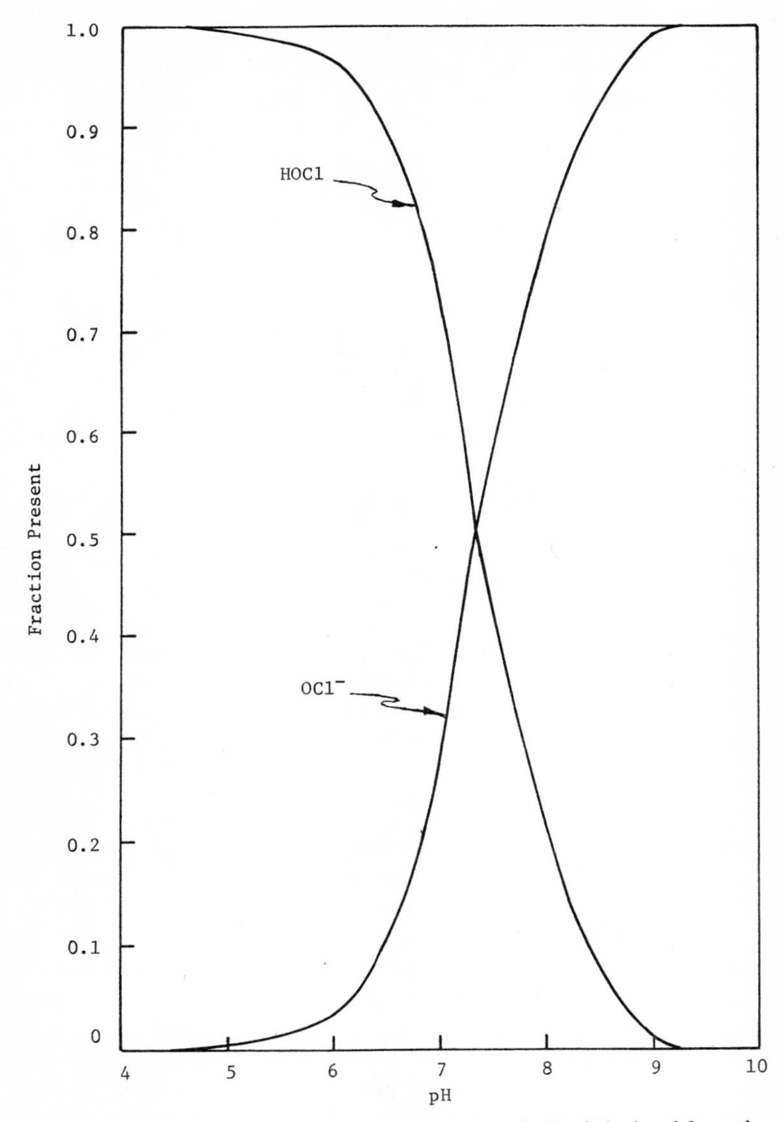

Fig. 9. Distribution of HOCl and OCl^- as a function of pH. Calculated from the appropriate equilibrium constants.

and ammonia and its derivatives. The break point is passed at the point where the addition of any more chlorine shows up as exactly the same increase in free available residual chlorine, i.e., HOCl and OCl^-. The curve up to point A in Fig. 10 represents the reaction of chlorine with

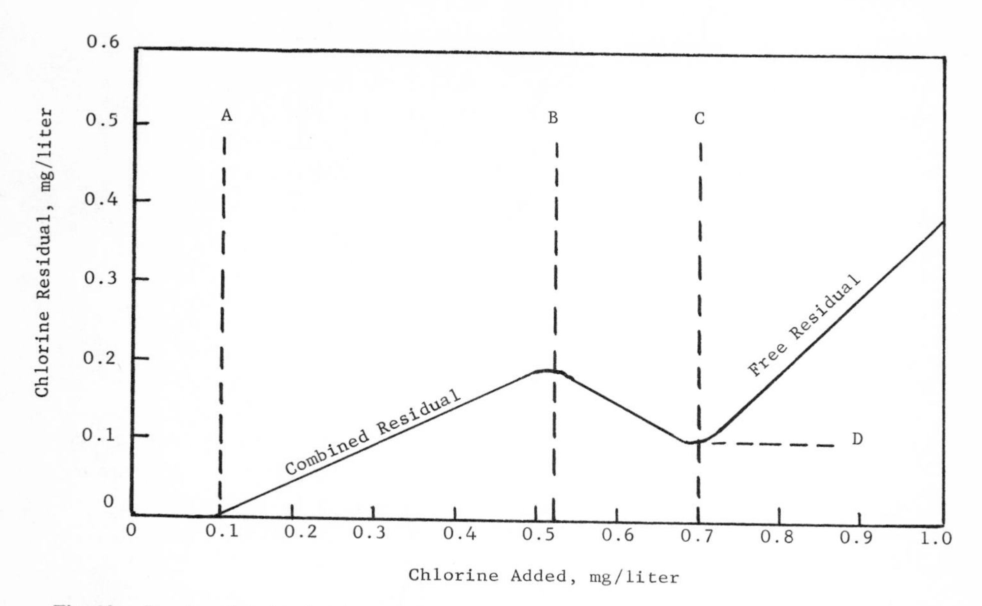

Fig. 10. Break point chlorination curve. This shows the effect of the addition of chlorine to a water that contains ammonia or its derivatives. The curve from 0 to A represents the chlorine demand of the sample. A to B represents combination with ammonia and its derivatives. B to C represents further oxidation of ammonia. Point C represents the "break point," and the curve for values of chlorine added greater than C represents the presence of additional HOCl and OCl^-, i.e., free residual chlorine. The chlorine residual equal to D represents the amount of total residual chlorine due to combined residual chlorine.

various reducing agents. From point A to point B, organic compounds and ammonia and its derivatives are oxidized, yielding combined available residual chlorine, i.e., chlorine that has some disinfecting power but much less than HOCl as it is already partially reduced. The curve in this range has a slope less than 1, as some of the chlorine is tied up and is not available as a disinfectant. The curve from B to C represents the oxidation of organic compounds and amines that can be oxidized no further at C, the break point. The combined residual at this point will be constant for all higher dosages of chlorine. Above C, the curve has a slope of 1, as each part of chlorine added yields a part of free available residual chlorine.

3. Disinfectants

Although chlorine is the most commonly used of all disinfectants, many others have been used. The most popular form of chlorine for larger plants is as the gas. Smaller plants may use either calcium hypochlorite or sodium hypochlorite. Some plants add chlorine, either as the gas or as the hypochlorite, but then add ammonia before transmission to the distribution system. This produces chloramines, which are less effective disinfectants than HOCl but which may be adequate in the well-disinfected water, as the lower oxidation potential will not attack the system as readily and may, therefore, provide a residual to the ends of the system.

Another substance that has been used is chlorine dioxide, a very strong oxidizing agent that is produced by the reaction of chlorine with sodium chlorite:

$$Cl_2 + 2NaClO_2 \rightleftharpoons 2ClO_2 + 2NaCl \tag{27}$$

Chlorine dioxide has the advantage of being effective at high pH values where most other disinfectants are ineffective. This offers advantages in applications such as softening plants (*36–38*).

Ozone has been used for many years in Europe and is an effective disinfectant, but it suffers the disadvantage of the difficulty of carrying a residual throughout the distribution system.

Iodine is a very effective disinfectant and has been used in many facilities. It is the disinfectant of choice by the armed services for field use but is generally too expensive for municipal use.

Other disinfectants that have been used or studied include ultraviolet light, radioactivity, and metal ions such as silver.

II. Advanced Treatment of Waste Waters

A. General

1. Purpose

The accelerating trend toward urbanization has accentuated the problems of waste disposal. The majority of cities have developed along watercourses and have used these convenient transmission routes for removal of a large portion of the waste products of the community. The increasing demand for water has required that such water supplies be utilized by downstream consumers. In the case of Chanute, Kansas, the city found it necessary to recycle its own sewage plant effluent through a stabilization pond to the influent of the water treatment plant for redistribution to the consumers. This was continued for a 5-month period during the severe drought in the fall of 1956 and winter of 1957. The mean cycle time was about 20 days and the water was reused from 8 to 15 times. The color gradually increased along with tastes and odors (*39*).

It has been estimated that a minimum of 40% of the population of the

United States is reusing water that has been returned to the raw water source as the effluent from a sewage treatment plant further upstream. In many cases there is little opportunity for natural purification between the outfall of one city's sewage treatment plant and the raw water intake for the next city's water treatment plant.

In addition to the problem of better water treatment for today's population, there must be concern for the future in nondegradation of existing streams, rivers, lakes, and underground aquifers because of future increased need for pure water.

2. Present Practice

The municipal waste treatment plants of today are designed to remove the solids and the bacteriologically degradable organics from the plant influent. Current laws in many states demand up to 90% removal of biochemical oxygen demand (BOD), which requires a biological process, so-called secondary treatment (see Chap. 9). The national trend is toward at least this level of treatment. The buildup of refractory organics, i.e., those not easily broken down biologically, and the increase in mineral content with reuse are not affected by this treatment. The technology is available today to reclaim or renovate these waters, but the cost is still too high to compete with alternate sources for water.

Alternate sources will soon become more expensive, however. To avoid limiting the growth of many areas of this country, more efficient methods of water renovation will have to be developed. The following sections discuss advanced waste treatment processes that show promise in solving this problem (*40*).

B. Distillation

One certain method of bacteriologically purifying and separating water from solids dissolved or suspended from the effluent from any waste treatment plant is distillation. It is equally certain that a simple, one-stage evaporation will not be economical. There have been several engineering modifications that offer economic promise through conservation of the heat (*41*).

Multiple-effect evaporation is one of the oldest such processes. It is based on providing steam from a boiler to heat the first stage. The pure vapor from the first stage is used to heat the second stage. Each pound of steam feed to the first stage should produce slightly decreasing amounts of steam distillate and pure condensate in succeeding stages. Up to 12 stages have been used in seawater conversion (*41a*).

Multistage flash evaporation is another possibility. It is based upon flash evaporation in a number of stages at successively lower temperatures with the heat of condensation used to heat the feed stream. Steam from a boiler is used to increase the temperature of the feed for the first stage (*41a*).

Vapor recompression is another possible modification of the simple evaporation process that offers economic promise. In this still, only one stage is used. The vapor is compressed and utilized to heat the evaporator. The heat of compression increases the temperature of the vapor slightly. The vapor which then condenses in the evaporator heating coils transfers a large portion of its heat to the evaporating liquid and passes to a heat exchanger where it transfers the rest of its excess heat to the feed. Modifications of the approach are rotary and wiped-film stills which increase the heat transfer efficiency.

The oldest type of evaporation is the solar still, in which there is current interest. This utilizes energy from the sun to accomplish the heating. In principle it is so designed that the sun's rays pass through a transparent film which also serves as the condenser. The latent heat of condensation is lost to the atmosphere. Modifications have included techniques for utilizing part of the heat of condensation for preheating the feed.

All of these methods show promise of producing pure water from sewage plant effluent for several cents per thousand gallons less than it can be produced from seawater because of the decreased concentration of the impurities which allows more stages or lower evaporation temperatures (*42*). The problem inherent in the use of solutions containing organic compounds and ammonia or its derivatives is the carry-over of tastes and odors with the distillate. One of the major advantages from an aesthetic or public acceptance standpoint is the general understanding of the sterilizing effect of the operation. A major disadvantage of current techniques is the problem of scaling or fouling of the heat exchange surfaces (*43*).

C. *Freezing and Hydrating*

Both of these processes are designed to purify waste water by separating the water as a solid from the impurities. They offer advantages in energy requirements as the heat of fusion of water is much lower than the heat of vaporization. In both processes the solid is separated from the aqueous solution, washed, and melted to produce pure water. In each case the pretreatment must be more extensive than for distillation but comparable to that for many other processes. All suspended matter must be removed, as must gases, or they will be trapped in the solid phase and carried through the process.

Both processes follow essentially the same steps. The feed is filtered, deaerated, and precooled by the cold product as it exits. In a cold reactor it contacts an immiscible gas, such as butane for the freezing process, or propane or a halogenated methane or ethane in the hydrating process. Ice or solid hydrate forms and is separated from the solution. It is then washed with pure water and melted to produce pure water. The residual concentrate is recycled for cooling. The gas is recondensed and returned to the reactor(*41*, *41a*).

In the hydrate process the solid formed is a clathrate containing a gas molecule enclosed in a cage of 10–18 water molecules. This clathrate has a melting point well above the melting point of water(*41*, *41a*).

The advantages of freezing and hydrating are(*44*): (a) The scaling associated with distillation is eliminated; (b) the energies involved are lower than for distillation, and (c) the equipment is simple. The disadvantages are(*44*): (a) contamination of the water with the refrigerant; (b) disposal of concentrated residue; and (c) lack of knowledge about effect on microorganisms.

D. Reverse Osmosis

Osmosis is the diffusion of a solvent through a selective membrane from a solution of higher solvent concentration to one of lower concentration. The osmotic pressure is the hydrostatic head built up by the downstream side that is required to counterbalance the tendency to flow. If this head is exceeded by an externally applied pressure, the solvent will flow from the solution that is lower in concentration of solvent to that which is higher in concentration of solvent, i.e., from the solution more concentrated in impurities to that less concentrated. This is "reverse osmosis." Pure water then can be extracted from a concentrated waste stream. One disadvantage is the lack of technological background compared with other processes of advanced waste treatment; these other processes have a long history of application in other areas, such as desalination of seawater and chemical processing, whereas reverse osmosis has been developed only in the last 15 years(*41*, *41a*, *45*, *45a*).

The membranes currently used are composed of cellulose acetate. Special techniques are required to give optimum properties in terms of flux rates, selectivity, and ease of preparation. Other polymers, including polyvinylene carbonate, B-glucan acetate, and various methacrylates, have been studied, but none has shown better overall performance than cellulose acetate. Polyelectrolyte membranes from poly(sodium styrene sulfonate) and poly(vinylbenzyl trimethyl ammonium chloride) have shown promise for removal of high molecular weight organic molecules and some have been shown to have high flux rates and to give good salt rejection.

Since the rejected organic macromolecules and high salt concentration tend to block or form a scale on the surface of the membranes, techniques to avoid this have been developed. These include turbulent flow and thin channel flow across the face of the membrane.

The membranes are mounted so that the concentrated waste stream passes on one side under pressure and the product water is removed on the other side. Membrane configurations have included (a) plate and frame, (b) spiral-wound module, (c) tubular module with directly cast membrane, (d) tubular module with separate membrane insert, (e) hollow fiber module, and (f) thin channel.

Relatively high flux rates have been obtained – up to several hundred gallons per day per square foot of membrane surface – at high salt rejection rates. Water can be produced with low TDS and low organic content at costs competitive with other methods of tertiary treatment. The advantages offered by reverse osmosis are (a) simplicity of design, (b) low energy requirements, since no change of state occurs, (c) organic and inorganic removal, (d) low maintenance, and (e) simplicity of operation.

The disadvantage is the present high cost of processing, principally due to inadequate membranes. Improvements in membrane technology, however, may make this method more attractive economically.

E. Electrodialysis

Electrodialysis has been used in chemical laboratories for many years as a means of concentrating charged particles in solution. Colloidal dispersions have been stabilized by the removal of the ionic components that would tend to agglomerate the colloidal particles by charge neutralization. The same principle can be employed in waste water purification (*41*).

In principle, a potential is imposed between two electrodes immersed in the solution to be purified. The positively charged ions migrate to the cathode, or negatively charged electrode, and the negatively charged ions migrate to the anode. The solution, then, midway between the two, has a lower concentration of ions than the original solution. In practice, the electrodes are separated by a series of compartments with sides composed of semipermeable membranes, alternating between cation permeable and anion permeable. The solution to be purified is passed into the compartments, continuously, with the cations attracted one way and the anions the other. After a cation passes through a cation-permeable membrane it next encounters an anion-permeable membrane and is thereby prevented from further migration. Anions passing the other way are similarly concentrated in the same compartment. Therefore, alternate compartments have concentrated ions and purified solution which can be collected for use or disposal.

There are several plants currently producing potable water by electrodialysis of brackish waters (*42, 45b*). The major advantage of electrodialysis is the ability to remove inorganic ions, and the major disadvantage is the failure to remove organic compounds. It may be possible to modify the organic molecules so that they are charged, but even then some are so large that they foul the membranes irreversibly. Pretreatment by coagulation-filtration or carbon adsorption may be necessary in order to eliminate this problem and to remove solid impurities that may also foul the membranes (*46*).

Even though the total dissolved solids may be reduced by 40–50%, the level of NH_4^+ may still be too high. This can be overcome by using an oxidized effluent from secondary treatment that contains the nitrogen as NO_3^-.

Electrodialysis shows promise as a demineralizing step in water renovation after the use of adsorption or coagulation and filtration for insoluble solids and organic removal. These additional steps add materially to the overall costs.

F. Coagulation

The use of coagulation followed by flocculation, sedimentation, and filtration is an obvious step following secondary treatment due to its established merit in the production of potable water. The mechanisms and controlling factors have been discussed in detail in Sec. I. In waste water treatment the same principles hold. In many cases it has been possible to use the synthetic polyelectrolytes as flocculants to improve sedimentation in the primary settling units in existing treatment plants, thereby increasing the level of solids removal.

The use of aluminum or iron salts has shown great promise for the removal of phosphates. This is an important step in reduction of eutrophication, as phosphorus is an essential element in algal growth (*47*).

Improvements in coagulants and in technology should make coagulation a major factor in tertiary treatment.

G. Adsorption

Adsorption of organic impurities, such as hydrocarbons and those causing tastes and odors, by activated carbon has been common practice in the waterworks industry for many years. Similarly, purification of various process streams in the chemical industry by carbon adsorption has been an integral part of many processes. In principle, the process is very simple and effective, as it is only necessary to bring the water to be treated into contact with the solid adsorbent and then physically separate the adsorbent from the purified water. In the case of packed bed or expanded bed adsorption systems, the water is passed through the adsorbent.

In order for adsorption to be economical the adsorbent must be very inexpensive or capable of efficient regeneration. Many inexpensive materials have been tried as adsorbents, but none has been shown to compare economically with activated carbon. The development of continuous thermal regeneration on a commercial basis has decreased the cost of carbon adsorption to the point that it is considered one of the most economical approaches to tertiary treatment. It has been shown repeatedly that the efficiency can be increased by improving the quality of the stream to be treated. This can be accomplished, in many cases, by coagulation, flocculation, and sedimentation prior to adsorption. This reduces physical blinding of the carbon bed and rapid buildup of pressure losses (*48–52*).

Many of the aspects of carbon adsorption have been studied extensively, so that it is frequently considered as a polishing step in many water renovation schemes; but it may be the major purifying step, as in the 2-mgd plant at the South Tahoe Public Utility District in California for treating municipal secondary effluent (*53*).

One disadvantage is the 2–3% continuous leakage that may be due to organics adsorbed on colloidal particles which pass through the adsorption unit.

Recently, two papers (*53a,53b*) gave details on the combined use of coagulation and carbon adsorption to treat raw sewage resulting in a very highly purified effluent.

H. Foam Fractionation

The separation of substances from water by flotation has been practiced for many years in the mining industry. The techniques and knowledge developed are of value in application to waste water treatment. The foam produced in many waste water treatment plants may be utilized as a means of purification simply by physically separating the foam from the liquid phase. Both solids and soluble organic materials may be separated by this technique. The efficiency of organic removal is a function of the surface activity of the organics involved. For example, the alkylbenzene sulfonates (ABS), the surfactants most commonly used in household detergents, are 80–90% removed by flotation. The 10-mgd Whittier Narrows water reclamation plant routinely produces an effluent containing less than 1.8 mg/liter of ABS. Foam separation, using the "soft" detergents, may be used to increase organic removal from the 90% level for biological treatment to 93–95% (*54–56*).

The details of separations utilizing gas bubbles are presented in Chap. 11. The authors point out that foam fractionation can separate both surface-active solutes and soluble inorganic or organic substances that are collected or attracted to the surface-active agent. In this way microorganisms, ions, colloids, and soluble molecules can be separated from solution.

I. Solvent Extraction

Solvent extraction may be of interest in certain aspects of waste water treatment such as removal of specific contaminants, but it does not seem to offer promise as a general method because of very low efficiencies and problems in contamination of the effluent with traces of the solvent.

Phenols have been extracted from domestic sewage with ketones (*57*) and with aromatic hydrocarbons (*58*).

J. Ion Exchange

Several of the purification methods discussed above, such as coagulation, adsorption, flotation, and solvent extraction, are practiced for the removal of organic impurities and do not reduce the inorganic content of the waste water significantly. While distillation would succeed in eliminating inorganic substances, ammonia may be carried into the purified stream.

In these cases, one possible technique for inorganic reduction is ion exchange. Mixed bed or serial treatment by cation and anion exchange resins would provide essentially complete deionization. As in the field of municipal water treatment, discussed previously, it is customary to treat only that portion of the stream required to produce the desired ion level and blend with the required amount of the untreated stream.

Many water renovation schemes have included ion exchange as one of the steps.

K. Virus Removal

Since the reuse of water will have to increase, it is essential that all pathogenic organisms be removed or inactivated in the treatment process. Current primary and secondary treatment of domestic sewage may remove a major part of the bacteria, but the extent of virus removal is not known accurately for many processes due to the problems involved in viral assay.

The major protection utilized against both bacteria and viruses has been disinfection with chlorine. It has been shown that chlorination of effluent from an activated sludge treatment plant reduces the bacterial population to less than 1% of its original value. Such data are not generally available for viruses, but it has been shown that not all viruses are removed or inactivated by chlorination of secondary effluent(*59*).

Although there has been increasing concern about the presence of viruses in public water supplies, it has been shown that the only enteric virus that is waterborne is infectious hepatitis(*60*). This particular virus has not been successfully cultured in the laboratory, so very little is known about the effect of various treatment processes.

A study by Wentworth et al.(*61*) showed that excess lime–soda softening was as much as 99.983% effective in the inactivation of poliovirus. Straight lime softening was about 70% effective, but pH increase alone was ineffective.

Merrell and Ward(*62*) state that no viruses were present in the swimming area of the Santee, California water reclamation project. This water was domestic sewage treated by a modified activated sludge process, then passed through an oxidation pond and percolation beds. Before use in the swimming area it was coagulated, filtered, and heavily chlorinated. Very low levels of viruses were found after the percolation step.

Berg et al.(*63*) showed that poliovirus can be 70–99.86% removed or inactivated from secondary effluent by lime flocculation and 98.6–99.995% removed or inactivated if the lime flocculation is followed by sand filtration.

III. Projections for the Future

A. Improved Coagulants

In both water and waste water treatment, coagulation plays a major role. The research activity in this area probably exceeds that in any other comparable area. This will certainly lead to major improvements. Among the most important of these should be the development of new coagulants with properties tailored to specific functions. The principal type should be the organic polymers, as they provide much more possibility of structural modification to give the desired properties. The cationic polymers show the greatest promise, as most of the impurities to be coagulated are negatively charged. The current problem in the production of cationic polyelectrolytes is that the monomers are very toxic and the production of a monomer-free polymer is difficult. This problem will be solved by better separation techniques, better methods of polymerization, and new monomers.

There is also the probability that inorganic polymers will be developed that will have general application at relatively low unit cost.

Another technique that is currently used in other fields and will probably be adapted to water and waste water coagulation is the blending of coagulants to achieve the desired properties. The development of this approach will depend on a better knowledge of the mechanisms of coagulation and the properties of coagulants.

B. Flotation or Foam Separation

The studies of flotation or foam separation in waste water purification show promise that may be translated to useful techniques in potable water treatment. Current studies of bacteria, algae, turbidity, and color removal by foaming processes may lead to methods capable of competing economically with the coagulation-flocculation-sedimentation processes. The engineering aspects require some development, but these would be forthcoming if the process economics warrant.

The major advantages of flotation over sedimentation are the relatively concentrated sludge, which simplifies disposal, and the ease of separating the sludge from the purified water.

C. High Rate Filtration

There will certainly be major improvements in filter design, construction, and operation. Among these should be improved media, allowing much higher filter rates, and modified distribution of media in the filter bed. Data supporting the efficiency of such filters and effective monitoring devices will lead to approval of high filter rates on potable waters. Many studies currently under way are directed toward improved understanding of the mechanisms of filtration.

D. Reuse

One of the most important developments in many water-poor areas will be the acceptance of recycled water as a source of potable water. The advancing technology of waste water treatment should provide treated effluent of better quality and lower cost than many currently available sources. The integration of water and waste water treatment plants into one entity will become a common practice in water-poor areas.

E. Others

1. Dual Systems

In areas where a limited supply of good water is available, such water may be used for human consumption, with another system being installed to provide treated sewage plant effluent or other less acceptable supply for all other purposes. The additional expense of essentially parallel systems may be justified by the consumer on esthetic grounds.

2. Transmission

There should be an increase in the volumes of water involved in inter-basin transfers, as there are adequate supplies of very high quality water in many undeveloped areas. The purchase of these supplies by one political entity from another will provide an economical solution to many water shortages. There are, of course, present-day examples of this solution, such as the Colorado River and Feather River projects in California.

3. Computerized Plant Control

The advances in electronics and in continuous monitoring of various chemical and physical parameters make the use of computerized plant control possible today. Plants will be constructed that will require a minimum of supervision, as all decisions for treatment and distribution can be controlled by properly programmed computers.

4. Specific Adsorbants

As our water quality requirements increase and as the recognition of the effects of specific contaminants increases, it will become necessary to remove specific substances from solution. This may best be accomplished by the use of adsorbents designed to attract a specific substance or class of substances. Currently, activated carbon is used in many water and waste water treatment processes for the removal of organics, with varying degrees of success. Improvements in the economics of the reactivation of carbon and development of other adsorbents will lead to increased use.

REFERENCES

1. K. A. MacKichan and J. C. Kammerer, *Circ. No. 456*, U.S. Geological Survey, Washington, D.C., 1961.
2. E. Nordell, *Water Treatment for Industrial and Other Uses*, Reinhold, New York, 1961.
3. *Public Health Service Drinking Water Standards, 1962, Public Health Serv. Publ. No. 956*, U.S. Dept. Health, Education, and Welfare, Cincinnati, Ohio, 1962.
4. C. N. Dufor and E. J. Becker, *J. Am. Water Works Assoc.*, **56**, 237 (1964).
5. V. K. LaMer and T. W. Healy, *J. Phys. Chem.*, **67**, 2417 (1963).
6. I. S. Hyatt, U.S. Pat. 293,740 (1884).
7. G. W. Fuller, *Water Purification at Louisville*, Van Nostrand, New York, 1898.
8. E. J. Theriault and W. M. Clark, *Public Health Rept. (U.S.)*, **38**, 181 (1923).
9. L. B. Miller, *Public Health Rept. (U.S.)*, **38**, 1995 (1923).
10. L. B. Miller, *Public Health Rept. (U.S.)*, **40**, 351 (1925).
11. L. B. Miller, *Public Health Rept. (U.S.)*, **40**, 1413 (1925).
12. L. B. Miller, *Public Health Rept. (U.S.)*, **40**, 1472 (1925).
13. A. P. Black, *Water Sewage Works*, Ref. No. (1961).
14. W. Stumm and C. R. O'Melia, *J. Am. Water Works Assoc.*, **60**, 514 (1968).
15. J. H. Sullivan and J. E. Singley, *J. Am. Water Works Assoc.*, **60**, 1280 (1968).
16. J. E. Singley and A. P. Black, *J. Am. Water Works Assoc.*, **59**, 1549 (1967).
17. J. E. Singley and J. H. Sullivan, *J. Am. Water Works Assoc.*, **61**, 190 (1969).
18. A. P. Black and M. Vilaret, *J. Am. Water Works Assoc.*, **61**, 209 (1969).
19. Anon., *J. Am. Water Works Assoc.*, **60**, 1094 (1968).
20. Anon., *J. Am. Water Works Assoc.*, **59**, 1291 (1967).
21. A. P. Black and R. F. Christman, *J. Am. Water Works Assoc.*, **55**, 897 (1963).
22. J. E. Singley, R. H. Haris, and J. S. Maulding, *J. Am. Water Works Assoc.*, **58**, 455 (1966).

22a. E. S. Hall and R. F. Packham, *J. Am. Water Works Assoc.*, **57**, 1149 (1965).

23. L. L. Hedgepeth, N. C. Olsen, and W. C. Olsen, *J. Am. Water Works Assoc.*, **20**, 467 (1928).
24. A. P. Black, J. E. Singley, G. P. Whittle, and J. S. Maulding, *J. Am. Water Works Assoc.*, **55**, 1347 (1963).
25. Anon., *J. Am. Water Works Assoc.*, **60**, 1317 (1968).
26. T. E. Larson, *J. Am. Water Works Assoc.*, **43**, 649 (1951).
27. A. P. Black, *J. Am. Water Works Assoc.*, **58**, 97 (1966).
28. R. Kunin, *Elements of Ion Exchange*, Reinhold, New York, 1960.

29. C. R. O'Melia and W. Strumm, *J. Am. Water Works Assoc.*, **59**, 1393 (1967).
30. C. R. O'Melia and D. K. Crapps, *J. Am. Water Works Assoc.*, **56**, 1326 (1964).
31. J. L. Cleasby and E. R. Bauman, *J. Am. Water Works Assoc.*, **54**, 579 (1962).
32. H. E. Hudson, *J. Am. Water Works Assoc.*, **61**, 3 (1969).
33. J. R. Baylis, *J. Am. Water Works Assoc.*, **10**, 365 (1923).
34. W. F. Langelier, *J. Am. Water Works Assoc.*, **28**, 1500 (1936).
35. T. E. Larson and A. M. Buswell, *J. Am. Water Works Assoc.*, **34**, 1667 (1942).
36. R. S. Ingols and G. M. Ridenour, *J. Am. Water Works Assoc.*, **40**, 1207 (1948).
37. A. T. Palin, *J. Inst. Water Engrs.*, **2**, 61 (1948).
38. J. A. Myhrstad and J. E. Samdal, *J. Am. Water Works Assoc.*, **61**, 205 (1969).
39. D. F. Metzler, R. L. Culp, H. A. Stoltenberg, R. L. Woodward, G. Walton, S. L. Chang, N. A. Clarke, C. M. Palmer, and F. M. Middleton, *J. Am. Water Works Assoc.*, **51**, 1021 (1958).
40. D. G. Stephan, *Civil Eng.* (New York), **35**, 46 (1965).
41. *Public Health Serv. Publ. No. AWTR-1*, U.S. Dept. Health, Education, and Welfare, Cincinnati, Ohio, 1962.
41a. W. A. Homer, *J. Am. Water Works Assoc.*, **60**, 869 (1968).
42. *Federal Water Pollution Control Admin. Publ. No. WP-AWTR-19*, U.S. Dept. Health, Education, and Welfare, Cincinnati, Ohio, 1968.
43. J. N. Neale, *Public Health Serv. Publ. No. 999-WP-9*, U.S. Dept. Health, Education, and Welfare, Cincinnati, Ohio, 1964.
44. A. J. Barduhn, *Public Health Serv. Publ. No. 999-WP-4*, U.S. Dept. Health, Education, and Welfare, Cincinnati, Ohio, 1963.
45. H. Beder, Master's Thesis, University of Florida, Gainesville, 1968.
45a. R. Eliassen, *J. Am. Water Works Assoc.*, **61**, 572 (1969).
45b. *J. Am. Water Works Assoc.*, **58**, 1231 (1966).
46. J. D. Smith and J. L. Eisenmann, *Water Pollution Control Res. Series Publ. No. WP-20-AWTR-18*, U.S. Dept. Health, Education, and Welfare, Cincinnati, Ohio, 1967.
47. D. F. Bishop, L. S. Marshall, T. P. O'Farrell, R. B. Dean, B. O'Connor, R. A. Dobbs, S. H. Griggs, and R. V. Villiers, *J. Water Pollution Control Federation*, **39**, 188 (1967).
48. C. B. Hopkins, W. J. Weber, and R. Bloom, *Federal Water Pollution Control Admin. Publ. No. TWRC-2*, U.S. Dept. Health, Education, and Welfare, Cincinnati, Ohio, 1968.
49. R. S. Joyce and V. A. Sukenik, *Public Health Serv. Publ. No. 999-WP-28*, U.S. Dept. Health, Education, and Welfare, Cincinnati, Ohio, 1965.
50. J. C. Morris and W. J. Weber, *Public Health Serv. Publ. No. 999-WP-11*, U.S. Dept. Health, Education, and Welfare, Cincinnati, Ohio, 1964.
51. J. N. Williamson, A. H. Heit, and C. Calmon, *Public Health Serv. Publ. No. 999-WP-14*, U.S. Dept. Health, Education, and Welfare, Cincinnati, Ohio, 1964.
52. J. C. Morris and W. J. Weber, *Public Health Serv. Publ. No. 999-WP-33*, U.S. Dept. Health, Education, and Welfare, Cincinnati, Ohio, 1966.
53. R. L. Culp, *J. Am. Water Works Assoc.*, **60**, 85 (1968).
53a. M. M. Zuckerman and A. H. Molof, *Natl. Symp. Sanit. Eng. Res. Develop. and Design, 2nd, Ithaca, N.Y., July, 1969*, American Society of Civil Engineers, pp. 83–84.
53b. W. J. Weber, Jr., C. B. Hopkins, and R. Bloom, *Ann. Conf. Water Pollution Control Federation, 42nd, Dallas, Texas, October, 1969.*
54. E. Rubin, R. Everett, J. J. Weinstock, and H. M. Shoen, *Public Health Serv. Publ. No. 999-WP-5*, U.S. Dept. Health, Education, and Welfare, Cincinnati, Ohio, 1963.
55. I. A. Eldib, *J. Water Pollution Control Federation*, **33**, 914 (1961).
56. C. A. Brunner and D. G. Stephan, *Ind. Eng. Chem.*, **57**, 40 (1965).

57. *Metallges. Gen.*, **941**, 544 (1956); through *Chem. Abstr.*, **52**, 20 (1958).
58. D. P. Manka, U.S. Pat. 2,812,305 (1957).
59. O. J. Sproul, L. R. Larochelle, D. F. Wentworth, and R. T. Thorup, *Water Reuse, Chem. Eng. Progr. Symp. Ser. 63*, American Institute of Chemical Engineers, New York, 1967, p. 130.
60. S. R. Weibel, F. R. Dixon, R. B. Weidner, and L. J. McCabe, *J. Am. Water Works Assoc.*, **56**, 947 (1964).
61. D. F. Wentworth, R. T. Thorup, and O. J. Sproul, *J. Am. Water Works Assoc.*, **60**, 939 (1968).
62. J. C. Merrell and P. C. Ward, *J. Am. Water Works Assoc.*, **60**, 145 (1968).
63. G. Berg, R. B. Dean, and D. R. Dahling, *J. Am. Water Works Assoc.*, **60**, 193 (1968).

Chapter 9 Biological Waste Treatment

Emil J. Genetelli
DEPARTMENT OF ENVIRONMENTAL SCIENCES
COLLEGE OF AGRICULTURE AND ENVIRONMENTAL SCIENCE
RUTGERS, THE STATE UNIVERSITY
NEW BRUNSWICK, NEW JERSEY

I. Introduction

It is recognized by most professionals in the field of waste water treatment that the most efficient method for eliminating or removing organic material from waste water is by utilizing biological treatment systems. There are three basic systems most often utilized for this removal: activated sludge, trickling filters, and aerobic oxidation lagoons.

While these three systems may differ somewhat regarding the detention times, oxygen requirements, and mode of utilization of biological slimes, the essential biochemistry that occurs within each system is identical.

II. Principles of Biological Waste Treatment Processes

In general, the purpose of the biological treatment unit is to remove organic material either by oxidation to carbon dioxide, water, and other derivatives, or by conversion of the organic material into a settleable form which can be removed by gravity sedimentation. The production of carbon dioxide, water, and ammonia is referred to as the respiration stage, while the conversion of the organic material into new bacterial cells which can be settled out is referred to as synthesis.

Since these are aerobic biological treatment processes, the final hydrogen acceptor for the oxidation of organic material is oxygen. During this hydrogen transfer the liberation of energy from the organic molecule is utilized for synthesis and for energy requirements of cellular substances.

The quantity of oxygen that is necessary to oxidize the organic material depends mainly upon the biological oxygen demand (BOD) satisfied during the biological treatment process.

Various treatment methods have been estimated as satisfying the following percentages of applied BOD; conventional activated sludge, 90–95%; high rate trickling filters, 65–85%; and low rate trickling filters, 80–90% (*1*).

The removal of organic material is accompanied by oxidation and synthesis of cells. Gellman and Heukelekian (*2*), Helmers et al. (*3*), and Placak and Ruchhoft (*4*) have shown that the amount of new cell material produced per pound of BOD added varies with the chemical composition of the substrate.

Hoover and Porges (*5*) found 52% by weight of cell yield from the oxidation of skimmed milk. Gellman and Heukelekian (*2*) reported a yield of 0.5 lb of volatile suspended solids per pound of BOD fed to the system for several industrial wastes. Helmers et al. (*3*) reported that solids production varied with BOD removal. Okun(*6*) found that the total quantity of new cell material formed was a function of the loading applied to the system. At high loading ratios high growth rates were obtained, while at low loadings sludge destruction was evident.

Table 1 shows the effect of various organic substrates on the relative proportions of oxidation and synthesis in 24-hr, batch-fed activated sludge systems (*4*).

The variabilities among the yields reported can be explained by visualizing that if the organic loading to the unit is quickly assimilated by the microorganisms then the organisms metabolize themselves, thereby oxidizing cellular material normally contributed to sludge yield. The energy yields of different compounds are not the same; consequently, more or less of a particular substrate may be used to satisfy the energy requirements of the system. Thus, the removal of organic material from liquid waste is achieved by complete destruction (oxidation), which yields energy, and by synthesis, which uses the energy produced during the oxidation of organic matter.

The object of this chapter is to briefly review the three basic biological waste treatment systems and to deal specifically with their methods of operation and the results which we have so far been able to achieve.

TABLE 1
Division of Substrate between Oxidation and Synthesis[a,b]

Class of compound	Per cent		
	Range present	Oxidized mean	Converted to sludge
Carbohydrates	5–25	13	65–85
Alcohols	25–38	30	52–66
Amino acids	22–58	42	32–68
Organic acids	30–80	50	10–60

[a]Data from Ref. *4*.
[b]24-hr, batch-fed activated sludge system.

A. Primary Treatment

Before undergoing secondary biological treatment a waste water is usually subjected to a preliminary form of treatment, referred to as "primary," in order to remove suspended and other insoluble matter.

These systems usually include rough screens or racks, constant velocity grit removal tanks, and primary settling tanks.

1. Coarse Screens or Racks

Racks are usually made of long, parallel-shaped bars placed on a slope of approximately 45° from the horizontal and 2–4 in. apart. Their purpose is to remove larger particles of floating or suspended matter from the waste stream. Quantities of screenings removed from domestic waste waters vary from 1 to 15 ft^3 of screenings per million gallons, depending upon the size of openings. There is no measurable reduction in BOD or suspended solids after screening, but the procedure prevents the clogging of treatment equipment, overcomes accumulations of unsightly deposits, and intercepts esthetically undesirable floating matter.

2. Grit Chambers

A grit chamber can be defined as a modified settling basin in which the horizontal velocity of flow is so controlled that only heavier solids such as grit and sand are removed while the lighter organic solids are carried forward in suspension. Grit removal is extremely important in protecting working equipment such as pumps and in preventing accumulations of undigestible solids in sludge disposal systems. From 1 to 12 ft^3/1 million gal of grit can be collected from systems, depending upon the type of sewers, weather conditions, and the age of the systems.

3. Primary Settling Tanks

Primary tanks are utilized in all trickling filter plants and in some activated sludge plants. Their purpose is to reduce the content of settleable solids so as to reduce the load on subsequent treatment systems and to prevent the formation of sludge deposits. They may be either manually or mechanically cleaned, have detention times of 2 hr or less, and can be responsible for as much as 35% of the BOD removal and 60% of the suspended solids removal.

The necessity for primary treatment prior to activated sludge systems is still heatedly discussed. On the one hand, the ease of removal of the primary sludge removed (approximately 400 ft^3/1 million gal at 4–6% and its subsequent effect on secondary treatment make it an attractive addition to a large system. On the other hand, the cost of disposing of the primary sludge removed (approximately 400 ft^3/1 million gal at 4–6% solids) by anaerobic systems has caused some investigators to advocate its discontinuance.

There is no doubt that it is necessary to use primary treatment prior to trickling filtration systems to avoid clogging of the filter media.

B. Activated Sludge

The activated sludge process of waste water purification is one of the most common processes for the secondary treatment of wastes. The activated sludge consists of a gelatinous matrix in which filamentous and unicellular bacteria are imbedded and on which protozoa crawl and feed. The bacterial genera which predominate depend on the characteristics of organic matter in the waste water, e.g., *Pseudomonas* for hydrocarbon and carbohydrate wastes and *Alcaligenes*, *Bacillus*, and *Flavobacterium* for proteinaceous wastes (*4a*). The process consists of mixing activated sludge, recirculated from a final settling tank, with incoming raw or primary sewage to form a mixed liquor, which is subsequently aerated and from which activated sludge is later settled. When a plant is first started it can be seeded with an activated sludge from a currently operating plant. If, however, no seed sludge is available, then one can be built up over a short period of time (4–6 weeks) by simply continually aerating, settling, and returning the residue of the sewage.

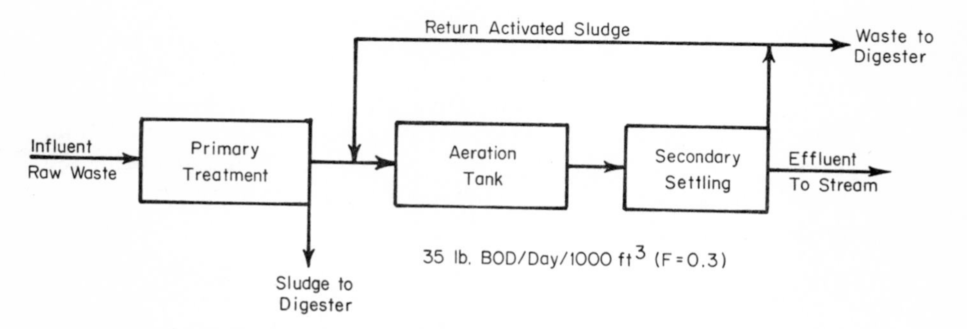

Fig. 1. Flow diagram of a conventional activated sludge plant. *F* refers to sludge loading ratio.

1. Description of Conventional Process

A flow diagram of a conventional activated sludge plant is shown in Fig. 1.

In the process, primary treated sewage is mixed with a portion of returned activated sludge and aerated for 4–6 hr. The process can be said to consist of the following steps:

1. Mixing of the activated sludge with the sewage to be treated.
2. Aeration and agitation of this mixed liquor for the required length of time.
3. Separation of the activated sludge from the effluent and the subse-

quent return of a portion of the settled sludge to be mixed with the incoming sewage.

2. THEORY OF ACTIVATED SLUDGE OPERATION

The mechanism of removal of organic material from sewage by activated sludge can be generalized in the following way:

$$\text{organic material} + \text{bacteria} \xrightarrow[O_2]{\text{sufficient}} CO_2 + H_2O + NH_3 + \text{energy} \quad (1)$$

The release of energy results in the formation of new cellular material; this formation is commonly referred to as synthesis, while the production of carbon dioxide, water, and ammonia is referred to as respiration. Thus, we can then say that the oxidation of organic material results in respiration and synthesis.

The bacteria necessary to utilize the organic material are present in the activated sludge flocs. These flocs are gelatinous matrices in which unicellular and filamentous bacteria are present. Protozoa and metazoa are usually present on the surface of the flocs.

When sewage is first mixed with activated sludge a portion of material is stored away (adsorbed) until the bacteria find it necessary to use it as food. The remaining portion of organic material is then oxidized and results in synthesis and respiration. Thus, the total removal of organic material is accomplished in two major parts, adsorption and oxidation. Adsorption occurs within the first section of the aeration tank during the first hour of aeration. The total removal of organic material by the activated sludge process is from 90 to 95%.

3. ORGANIC LOADING PARAMETERS

The main treatment units involved in the activated sludge process are shown in Fig. 1. They include the aeration tank and the sludge separation or secondary settling tank. Primary effluent, or the untreated waste water containing colloidal and soluble organic material, enters the aeration tank where it is attacked and stabilized by the mixed flora and fauna known as activated sludge. This activated sludge, when combined with the influent waste water, is known collectively as "mixed liquor," and the sludge solids are designated as either the "mixed liquor suspended solids" (MLSS) or "mixed liquor volatile suspended solids" (MLVSS). The MLVSS is the volatile portion of the MLSS and is usually related by: MLVSS = 0.80 (MLSS). Most investigators feel that the use of MLVSS more closely represents the total active (biological) mass in the system. This concept is discussed more fully in Sec. II.B.7. The overflow from the secondary settling tank leaves the process as effluent. Since there is usually a net production of biological cellular material by the aerobic treatment process, some sludge must be removed from the system. It is either removed intentionally by separation of sludge or occurs unintentionally by the loss of solids in the process effluent.

The nutrient substrate level, or food value of a waste water, is measured in terms of the biochemical oxygen demand (BOD) of the constituents present in the waste (see Chap. 15). The relationship existing between the quantity of substrate and the BOD removal efficiency of the process has led to the development of many concepts proposed for use as loading parameters.

Loadings have been expressed in terms of pounds of BOD/1000 ft^3 of aeration tank volume. In the literature, however, constant BOD removal efficiencies (87–90%) have been reported (*7,8*) at loadings of 30–120 lb/1000 ft^3, thus indicating the ineffectiveness of this ratio as a general loading parameter. Parameters based on tank volumes do not take into account the amount of solids under aeration or specify the length of the aeration period.

Gould (*9*), in 1939, suggested the term "sludge age" as a measure of the quantity of organic material entering the process. This term was defined as the average length of time a particle of suspended solids remains under aeration. It was expressed as the ratio of the pounds of mixed liquor solids under aeration to the pounds of suspended solids entering the system per day, and thus had a unit of days.

When dealing with sewage, sludge age is adequate since the suspended solids concentration in milligrams per liter usually is about equal to the BOD concentration in milligrams per liter. The parameter is modified for relatively soluble organic wastes by substitution of the pounds of BOD entering the system per day in place of the pounds of suspended solids. This substitution makes the sludge age equivalent to the ratio of the pounds of mixed liquor solids under aeration per pound of BOD entering the process per day.

Harris et al. (*10*), in 1926, were actually the first to propose, if not actually formulate, a loading parameter combining the three major factors applicable to the activated sludge process. The factors evaluated were the BOD of the applied waste, the quantity of sludge present in the aeration tank, and the period of aeration. Since then, ample evidence has been presented to demonstrate that the effectiveness of the activated sludge process and the amount of aid required for its operation are primarily dependent upon the daily BOD input/mixed liquor volatile solids ratio.

In the activated sludge process, it is difficult to obtain an exact measure of the amount of active cell material held in a system. However, because of the simplicity of the test procedures involved, it has become standard practice to measure the weight of suspended "solids" (MLSS) contained in the mixed liquor and assume that there is a relationship between the amount of solids in the mixed liquor and the number of active organisms present.

The National Research Council Sub-Committee on Military Sewage Treatment (*11*) established a loading parameter expressed as pounds of BOD applied daily per 1000 lb of suspended solids in the aeration tank.

Okun (*6*) formulated a similar parameter based on volatile suspended solids, and Heukelekian et al. (*12*) found that volatile solids content was a better measure of excess activated sludge produced than suspended solids content.

The Water Pollution Control Federation (*13*), after reviewing the literature, has recommended the "sludge loading ratio" (SLR) as the loading parameter to be used for activated sludge. The SLR is expressed as pounds of BOD applied per day per pound of MLVSS and thus amounts to a food-to-organisms ratio. (The symbol F is also used for SLR, as in Figs. 1–4 and 6.)

The concentration of suspended solids in the mixed liquor depends upon the waste being treated and the aeration capacity of the plant. Normally, the solids concentration is between 1000 and 2000 mg/liter. Calculations of the required solids concentration necessary to obtain efficient treatment under specific conditions can be made by utilizing the SLR.

The equation for the calculation of the SLR is:

$$\text{SLR} = \frac{24\,La}{Sat\,(1+R)} \tag{2}$$

Where La is the BOD in milligrams per liter, t is the detention time in hours, Sa is the concentration of MLVSS in milligrams per liter, and R is the recycle ratio.

It is usually expected that the sludge loading ratio should never exceed a value of 0.3/day in normal sewage plant operation utilizing conventional activated sludge.

If we recognize that the area of contact surface and the opportunity for contact are two of the most important factors in the activated sludge process, the sludge loading ratio can be readily accepted as a general parameter of loading intensity. It can be considered as (a) the weight of removable substrate applied in a unit of time to (b) a unit of contact surface for (c) a unit of contact time. The first factor (a) is readily determined by analytical procedures (BOD) and measurements of sewage flow. The second factor (b) can be evaluated only indirectly by the volatile suspended solids concentration in the mixed liquor, while the third factor

(c) is usually represented as a statistical average for activated sludge plants (*14*).

Even though much research has gone into the development of this loading parameter, situations are encountered in plant practice that indicate the ineffectiveness of this ratio.

The quality of the material making up the BOD is extremely important in plant performance and also for design purposes. BOD is not a simple entity, but must be evaluated in light of the composition of the substrate contributing the BOD (*15*). Some domestic sewage treatment plants experience process trouble at one SLR but other plants at the same SLR operate extremely well.

4. Sludge Volume Index

Knowledge of the volume of sludge present or of the solids (on a dry basis) in the aeration tank is not sufficient for good process control. A combination of these two is, however, an essential feature of good process control. This combination, called the sludge volume index (SVI), is the volume in milliliters occupied by 1 g (dry weight) of sludge after 30 min of settling, and is sometimes referred to as the Mohlman index (*16*). The test is performed by settling a 1-liter sample of mixed liquor for 30 min in a 1000-ml graduated cylinder. The volume occupied by the sludge is reported as per cent or in milliliters; the suspended solids are determined and reported in per cent by weight or in milligrams per liter. The SVI can be expressed as:

$$\text{SVI} = \frac{\%\ \text{volume of sludge settled in 30 min}}{\%\ \text{suspended solids}} \tag{3}$$

Although the 30-min period has been adopted as a standard, variations in sludge settling rates taking place in this time, due to the effect of different mixed liquor solids concentrations, have been reported (*17*). A modification of this test has been developed by Heukelekian and Isenberg which somewhat eliminates these variations and is usually used for all the index calculations in experimental work (*18*). This modification calls for an adjustment of the mixed liquor solids concentration to between 1000 and 1500 mg/liter before the SVI determination is made.

A well-settling sludge may have a Mohlman index between 50 and 100, but an index of 200 is indicative of a sludge with poor settling characteristics.

Knowledge of the sludge volume index and the SLR is necessary for good process control since there is a critical value of sludge volume index below which the volume of settled sludge in the final tanks will exceed the return sludge rate. If the SVI rises to 200 then the return sludge rate must be increased to maintain a constant solids concentration under aeration. If the concentration of mixed solids decreases, the sludge loading ratio will increase, thereby increasing the bulking tendency of the sludge and compounding the problem.

5. Sludge Density Index

The Water Pollution Control Federation(*13*) has recommended the use of the sludge density index (SDI) rather than the sludge volume index (SVI). This can be defined as the weight of a specific volume of sludge after it has settled for 30 min. It is calculated as follows:

$$\mathrm{SDI} = \frac{100}{\mathrm{SVI}} \tag{4}$$

SDI values are more amenable to graphical expression than are SVI values. Another reason for their use is to satisfy the recommendations given in Ref. *13* and to encourage standardization of this nomenclature for usage in the field. This means that a sludge with good settling characteristics has an SDI of between 2.0 and 1.0, while an SDI of 0.5 indicates a "bulky" or "nonsettleable" sludge.

6. Activated Sludge Bulking

During normal operation, mixed liquor flows into the final settling tanks from the aeration tanks. The activated sludge forms flocs, settles, and the effluent flows over the weirs of the final tank. At times the activated sludge does not settle well; its volume becomes greater in comparison to its density and the return sludge pumps cannot keep up with the large volumes of light sludge settling in the final tanks. If this condition persists, the sludge in the final tanks will spill over the weirs and the BOD of the final effluent will increase. This phenomenon, known as "sludge bulking," is a major problem of the activated sludge process. Quite frequently, bulking occurs unexpectedly when the plant seems to be operating at its peak efficiency and producing an excellent effluent(*19*).

There have been many theories advanced as to the cause of bulking, none of them completely satisfactory. It was observed by many early workers that the organism *Sphaerotilus natans* was often present in large numbers when bulking occurred (*20–25*). *S. natans* is a sheathed, filamentous bacterium and it was reasoned that its growth in excessive numbers caused the sludge to be less dense, and hence caused it to bulk. Ingols and Heukelekian(*22*) stated that bulking "produced by carbohydrates is a direct response of *Sphaerotilus* to a relatively long contact with an available energy food." Babbitt and Baumann(*26*), on the other hand, have stated that "these organisms are a result, not a cause of bulking." A causal relationship between the presence of *Sphaerotilus* and bulking has yet to be established, although Finstein(*27*) has shown a correlation between the number of measurable filaments in a floc and high SVI values. This suggests that the filaments exert control over the sludge volume index. The author has also shown a definite link between *Sphaerotilus* growth and SVI increase(*28*).

Bulking has been associated with such characteristics of the raw waste as septicity, heavy organic load, trade wastes, mineral oil, and excessive carbonaceous content. In the treatment plant, overaeration, underaeration, poor mixing, short circuiting, and too high or too low mixed liquor solids have been listed as causes of bulking. Other causes within the plant have been listed as septic return sludge and excessive detention periods.

Although the characteristics of the raw waste are, without question, associated with the causes of bulking, the experience of this writer indicates that in-plant causes of bulking are more significant than those attributed to raw waste characteristics.

The many different measures used to control bulking reflect the incomplete knowledge of its causes. Reducing the suspended solids in the aeration tank, increasing the quantity of air, increasing the solids in the aeration tanks, use of inert materials, use of iron compounds, chlorination of the return sludge, reaeration of return sludge, and addition of lime to the mixed liquor are some of the methods which have been recommended at various times(*19*).

Since most treatment operations are concerned with maintaining a constant SLR or sludge age, it becomes apparent that perhaps there are inconsistencies contained within the formulation of the parameter.

Recent work by the author has shown that differences inherent within the SLR factor are related primarily to solids concentration values alone. The MLVSS value which is maintained is the single most important operational parameter in activated sludge operation(*22*).

7. Active Mass Approximation and Loading Parameters

The food-to-organisms ratio designated as the sludge loading ratio (SLR), mentioned earlier, is the parameter currently used in waste water treatment practice. The SLR corresponds to the symbol F which is sometimes used in mathematical formulations of biological treatment processes (*22, 29, 30*). and utilized herein later in Figs. 1–4 and 6. This symbolic representation of a biological system depends on an understanding of the fundamental biological concepts, within the activated sludge process. Without these concepts, it is extremely easy to represent an analysis that is mathematically correct but biologically incorrect.

The constant assumption made in most mathematical formulations is that the suspended solids or volatile suspended solids contained in the mixed liquor represent the total number of bacteria present and are therefore related to the total enzymatic activity of the system. This assumption is not entirely correct(*31*, *32*). For any given system it may be possible to use some solids parameter as an index of active mass, but it must be used with extreme caution. In an unchanging ecological system the solids concentration may be adequate, but as a general estimation of active mass it is sorely lacking.

Hartmann(*33*) in 1960 reported that the weight of nitrogen in the sludge was a better estimation of the quantity of bacteria and was hence the optimum loading parameter for activated sludge systems. Hartmann designated his loadings as BOD per milligram sludge nitrogen.

More recently, interest has been expressed in the possibility of utilizing pounds of BOD applied per day per unit weight of deoxyribonucleic acid (DNA) as a loading parameter(*8*). This parameter would appear to provide a more accurate measure of the food-to-organisms ratio since the amount of DNA present gives a fairly accurate measure of the amount of cell material present in the system.

Recently, it was shown that a parameter based upon DNA is more responsive than any others currently employed in predicting certain types of operational upsets(*28*).

C. Anaerobic Digestion

When primary treatment is employed prior to secondary systems, a sludge is produced which is usually disposed of by anaerobic digestion. Excess activated sludge may also be treated in such fashion. The main purpose of sludge digestion is to produce an innocuous residue which can be easily disposed of, and also to reduce the volume of sludge which must ultimately be removed. It is impossible to completely discuss anaerobic digestion in all its complexities, but those interested are referred to the Manhattan College Conference of 1957, which gives excellent coverage of the topic.

In general, the decomposition of complex organic matter is accompanied by production of intermediate and endproducts. Such compounds as methane, hydrogen, organic acids, and alcohol are the main products of decomposition of carbonaceous organic materials under anaerobic conditions. Similarly, degradation of proteins will result in compounds such as ammonia, amino acids, amides, peptones, hydrogen sulfide, indole, skatole, and mercaptans.

Sewage solids subjected to anaerobic decomposition pass through three stages: (1) a period of intensive acid production (acidification), (2) a period of acid digestion (liquifaction), and (3) a period of intensive digestion and stabilization (gasification). Each step is represented by the production of typical intermediate and end products. Under normal operating conditions all three stages occur simultaneously.

The term "liquifaction" as applied to digestion connotes the transformation of large solid particles into either a soluble or finely dissolved form. This process is brought about by hydrolysis utilizing extracellular enzymes. It is during this period that intermediate products of fermentation accumulate and gasification is at a minimum. When sludge is digested without a seed source (a sludge that has been digested under similar environmental conditions) this condition is greatly exaggerated. With seeded sludge, liquifaction is in balance with gasification and there is usually no undue accumulation of intermediate products. Although the terms "liquifaction" and "hydrolysis" are used interchangeably they are not strictly synonymous. Hydrolysis is a well-defined chemical term designating the addition of water to the molecule to break down complex substances into simpler ones. Liquifaction does not have such an exact connotation and refers merely to the transfer of substances from a solid sludge stage to a liquid phase.

The primary gases produced during the gasification stage are methane and carbon dioxide. These two gases normally form more than 95% of the gas evolved. The average heat value of the gas is approximately 700 BTU/ft^3. A good indicator of the degree of digestion is the percentage of methane gas contained in the total digester gas. Usually, when the methane production is low (under 65%) the digestion is poor. The maximum volume of gas that is generated from a heated anaerobic digester is approximately 10 ft^3/pound of volatile solids added to the tank, or approximately 0.7 ft^3/capita per day.

Methane production results from the breakdov'n of many compounds by numerous interdependent and interaction reactions which take place in an orderly and integrated fashion.

Methane organisms which produce methane do not utilize such substances as cellulose, glucose, proteins, amino acids or fats but they do utilize a restricted group of simple compounds consisting of lower fatty acids (formic, acetic, propionic, *n*-butyric, etc.). The transformation of complex organic materials contained in sludge to methane and carbon dioxide is brought about in two stages by two different groups of bacteria.

The complex organics are converted by a variety of common bacteria to volatile acids and alcohols without the production of methane (acid production). These products are then converted to methane (methane fermentation stage) by a restricted and specialized group of bacteria among which are: *Methanobacterium omelianskii*, which utilize primary and secondary alcohols; *Mbact. suboxydans*, which partially utilizes butyrate, valeric acid, and other four- and six-carbon fatty acids to produce acetic and propionic acid; and *Mbact. propionicum*, *Mbact. mayei*, and *Mbact. barkerii*, which utilize the simpler organic acids and alcohols, and produce methane and carbon dioxide.

The organisms responsible for active and thorough digestion of waste solids require an environment in which the pH is about 7. The optimum pH value for digestion varies slightly with the characteristics of water supply and the types of waste present. Insufficient seed or low temperature results in retardation of gasification and accumulation of acidic intermediate decomposition products.

In the normal digestion of domestic sewage sludge small amounts of carbon monoxide, as well as hydrogen sulfide, may be present. Oxygen should not be present in the digester gas. Its presence is indicative of an air leak in the digester or an error in the gas sampling procedures. Although some claims of improved digestion with aeration have been made these can usually be attributed to increased mixing of the tank contents with the air. The introduction of large quantities of air on a regular basis will adversely affect methane fermentation.

Some of the most important considerations affecting sludge digestion are: food supply, which is influenced by the type of primary sludge generated in the primary treatment system; time of digestion; utilization of seed sludge; temperature; mixing; pH; the volatile acids to alkalinity ratio; and the quantity and amount of chemicals added to the digestion system. The primary indices of digestive action are:

(1) Gas production—A good rule of thumb is that 10 ft^3 of gas should be produced per pound of volatile solids added to the tank. The gas produced should be approximately 70% methane and 30% carbon dioxide with only traces of miscellaneous gases.

(2) Volatile solids—The sludge produced should have a solids content of approximately 5% with a volatile content of roughly 50%.

(3) Volatile acids—The volatile acid content of the digested sludge should be approximately 500 mg/liter or less. Levels of higher acidity can

be corrected by utilizing chemicals such as lime. Care should be taken, however, not to utilize great amounts of chemicals but rather to change environmental conditions to secure better digestion.

(4) Sludge characteristics—The digested sludge should be black in color, easily dewatered and have a "tarry" odor that is not repellent.

In high rate digestion, rates of digestion can be substantially increased by (1) thorough mixing of the tank contents either mechanically or by gas recirculation; (2) optimum loading of the digesters by regulation of the solids in the feed sludge so that as dense a sludge as possible is fed to the digesters. Detention times can be decreased to as little as five days when adequate heating is provided.

The rate of activities of the organisms responsible for digestion is greatly influenced by temperature. The time required for digestion is indicated by the quantity of gas produced and the increased amount of volatile matter destroyed. The total amount of gas is not appreciably different at the end of digestion but is produced in a shorter time at higher temperatures. Increased gas production follows an increase in the amount of volatile matter destroyed. The best temperature for mesophillic digestion is about 85°C. Temperatures up to 95°F increase the rate of digestion slightly but may be more difficult to maintain throughout the year and may result in problems. The volume of gas may be greater in the presence of some organic industrial wastes, while other wastes may reduce the amount of gas. The carbon dioxide in the gas ranges from 15 to 35% and is affected by the degree of digestion and the types of trade waste present.

III. Activated Sludge Process Modifications

A. Tapered Aeration

In an aeration tank having a definite pattern of longitudinal flow, the impact of the high BOD of the influent entering the head end of the tank will create a relatively high oxygen demand in this point in the mixed liquor. As the oxygen demand of the waste is gradually decreased, the demand for oxygen becomes less and less. If one can envision a plug flow system passing down the length of the tank, then the tapered aeration process (Fig. 2) can be seen to take the gradually decreasing oxygen demand of the mixed liquor into account by making more oxygen avail-

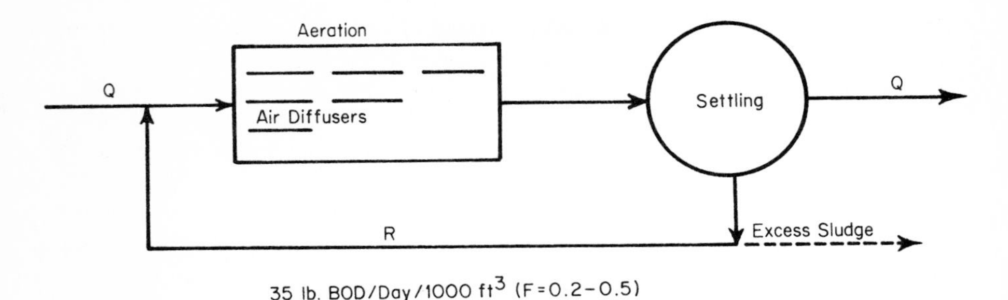

Fig. 2. Tapered aeration. Q, R, and F refer to flow through plant, recirculated effluent, and sludge loading ratio, respectively.

able at the head end of the tank by the use of more diffusers. As with the conventional activated sludge process, tapered aeration has a volumetric loading of about 35 lb BOD/day per 1000 ft^3 of aeration tank capacity, and an F value of 0.2–0.5, which is in the same order of magnitude as conventional activated sludge.

B. Step Aeration

Another activated sludge modification which is capable of handling shock loadings as well as evening out the oxygen demand in the mixed liquor entails the introduction of the waste flow at intervals throughout the length of the tank. This process is termed step aeration. This system (Fig. 3) has a volumetric loading of greater than 50 lb/day per 1000 ft^3 of aeration tank capacity. The F ratio, however, is the same magnitude as that used in conventional activated sludge plant operation.

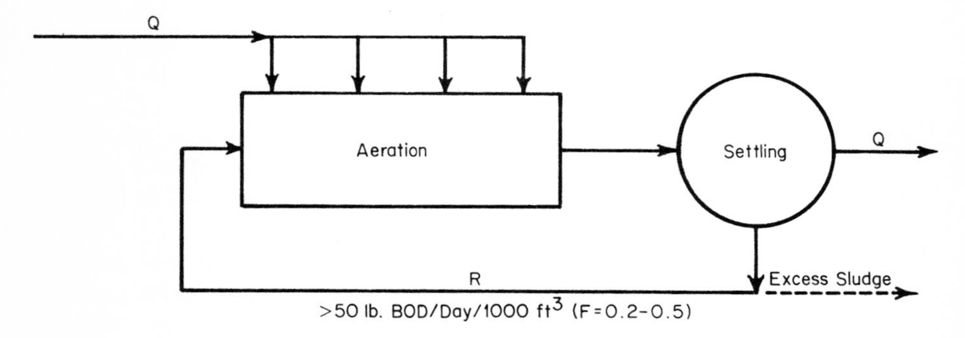

Fig. 3. Step aeration (step loading). Q, R, and F refer to flow through plant, recirculated effluent, and sludge loading ratio, respectively.

C. Contact Stabilization or Sludge Reaeration

Contact stabilization is yet another modified process that permits up to twice the volumetric loading of the conventional process. In contact stabilization, the volumetric loading for the system is in the order of 70 lb BOD/day per 1000 ft^3 of aeration tank capacity. This system is shown in Fig. 4. The mixed liquor, displaced to the settling unit, is settled out and pumped to the reaeration unit where it is aerated without further waste water addition. The organisms thus enter a declining phase and, when finally discharged into the aeration tank, they have the capability of removing large amounts of substrate BOD by the anabolic processes involved in the assimilation and storage of substrate previously discussed.

The actual contact time between waste water and mixed liquor is maintained between 30 min and 1 hr by the design of the system. Unlike the step aeration process, in which this contact time can be varied by the operator regulating the point(s) of waste entry, the contact stabilization process does not give extremely great flexibility in load assimilation. It is primarily used in the design of "package" treatment systems as well as in industrial waste applications.

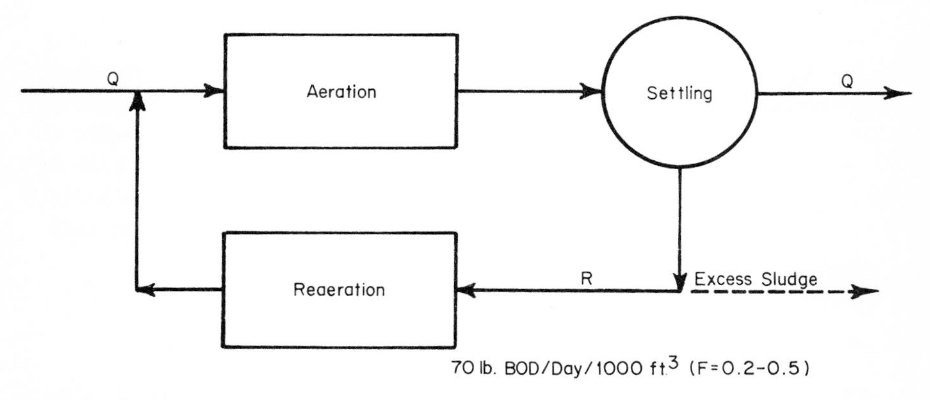

Fig. 4. Contact stabilization, Q, R, and F refer to flow through plant, recirculated effluent, and sludge loading ratio, respectively.

D. Hatfield Process

The Hatfield process, shown in Fig. 5, is a system that is not widely used in the United States. This process differs from contact stabilization in that anaerobic digester supernatant or, in some cases, digested sludge is fed to the reaeration tank. Proponents of this process feel that in a waste flow containing large amounts of highly carbonaceous industrial material,

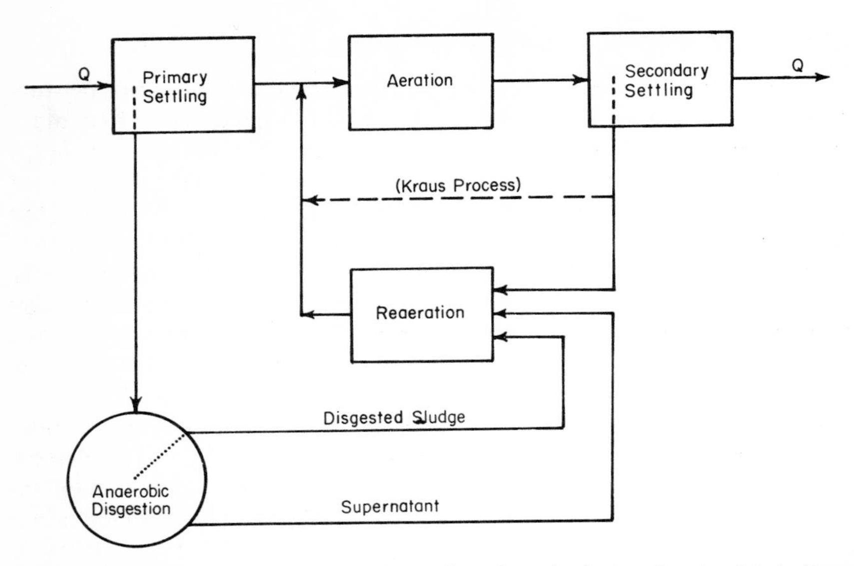

Fig. 5. Hatfield process. Q and R refer to flow through plant and recirculated effluent, respectively.

the supplying of anaerobic digester effluent to the reaeration unit fortifies the active sludge solids with amino acids and other nitrogenous substances. In addition, there is some thought that the addition of a heavier type of solid to the aeration tank would act to prevent the nonsettleability or bulking of the activated sludge. As with the contact stabilization process, the Hatfield process has the advantage of being able to maintain a large weight of organisms under aeration in a relatively small aeration system.

E. Kraus Process

Also shown in Fig. 5 is the Kraus process modification. In this process some of the return sludge bypasses the reaeration unit and is delivered directly to the mixed liquor aeration tank. Therefore, in certain aspects, the Kraus process is a hybrid between the conventional activated sludge process and the Hatfield system.

F. Short-Term or Modified Aeration

The short-term aeration processes have extremely high loading factors, varying from about 0.5 up to 5 lb BOD/day per lb MLVSS. Modified aeration has a loading factor range of about 2–5 lb BOD/day per lb MLVSS. The volumetric loading for modified aeration is about 100 lb BOD/day per 1000 ft^3 of aeration tank capacity.

The short-term aeration processes offer considerable economy of construction due to the very small aeration tank capacities that are required. However, it will be noted that aeration systems are able to contain only relatively low organism weights and that the effluent quality will suffer accordingly, since the lower the organism weight in a system, the greater will be the amount of unused BOD in the process effluent. However, this effect is not directly proportional to the weight of organisms contained in the system, and effluent quality reduction becomes serious only when the weight of organisms is extremely small, as is the case for the modified aeration and supraactivation systems.

Short-term aeration systems produce a relatively large amount of net growth of MLVSS. Sludge disposal becomes a major problem if the sludge is intentionally removed from the system. If sludge is allowed to remain within the system the BOD removal efficiency would range between 50 and 75% since excess sludge would pass out in the process effluent. Removing the sludge intentionally, although creating a sludge handling problem, would increase the BOD removal efficiency to as high as 90%.

G. Extended Aeration

The extended aeration process (Fig. 6) is characterized by a low loading spectrum. The objective of this process is to oxidize the biological solids produced by synthesis from the removal of BOD. Extended aeration typically operates at loading factors ranging from about 0.05 to 0.2 lb BOD/day per lb MLVSS, and the volumetric loading is generally about 20 lb BOD/day per 1000 ft^3 of aeration tank capacity. One unique feature of the extended aeration process is the system's ability, because of its relatively large aeration tank volume, to contain a relatively large weight

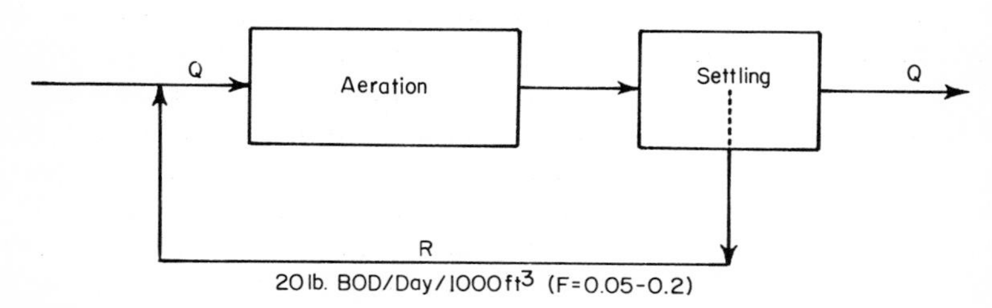

Fig. 6. Extended aeration. Q, R, and F refer to flow through plant, recirculated effluent, and sludge loading ratio, respectively.

of volatile sludge. The low volumetric loading rate, together with the large weight of organisms, combine to give the typically low loading factors.

As usually operated, the extended aeration process has no intentional wasting of sludge from the system because, theoretically, no excess activated sludge is produced. Practically, however, net growth is produced and a large amount of it is wasted from the system, unintentionally, in the effluent. This escape of solids results in lowered efficiencies in the range of 75–85%. If sludge is removed intentionally there are indications that removal efficiencies can be improved to those of conventional activated sludge.

Because of the relatively large concentration of microorganisms carried within the aeration tank and the extended length of aeration time, endogenous respiration plays a major role in sludge quality. The volatile portion of the sludge remaining is not degraded at the same rate as normal activated sludge and thereby results in a lower BOD exertion per unit weight of solids. Effluents from this process, therefore, contain higher suspended solids contents but relatively less BOD than conventional effluents containing equivalent solids concentrations. For this reason effluents from extended aeration systems often meet regulatory agency requirements for BOD levels but contain unsatisfactory levels of suspended solids.

IV. Trickling Filters

The second major biological treatment system is called the trickling filter. In essence, the name is a misnomer since the biological unit neither filters nor does it trickle. The major difference between the trickling filter and the activated sludge system, as far as its ecological patterns are concerned, is that the trickling filter utilizes a succession of biological communities established at different levels within the trickling filter and associated with correspondingly different degrees of purification. The activated sludge system, however, has the same biological community within the floc at any one time, and this community is associated both with the raw untreated waste entering the basin and with the purified effluent.

Within the trickling filter the interrelationships and activities of the different members of the biological community are similar to those outlined for the activated sludge system, although they are limited to the extent that stratification occurs. In general, there seems to be an underlying consensus of opinion in the field of pollution control stating that trickling filters are more easily adapted or utilized to treat shock loadings of waste sources. The concept of shock loadings and their relationship to both trickling filters and activated sludges is discussed in a later portion of this chapter. There are, however, basic physical differences as well as operational differences between the two biological treatment systems.

A. Description

A trickling filter is a bed of coarse, rough, hard, impervious material over which the sewage is sprayed or otherwise distributed (by a distributor) through the air. A biological slime which grows on the filter packing is responsible for the biological reactions. The sewage then flows downward through the filter in contact with the air. The filter is usually 3–12 ft deep and is provided with an underdrainage system to remove the filter effluent and provide ventilation.

The main function of a trickling filter is to remove unstable, organic, pollutional materials in the form of dissolved and finely divided organic solids and to oxidize these solids biologically to form more stable materials.

There are many variations in flow pattern for trickling filtration plants. The most common patterns are given in Figs. 7(a)–(d). No universally applicable flow system exists, and it is not unusual for one plant to operate on many different patterns during the course of a year.

B. Theory of Filter Operation

The trickling filter process depends upon the biochemical oxidation of complex organic material in sewage. Soon after a filter is placed in operation, the surface of the filter medium becomes covered with zoogleal slime, a viscous, jelly-like substance containing bacteria and other biota. Under favorable environmental conditions, the slime adsorbs and utilizes suspended, colloidal, and dissolved organic matter from the sewage, which passes in a relatively thin film over the slime's surface. Eventually a population equilibrium is reached. As biota die, they are discharged from the filter together with the more or less partly decomposed organic matter. This "sloughing off" of material may occur periodically or continuously depending upon the type of filter. A low or standard rate filter sloughs periodically, while a high rate filter sloughs continuously. Generally, secondary settling is provided to retain the settleable solids sloughed from the filter.

The essential features necessary to the process are: (1) Surface area

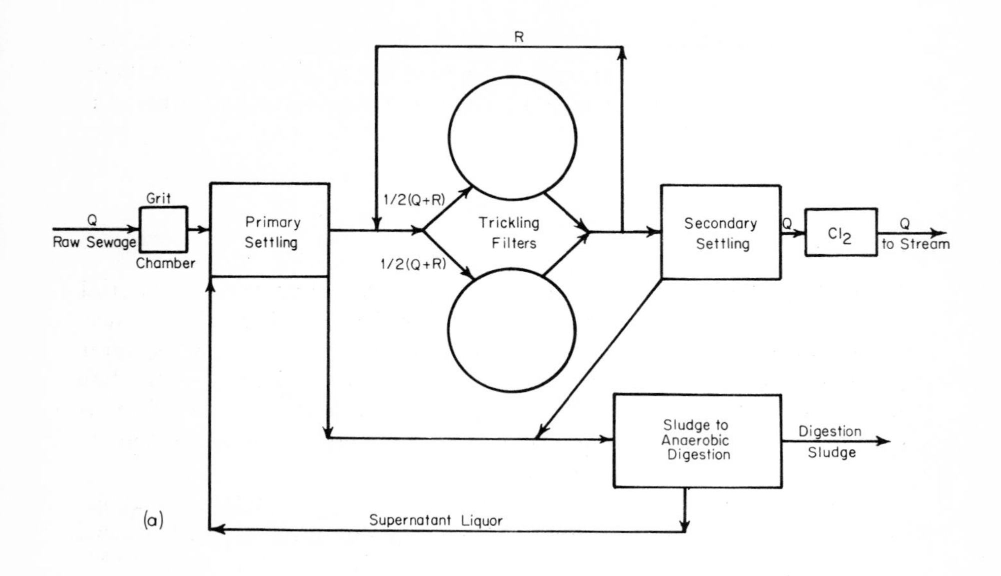
R
Grit
Q
Raw Sewage
Chamber
Primary Settling
1/2(Q+R)
1/2(Q+R)
Trickling Filters
Secondary Settling
Q
Cl2
Q
to Stream
Sludge to Anaerobic Digestion
Digestion Sludge
(a)
Supernatant Liquor

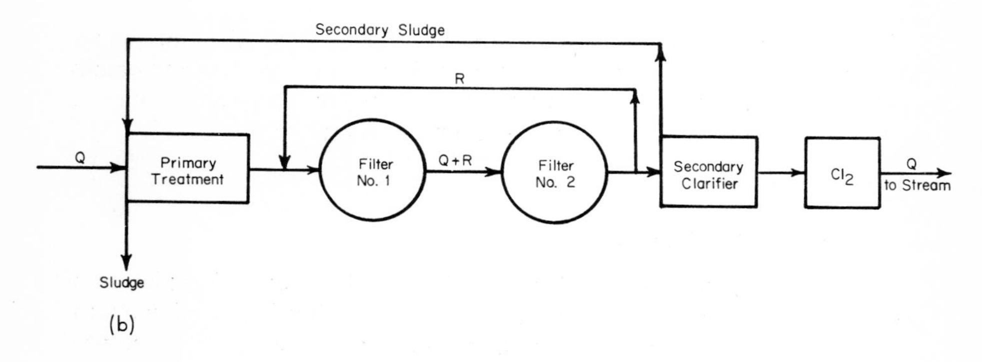
Secondary Sludge
R
Q
Primary Treatment
Filter No. 1
Q+R
Filter No. 2
Secondary Clarifier
Cl2
Q
to Stream
Sludge
(b)

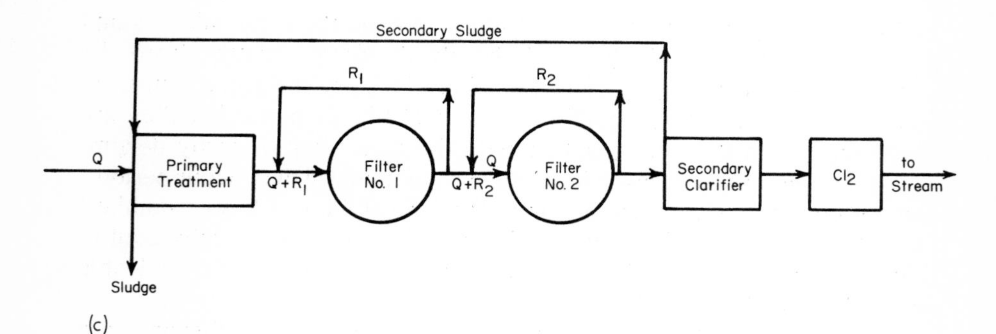
Secondary Sludge
R1
R2
Q
Primary Treatment
Q+R1
Filter No. 1
Q+R2
Q
Filter No. 2
Secondary Clarifier
Cl2
to Stream
Sludge
(c)

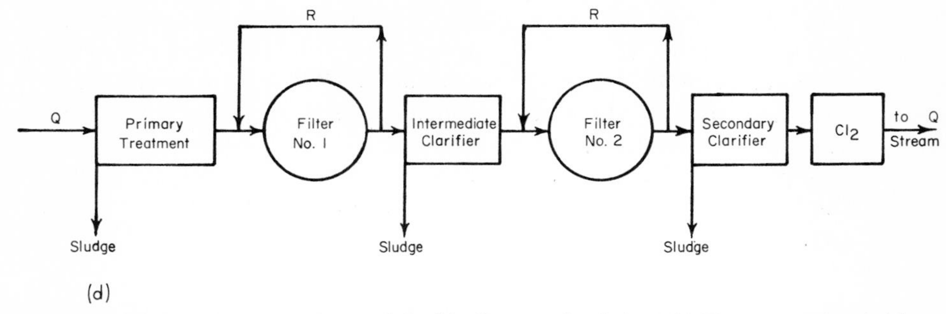

Fig. 7. (a) Trickling filters in parallel with direct recirculation. (b) Two-stage plant with direct recirculation to primary stage. (c) Two-stage plant with direct recirculation within each stage. (d) Two-stage plant with recirculation and clarification between stages. Q and R refer to flow through plant and recirculated effluent, respectively.

must be provided for biological growth, and the surface area/volume ratio must be large. (2) Free oxygen must be available so that aerobic conditions exist. (3) Waste must be amenable to biological treatment; sewage is no problem, but industrial wastes may be.

C. BOD and Suspended Solids Removal

Trickling filters are capable of providing adequate treatment of wastes susceptible to aerobic biologic processes where the production of a plant effluent of 20–30 mg/liter BOD is acceptable. Table 2 shows some generally expected BOD and suspended solids removals.

TABLE 2
Comparison of BOD and Suspended Solids (SS) Removal for Various Processes

Type of plant	Expected BOD removal, %	Expected SS removal, %
Primary treatment	15–35	45–65
Primary and trickling filtration treatment	80–90	80–95

BOD removals are affected by:

1. Quantity and quality of waste. A definite ratio exists between the quantity of a waste applied to a trickling filter and the BOD removal

efficiency. In general, the greater the quantity of waste applied, the lower the BOD removal efficiency. In order to obtain a reasonable degree of treatment the waste should be free from constituents that are toxic to the filter organisms, such as cyanide, copper, chromium, and other heavy metals.

2. Temperature. Greater BOD removals should be expected during the summer months due to the increased activity of the microorganisms.

D. Construction Features

The shape of a filter is related to the type of distributor used. As most plants constructed since 1935 have rotary distributors, the filters are usually circular. Plants utilizing fixed nozzle distributors usually have rectangular filters.

Some filters have been built without retaining walls for the media. However, such construction is seldom economical. The majority of filters have circumferentially reinforced concrete walls, usually 8–12 in. thick.

The choice of the filter medium is often governed by the material locally available and the cost of transporting it. Field stone, gravel, blast furnace slag, redwood blocks, and synthetic inert materials have all been used. The medium should be hard, clean, free of dust, insoluble in sewage, and approximately cubical in shape to obtain a large surface area/volume ratio. Ninety-five per cent or more of the medium should pass a $4\frac{1}{2}$-in. square screen but be retained on a 2-in. square screen. According to New Jersey regulations, the depth of filtering medium at any point in the filter should not be less than 6 ft or greater than 9 ft for standard rate filters. The minimum depth of medium for high rate filters is set at 5 ft.

The medium must serve two primary purposes: (1) It must provide surface area for slime growth, and (2) it must leave sufficient voids for free circulation of air.

The underdrainage system in the filter serves two primary purposes: (1) to carry sewage effluent passing through the filter away for further treatment, and (2) to provide ventilation and maintenance of aerobic conditions.

The entire underdrainage system should be designed to permit free passage of air, and provisions should be made for flushing out of the lateral channels. The inlet openings should have an unsubmerged gross combined area equal to at least 5% of the surface area of the filter.

The floor of the filter must be strong enough to support the underdrainage system, the filter medium, and the water load if the filter is to be flooded.

E. Distribution Systems

Sewage is applied to the filters by either fixed nozzles or rotary arm distributors. Fixed nozzle filters are usually rectangular in shape and utilize deflector plates along with the orifices, as shown in Fig. 8.

Rotary distributors are the most commonly employed. The arms are rotated by the reaction of the sewage discharge or, in some cases, are motor-driven. The arms are usually set 6 in. above the filter bed.

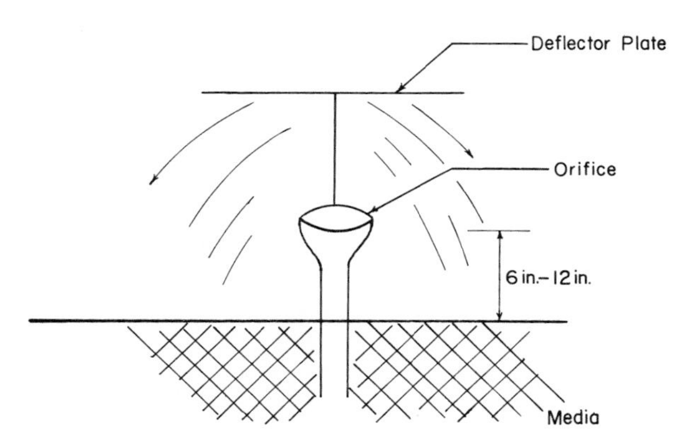

Fig. 8. Fixed nozzle distributor.

F. Loadings

Trickling filters are classified according to the applied hydraulic and organic loadings. The hydraulic load is the total volume of sewage (including recirculation) applied to the filter per day per square feet of surface area [gal/day (gpd) per ft^2 or million gpd/acre]. The organic load is the pounds of 5-day BOD applied to the filter per day per cubic foot of filter medium (lb BOD/day per ft^3 medium) or (lb/day per 1000 ft^3), Table 3 shows a differentiation between standard rate and high rate filters.

Some filters are called "roughing filters." These are usually high rate filters receiving high organic loadings. Although they may give a high pound per unit volume of organic load removal, their settled effluent still contains substantial BOD. They are used to provide intermediate treatment or as the first of a multistage biological treatment.

Most high rate filters utilize recirculation, which is the recycling of filter effluent through the filter. In such cases the ratio of recycled flow to sewage flow is known as the recirculation ratio. Among the useful purposes of recirculation are: (1) reducing "out of service" periods (due to

TABLE 3
Classification of Trickling Filters[a]

Type of filter	Hydraulic load, gpd/ft²	Organic load, lb BOD/day per 1000 ft³	Appli-cation	Slough-ing	BOD removal through filter, %
Standard rate	25–100	5–25	Intermittent	Largely periodic	80–85
High rate	200–1000	25–300	Continuous	Continuous	65–80
Roughing filters	500	300	Continuous	Continuous	25–65

[a]Data obtained in part from Ref. *11*, by courtesy of the Water Pollution Control Federation.

shock loadings; see Sec. V) to a minimum by adjusting recirculation to influent flow, (2) keeping self-propelled distributors turning by adjusting recirculation to influent flow, (3) lowering film thickness and fly breeding by film sloughing, and (4) improving the quality of the effluent, at constant efficiency of treatment, by reducing the concentration of applied sewage. Some high rate filters are given special names which signify specific patented recirculation or distribution schemes.

G. Biofilter

The biofilter (Fig. 9) employs recirculation and a high rate of application to a shallow trickling filter. The recirculation in this case involves bringing the effluent of the filter or the secondary settling tank back to the

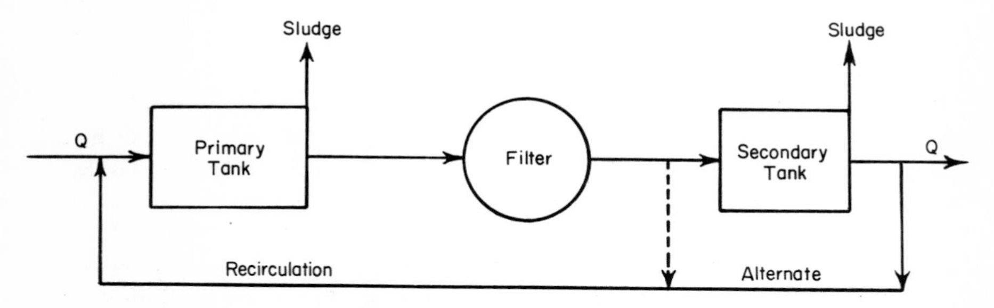

Fig. 9. Biofilter. Q refers to flow through plant (*48*).

primary sedimentation basin. This requires designing the primary settling tank to accommodate not only the average daily sewage flow but also the recirculated flow, which may be as much as 10 times the average flow.

H. *Accelofilter*

This filter (Fig. 10) employs recirculation of the filter effluent directly back to the filter.

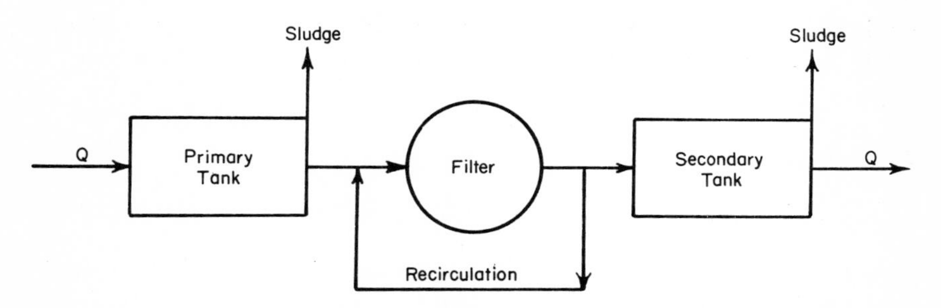

Fig. 10. Accelofilter. *Q* refers to flow through plant. (Patented)

I. *Aerofilter*

The aerofilter (Fig. 11) is a filter over which the sewage is distributed by maintaining a continuous rain-like application of the sewage over the filter bed. A disk distributor revolving at speeds as high as 380 rpm and set 20 in. above the surface of the filter is sometimes used on small beds.

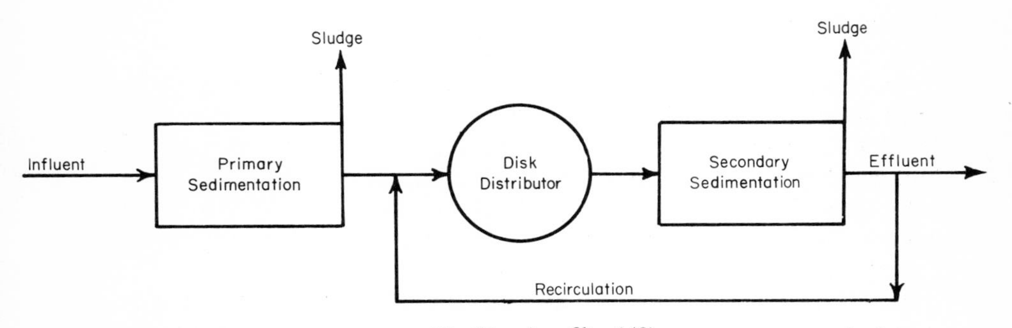

Fig. 11. Aerofilter (*49*).

For large beds 10 or more revolving distributor arms are used to obtain the uniform rain-like application. Aerofilters operate at hydraulic loadings greater than 300 gpd/ft^2 of surface area.

J. Industrial Waste Applications

For certain industrial applications sectional filters utilizing controlled quantities of forced air within each section can be utilized. Many states now make provisions for the use of synthetic media which allows greater flexibility in construction. These filters are often referred to as "Tower" filters and can receive greater organic loads at relatively high rates of removal efficiency.

K. Conclusions

Among the advantages of a trickling filter are:

1. Relatively high nitrifying effect, i.e., biological oxidation of amino nitrogen to nitrates (see Chap. 15).
2. Dependability to give a good effluent under wide variations in influent quality.
3. Relatively low operating and maintenance cost.
4. Ability to function under extreme weather conditions.

Disadvantages are:

1. The high head loss through the filter.
2. Odor and fly nuisance (low rate filters).
3. Large surface area required.
4. Relatively high construction costs.

V. Shock Loadings

An important consideration in the design of a biological sewage treatment plant should be the ability of the system to cope with immediate or rapidly occurring changes in the chemical or physical environment of the biological population. Such environmental changes may be broadly categorized as shock loads, the effects of which may vary from slight malfunctioning of the system to a complete cessation of metabolic activity resulting in the shutdown of the system.

Shock loading of systems may take any of several forms. The most common type is the quantitative shock, which involves a change in the concentration of the waste or the organic loading of the system. A rapid decrease in organic loading, termed a hydraulic shock, might result from an influx of storm waters which dilute the influent waste. In this same category is the rapid increase in loading which might result from the removal of a blockage in the sewer system. It is important to bear in mind that the chemical nature and structure of the waste are unchanged from those normally handled by the system, and thus the existing population requires no acclimation period for this type of shock.

A toxic shock is the result of materials which damage or inhibit the existing metabolic pathways or disrupt the physiological condition of the microbial population. The occurrence of heavy metals, cyanide compounds, or materials producing rapid pH changes would be considered toxic shocks. The treatability of such a waste depends on a prolonged acclimation period, and even then adaptation of the microorganisms to the new substrate is not guaranteed.

Qualitative shock loads are associated with a change in the structural configuration of the carbon source and do not of themselves entail any change in the organic loading of the system. This type of shock could occur when a new waste is introduced into the system, such as cannery or meat-packing wastes (*34*).

Trickling filters and the activated sludge process are the most commonly encountered biological treatment systems. Since quantitative shock loads involve an increase in the hydraulic or organic loading of the system, the apparent method of treatment would be to increase the capacity of the plant by having additional sedimentation tanks, filters, or aeration tanks on a standby basis. It is immediately apparent that this is an extremely uneconomical approach to the problem.

Inherently, a high rate trickling filter has the potential to cope with quantitative shocks since it employs recirculation to maintain the hydraulic loading on the system. Recirculation may be increased to dilute high organic loads or decreased to accommodate hydraulic shocks. By these adjustments the trickling filter effluent can be maintained at a fairly stable level irrespective of large variations in the influent (*35*).

The standard activated sludge process does not provide sufficient latitude in operation to handle large quantitative shocks. Sludge bulking (*34*) and inadequate treatment times caused by the shock result in the release of an unsatisfactory effluent. This inability of the activated sludge process to cope with shock loads has led to numerous modifications of the system in an attempt to alleviate this problem.

The step aeration system (Fig. 3) provides for the introduction of the primary effluent at several points along the course of the mixed liquor. All of the return sludge is introduced at the head of the plant (*36*). By varying the feed to the various passes, the BOD/suspended solids ratio may be completely controlled (*37*), thus evening out the oxygen demand in the mixed liquor. Step aeration provides great flexibility in the handling of quantitative shocks, but BOD removal is sacrificed.

Other variation of the activated sludge (AS) process is commonly known by the names sludge reaeration, contact stabilization, or biosorption (Fig. 12). By separating the adsorption and oxidation phases, double the volumetric loading of step aeration is facilitated. This ability to handle larger loadings than the standard process is the property which provides for adequate treatment of shock loads, but again, it is at the expense of BOD removal(*38*). It should also be noted that this system is not as flexible as step aeration.

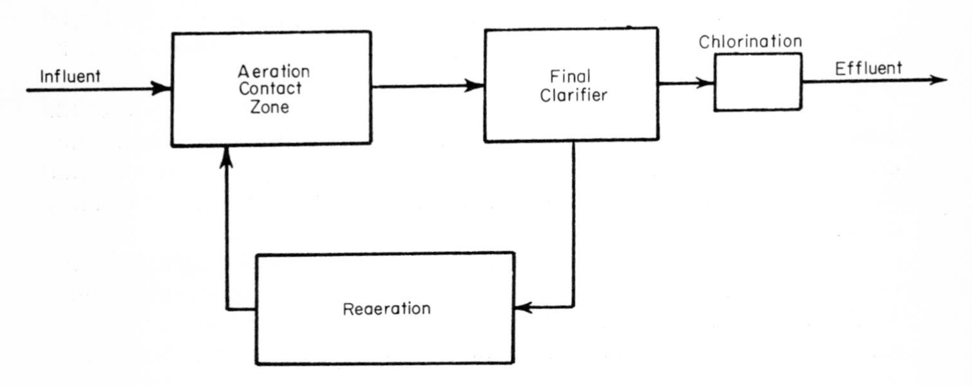

Fig. 12. Sludge reaeration or biosorption.

A somewhat different approach to the problem of quantitative shock loading has been the use of multistage biological treatment. To this end, systems have been designed with high rate biological filters in parallel with the AS process(*39*), or trickling filters in series with the AS process (*40,41*). Systems such as these take advantage of the trickling filter's ability to handle shock loads and the ability of AS to produce an effluent of high quality, and in this fashion eliminate some of the shortcomings of each.

Along these lines, AS systems have been used sequentially(*41*). Such serial operation of AS systems permits plant operation to be easily adjusted to shock overloads(*42*).

The activated sludge aeration system is a combined approach to shock loading of a quantitative type. It employs a step aeration system run in parallel with a conventional AS process, the step aeration system serving as a source of return sludge for both legs of the circuit. This combination provides the flexibility of step aeration and the high quality effluent of the AS process(*37*).

Quantitative shock loads may be successfully handled, in general, by processes designed for increased volumetric and/or organic loading without provision for duplication of system components. This ability has been provided by the use of recirculation, separation of the adsorption and oxidation phases of treatment, or by multistage biological treatment, in general, at the expense of BOD removal.

The treatment of qualitative or toxic shock loads poses very different problems since an adaptation phase is required for the biological population to adjust to the new substrate.

The utility of trickling filters is open to question when nonquantitative shocks are considered. One opinion is that they are more resistant to shocks of toxic and organic substances than AS, because of their relatively short contact time, and that they recover more quickly from the effects of deleterious substances or unfavorable conditions(*35*). Rhodes(*41*) maintains that substances which are detrimental to biological life (toxic shocks) have the same effect on microorganisms whether they live on filter media or in an AS floc. The utility of trickling filters is also questioned on the grounds that it is more time-consuming and costly to clean filter material and renew biological activity on a trickling filter than it is to renew the activity of the biological mass in the activated sludge process (*41*). Since the successful response of a system to a qualitative shock must be initiated and completed during the time of substrate-biological solids contact, activated sludge is better than a trickling filter from a biochemical standpoint(*34*). The treatability of a toxic shock depends on a prolonged acclimation period which is very much greater than the delay time(*34*) in the system, and hence neither method can adequately treat such a shock.

It has been shown, in practice, that the sludge reaeration or biosorption system (see Fig. 12) recovers from wide variations in pH of short duration in a short period of time and that long-term shocks (36 hr) require recovery times of 5 days(*43*). This is a measure of the system's ability to resist a toxic shock and has nothing to do with its ability to treat it. Similarly, it has been found that biological treatment systems can be acclimated to specific toxic shocks(*44*) not exceeding some maximum concentration, but this is in the realm of industrial waste treatment, including such patented systems as the "integrated waste treatment system" for the removal of cyanide and chromium from an acid medium(*45*). It seems that the best way to handle a toxic shock is prevention of it at the source, i.e., before the material enters the biological treatment system.

Biological treatment systems can and do satisfactorily treat qualitative shocks. Stage operation of activated sludge systems can be used since there is a highly active sludge available in the second stage which is not endangered, since the new substrate is caught up in the first stage(*41*).

Especially suited to the treatment of qualitative shocks are the Hatfield process (Fig. 5) and its modification, the Kraus process. In the Hatfield process the supernatant from an anaerobic digester or digested sludge is fed to the reaeration tank and subsequently to the aeration tank. This provides an extracellular source of nitrogen, which is essential to the successful response of the system to a shock high in carbonaceous materials. The Kraus process differs only in that some of the return sludge bypasses the reaeration tank and goes directly to the mixed liquor aeration tank (*38*). This is shown by the dotted line in Fig. 5.

That biological treatment can treat quantitative and qualitative shock loads is demonstrated by the number of plants using the methods already discussed (*40, 46, 47*). One of these plants is of special interest in that it was designed to optimize treatment of shock loads. This plant utilizes a Kraus nitrified sludge interchange process which is provided with a balanced nutrient through the continuous introduction of a mixture of activated sludge and waste digester liquor into the aerated sewage. The nutrient sludge is produced by a 24-hr aeration period in nitrification tanks. The aeration tanks are arranged so that they may be used as either two- or four-pass units. Air is introduced, at different depths, from each side of the tank to maintain turbulent mixing (*47*).

The processes employed at this plant allow great flexibility in the handling of shock loads, as they provide a step aeration for quantitative shocks and an extracellular source of nitrogen for qualitative shocks. It is a prime example of the ability of biological systems to treat shock loads.

VI. Summary

A. General Comments

We have attempted to show in this chapter some of the basic mechanisms available to the sanitary engineer for removing up to 95% of the organics present in a waste water. Of the more than 750 mg/liter of total solids which can be present in a domestic waste water, only about 25% are removed during conventional waste water treatment. It is relatively clear to many professionals that we cannot treat waste sufficiently to avoid nuisance conditions under our best, most modern, practices today. More faithful collection of sewage and industrial waste, with treatment efficiencies increased a few points, would not eliminate the need for research that would demonstrate the extent and character of the remedial measures that may be in order. Basic and applied research implying new concepts in sewage and waste water treatment may be able to solve our major problems and a greater margin of water resource can thus be made available to the environment.

If we are to achieve a more advanced waste treatment capability we must meet two important objectives: (1) It will be necessary to concentrate the impurities in municipal waste water into very small volumes and to dispose permanently of these concentrates; (2) it will become necessary to find an outlet for the relatively pure waters which will be produced from these advanced treatment systems.

Concentrated wastes removed by new waste treatment processes cannot, of course, be returned to surface waters; if this were done, stream pollution problems would not be diminished and downstream water users would have to remove the same contaminants over and over. The development of methods for permanently disposing of concentrated impurities is, therefore, an important part of the concept. Furthermore, if contaminants in waste waters are removed efficiently, it is evident that the purified water discharged will be of excellent quality. Such water, especially where concern over quality and/or quantity of water from conventional sources exists, must not be wasted. It would be folly to discharge such water into a stream less pure than the discharge itself. Therefore, the development of a successful advanced waste treatment technology could lead—at one step—to the alleviation of both water pollution and water supply problems.

B. Processes under Consideration

It is conceivable that almost any physical or chemical principle of separation is applicable to advanced waste treatment. Work has already been started on chemical methods to replace biological treatment for the removal of contaminants. It is recognized that these operations are foreign to the municipal waste treatment field and that they are unconventional by today's standards. Yet we are faced with an unconventional problem.

Excessive enrichment or eutrophication of receiving waters by nutrient-rich wastes is emerging as a major water pollution problem in many areas. It has been recognized for some time that ordinary domestic sewage is a rich source of the nutrients required by phytoplankton. Experience has shown that the degree of eutrophication and, hence, the severity of subsequent water quality problems are dependent largely upon the supply of inorganic nitrogen and phosphorus.

To remove 95–99% of the soluble phosphates from a sewage treatment plant effluent containing 5 ppm of phosphates, expressed as phosphorus, relatively high concentrations of coagulants are required. Approximately 6–10 times the amount of coagulant is required as is ordinarily used in the clarification of surface water supplies employed by a community as a source of potable water; the cost of the chemicals required is thus also 6–10 times the ordinary amount. It is apparent that the removal of soluble phosphates from sewage treatment plant effluent by means of a coagulant is a very costly process. There is also the problem of disposing of many tons of metallic (Al, Fe, and Ca) hydroxide sludge.

C. Conclusions

This paper has been offered to stimulate thought regarding the waste treatment field. Only some of the available methods have been discussed due to the complexity of the problem. A full discussion of all possible methods would necessitate at least one full volume.

Research on new treatment systems is proceeding rapidly, but no one can as yet predict its outcome. One thing is certain: We have come to the time when we must think not in terms of what we are accomplishing in the waste water treatment field, but of what we are *not* accomplishing.

REFERENCES

1. K. Imhoff and G. M. Fair, *Sewage Treatment,* 2nd ed., Wiley, New York, 1956.
2. I. Gellman and H. Heukelekian, *Sewage Ind. Wastes*, **25**, 1196 (1953).
3. E. N. Helmers, J. D. Frame, A. E. Greenberg, and C. N. Sawyer, *Proc. Ind. Wastes Conf., 7th, Purdue University, 1951*, p. 375.
4. O. R. Placak and C. C. Ruchhoft, *Public Health Rept. (U.S.)*, **62**, 697 (1947).
4a. R. E. McKinney, *Microbiology for Sanitary Engineers*, McGraw-Hill, New York, 1962, Chap. 21.
5. S. R. Hoover and N. Porges, *Sewage Ind. Wastes*, **24**, 306 (1952).
6. D. A. Okun, *Sewage Works J.*, **21**, 763 (1949).
7. S. A. Greenley, *Sewage Works J.*, **15**, 1062 (1943).
8. *News Quarterly*, Vol. XIII, No. 1, Sanitary Engineering Research Laboratory, Univ. California, Berkeley, 1963.
9. R. H. Gould, *Munic. Sanit.*, **10**, 185 (1939).
10. F. W. Harris, T. Cockburn, and T. Anderson, *Proc. Assoc. Managers of Sewage Disposal Works*, N.Y. Public Library, New York, 1926, p. 67.
11. Report of the National Research Council Sub-Committee on Military Sewage Treatment, *Sewage Works J.*, **18**, 794 (1946).
12. H. Heukelekian, H. E. Orford, and R. M. Manganelli, *Sewage Works J.*, **23**, 934 (1951).
13. Water Pollution Control Federation, *Units of Expression for Wastes and Waste Treatment, MOP No. 6*, Washington, D.C., 1958.
14. G. M. Fair and J. C. Geyer, *Water Supply and Waste Water Disposal*, Wiley, New York, 1966, p. 419.
15. E. J. Genetelli, Ph.D. Thesis, Rutgers Univ., New Brunswick, N.J., 1962.
16. F. W. Mohlman, *Sewage Works J.*, **6**, 121 (1943).

17. J. Finch and H. Ives, *Wastes Eng.*, **24**, 214 (1954).
18. H. Heukelekian and E. Isenberg, *Water Sewage Works*, **106**, 525 (1959).
19. *Manual for Sewage Plant Operators*, 3rd ed., Texas State Department of Health, Austin, 1964, p. 268.
20. R. W. Butcher, *Trans. Brit. Mycol. Soc.*, **17**, 112 (1932).
21. N. C. Dondero, *Advan. Appl. Microbiol.*, **3**, 77 (1961).
22. R. S. Ingols and H. Heukelekian, *Sewage Works J.*, **11**, 927 (1939).
23. E. H. Morgan and J. A. Black, *Sewage Works J.*, **1**, 46, (1928).
24. E. Naumann and J. Wanselin, *Bot. Notis.*, **1/2**, 141 (1927).
25. C. C. Ruchhoft and J. H. Watkins, *Sewage Works J.*, **1**, 52 (1928).
26. H. E. Babbitt and E. R. Baumann, *Sewage and Sewage Treatment*, 8th ed., Wiley, London, 1958.
27. M. S. Finstein, *J. Water Pollution Control Federation*, **39**, 33 (1967).
28. E. J. Genetelli, *ibid.*, **39**, R 32 (October, 1967).
29. W. Eckenfelder and D. J. O'Conner, *Biological Waste Treatment*, Pergamon, 1961.
30. J. McCabe and W. W. Eckenfelder, *Biological Treatment of Sewage and Industrial Wastes*, Vol. 1, Reinhold, New York, 1956.
31. R. E. McKinney, *J. Sanit. Eng. Div. Am. Soc. Civil Engrs.*, **88**, SA3, 87 (1962).
32. D. B. Smith, *Sewage Works J.*, **24**, 1077 (1952).
33. L. Hartmann, *Vom Wasser*, **27**, 107 (1960).
34. A. F. Gaudy, Jr., and R. S. Engelbrecht, *J. Water Pollution Control Federation*, **33**, 800 (1961).
35. E. H. Bryan, *Ind. Water Wastes*, **8**, 31 (1963).
36. L. Van Kleeck, *Waste Eng.*, **28**, 185 (1957).
37. M. H. Klegerman, *Ind. Water Wastes*, **7**, 97 (1962).
38. M. J. Stewart, *Water Sewage Works*, **111**, 246 (1964).
39. G. H. Hamlin, *Ind. Water Wastes*, **7**, 148 (1962).
40. A. W. Bannister and A. D. Cloud, *Water Waste Treat. J.*, **1**, 32 (1964).
41. I. H. Rhodes, *Inst. Sewage Purif. J. Proc.*, Part 2, 1961, p. 90.
42. T. Jaffe, *Water Sewage Works*, **103**, 428 (1956).
43. A. P. Courrette, *Water Works Waste Eng.*, **1**, 43, (1964).
44. G. H. McDermott, W. A. Moore, M. A. Post, and M. B. Eltinger, *J. Water Pollution Control Federation*, **35**, 227 (1963).
45. C. R. Bauerlain, L. Pieta, and J. Shackcor, *Waste Eng.*, **1**, 12 (1956).
46. J. O. Cessna, *Water Works Waste Eng.*, **1**, 50 (1964).
47. L. F. Rehm and F. N. Van Kirk, *Water Works Waste Eng.*, **1**, 32 (1964).
48. U.S. Pat. 2,168,208.
49. U.S. Pat. 2,141,979.